The Road Map to Repeatable Success

Using QFD to Implement Change

Systems engineering series

A. Terry Bahill, Editor-in-Chief

The Road Map to Repeatable Success: Using QFD to Implement Change
Barbara A. Bicknell, Bicknell Consulting, Inc.
Kris D. Bicknell, Bicknell Consulting, Inc.

Linear Systems Theory
Ferenc Szidarovszky, University of Arizona
A. Terry Bahill, University of Arizona

Engineering Modeling and Design
William L. Chapman, Hughes Aircraft Company
A. Terry Bahill, University of Arizona
A. Wayne Wymore, Systems Analysis and Design Systems

Model-Based Systems Engineering
A. Wayne Wymore, Systems Analysis and Design Systems

The Theory and Applications of Iteration Methods
Ioannis K. Argyros, Cameron University
Ferenc Szidarovszky, University of Arizona

System Integration
Jeffrey O. Grady, Jeffrey O. Grady System Engineering

The Road Map to Repeatable Success

Using QFD to Implement Change

Barbara A. Bicknell
Kris D. Bicknell

CRC Press
Boca Raton Ann Arbor London Tokyo

Library of Congress Cataloging-in-Publication Data

Bicknell, Barbara A.
The road map to repeatable success: using QFD to implement change / Barbara A. Bicknell, Kris D. Bicknell.
p. cm. -- (Systems engineering series)
Includes bibliographical references and index.
ISBN 0-8493-8019-7
1. Quality control. 2. Quality function deployment. 3. Production management--Quality control. I. Bicknell, Kris D. II. Title. III. Series.
TS156.B565 1995
658.5'62--dc20 94-17627
CIP

Direct all inquiries to CRC Press, Inc., 2000 Corporate Blvd., N.W., Boca Raton, Florida 33431.

International Standard Book Number 0-8493-8019-7
Library of Congress Card Number 94-17627
Printed in the United States of America 1 2 3 4 5 6 7 8 9 0
Printed on acid-free paper

Dedication

To each person who has tried to implement change,
not because it is new,
but because it is the right thing to do.

To each person who would rather do,
than complain;
because there is no profit in complaining.

To each person who would rather become part of the solution,
rather than remain part of the problem,
because that is what is meant by being part of a team.

To those with a vision
and the courage to follow it —
We salute you.

Acknowledgments

No man (or woman) is an island and the authors are no exception to this universal truth. Without the assistance and support of friends, family and clients this book would never have taken form. Although there have been times in the last 2 years when we have felt isolated and alone in struggling to create the language and images that convey our ideas in this book, we realize that this was only our perception and certainly not reality. To those individuals who had more faith in us at times than we did in ourselves we'd like to thank them for their encouragement, good humor and patience. We would also like to thank those clients and business associates who were kind enough, and brave enough, to let their success stories be told here. Unfortunately, there is not ample room here to individually describe the contributions made by all of these individuals, but the authors would like to pay tribute to some of them by mentioning their names:

Terry Bahill
Nick Batik
W. D. and Adalyne Bicknell
Sandra Chevalier
Al DeVries
Scott A. Flores
Timothy M. Gordon
Doris Lach
Dr. Richard Seebass
Dr. Jackie Sullivan
Jim Widenmyer
Judy Zabdyr

The authors would also like to thank the following companies and organizations for their support:

3M Austin Center
Cobe Laboratories
Die Cut Technologies
Martin Marietta Astronautics Group
Resources Trust Company
University of Colorado at Boulder College of Engineering and Applied Science
Westinghouse Electric Corporation

Bibliography

Magazine Articles:

Business Week, 1992 Special Bonus Issue.

Business Week, June 9, 1987.

Business Week, October 25, 1993.

"Leading the World-Class Company," Quality '92, Special Edition, *Fortune*, October 1992.

"Quality Function Deployment Improves Furniture Design," *Quality Digest*, September 1992.

"Reengineering: The How New Managing Tool," *Fortune*, August 23, 1993.

"Some Manufacturers Drop Efforts to Adopt Japanese Techniques," *Wall Street Journal*, May 7, 1993.

"The Customer-Driven Company," *People Who Make A Difference*, Supplement, *Fortune*, May 31, 1993.

"The Tough New Consumer," *Fortune*, Special Issue, Autumn/Winter 1993.

Books, Conference Proceedings:

Akao, Yoji, ed., *Quality Function Deployment*, Productivity Press, Cambridge, MA, 1990.

Akao, Yoji, ed., *Hoshin Kanri: Policy Deployment for Successful TQM*, Productivity Press, Cambridge, MA, 1991.

Blackburn, Joseph D., ed., *Time-Based Competition*, Business One/Irwin, Homewood, IL, 1991.

Blanchard, Benjamin S., *Systems Engineering Management*, John Wiley & Sons, Inc. New York, NY, 1991.

Carter, Donald E. and Baker, Barbara Stilwell, *Concurrent Engineering: The Product Development Environment for the 1990s*, Addison-Wesley Publishing, New York, NY, 1991,

Chapman, William L., Bahill, A. Terry, and Wymore, A. Wayne, *Engineering Modeling and Design*, CRC Press, Boca Raton, FL, 1992.

Danter, Rosabeth Moss, *The Change Masters*, Simon & Schuster, New York, NY, 1983.

Ernst & Young Quality Improvement Consulting Group, *Total Quality, An Executive's Guide for the 1990s*, Business One/Irwin, Homewood, IL, 1990.

Eureka, William and Tyan, Nancy E., *The Customer-Driven Company*, ASI Press, Dearborn, MI, 1988.

Giegold, William C., *Practical Management Skills for Engineers and Scientists*, Life Learning Publications, Belmont, CA, 1982.

Hall, Gene, Rosenthal, Jim, and Judy Wade, "How to Make Reengineering *Really* work," *Harvard Business Review*, November-December 1993.

Hammer, Michael and Champy, James, *Reengineering the Corporation*, HarperCollins, New York, NY, 1993.

Hammer, Michael, *Harvard Business Review*, July-August 1990.

Imai, Masaaki, *Kaizen: The Key to Japan's Competitive Success*, McGraw-Hill Publishing Company, New York, NY, 1986.

Kiemele, Mark J. and Schmidt, Stephen R., *Basic Statistics: Tools for Continuous Improvement*, 3rd Edition, Air Academy Press, Colorado Springs, CO, 1993.

King, Bob, *Better Designs in Half the Time*, Third Edition, Goal/QPC, Methuen, MA, 1989.

Lam, K.D., Watson, Frank D. and Schmidt, Stephen R., *Total Quality: A Textbook of Strategic Quality, Leadership and Planning*, Air Academy Press, Colorado Springs, CO, 1991.

Mansir, Brian E. and Schacht, Nicholas R., *An Introduction to the Continuous Improvement Process*, Educational Services Institute, Falls Church, VA,

MIL-STD-499B, "Systems Engineering," United States Department of Defense, May 15, 1991.

Mudge, Arthur E., CVS, *Value Engineering, A Systematic Approach*, J. Pohl Associates, Pittsburgh, PA, 1990.

Pahl, G. and Beitz, W., *Engineering Design, A Systematic Approach*, Springer-Verlag, New York, NY, 1988.

Pearce, John A., II and Robinson, Richard B., Jr., *Formulation, Implementation, and Control of Competitive Strategy*, Fourth Edition, Irwin, Homewood, IL, 1991.

Phadke, Madhav S., *Quality Engineering Using Robust Design*, Prentice Hall, Englewood Cliffs, NJ, 1989.

Podras, Sandy, *Systematic Problem-Solving and Decision-Making*, Crisp Publications, Inc., Los Altos, CA, 1989.

Proceedings of the 7th Annual NASA/Contractors Conference on Quality and Productivity, Summary Report, April 1991.

Pugh, Stuart, *Total Design: Integrated Methods for Successful Product Engineering*, Addison-Wesley Publishing Company, Wokingham, England, 1990.

Scherkenbach, William W., *The Deming Route to Quality and Productivity*, CEEPress Books, Washington, D.C., 1990.

Shina, Sammy G., *Concurrent Engineering and Design for Manufacturing of Electronics Products*, Van Norstrand Reinhold, New York, NY, 1991.

"Systems Engineering for the 21st Century," *Proceedings of the Second Annual International Symposium of the National Council on Systems Engineering*, Seattle, WA, 1992.

Thomas, Philip R., *Competitiveness Through Total Cycle Time*, McGraw-Hill Publishing Company, New York, NY, 1990.

Tomasko, Robert M., *Rethinking the Corporation*, AMACOM, New York, NY, 1993.

Wertz, James R. and Larson, Wiley J., eds., *Space Mission Analysis and Design*, Kluwer Academic Publishers, Boston, MA, 1991.

Winner, Robert I., Pennell, James P., Bertrand, Harold E. and Slusarczuk, Marko M. G., *The Role of Concurrent Engineering in Weapons System Acquisition*, IDA Report R-388, Institute for Defense Analysis, Alexandria, VA, 1988.

Preface

Only bet a lot when you know you're going to win.
Old Las Vegas Proverb

The best throw of the dice is to
thow them away.
Losing Gambler's Proverb

What does a book on Quality Function Deployment and management methodology have to do with gambling? Nothing — and everything. Gambling is not a uniquely American pastime; but Americans seem to embrace the practice with a fervor unlike any other modern culture — witness Las Vegas. At the time the authors were writing this book, Las Vegas was the fastest growing city in America; casinos imitating huge pyramids are built next to medieval castles; volcanoes roar into the night sky; pirates fight mock battles with the English on 18th century frigates sailing in the middle of a desert; and "Land of OZ" hotels boast over 5000 rooms, ushering visitors through an entrance under the jaws of a 70-foot MGM lion. The casinos lure millions every year to the desert with a "chance" at hitting it big. Gambling is a very, very successful American business. But this book is not about the *business* of gambling itself, and therefore, it has nothing to do with gambling.

The metaphorical throw of the dice

However, every business enterprise is a gamble. One out of every two new business ventures fails within the first 2 years of operation. That statistic in itself says that starting a new business involves risk: and engaging in any activity involving risk, no matter what the odds, is gambling. Businesses gamble every day. They gamble on whether or not the customer will buy a product or service at a specified price. They gamble by investing in new markets. They gamble by expanding. They gamble by "downsizing" to reduce costs. Granted, these are "calculated" risks and actions backed, sometimes, by enormous amounts of data — some of it even useful data. None-the-less, these businesses are taking risks. Often a business is hoping, like the first maxim above, that it's "betting" more when it's going to "win" — either with a new product or marketing strategy. Hopefully, unlike the high roller on the craps table betting on the outcome, a business has more control over the factors that affect the success of the "bet" than a serendipitous throw of the dice. But how does a business *know* it's done everything it can to increase its odds of success? And if it does succeed, how does it repeat that success on the next "throw" of the dice?

This book is about how to increase the odds of success for a business — not only on the first throw of the dice in the market, but on every throw. In addition, this book uses a tool to allow the analysis as to *why* the dice may be coming up 7 or 11 — and how to avoid crapping out! The main focus of the book is on a technique called QFD — Quality Function Deployment — as used to translate the voice of the customer into actionable steps in product and service delivery. But the authors believe that the power of QFD is unlocked when an organization adopts a wider management approach the authors call RSM™. Using these techniques, this book discusses reducing (or at least intelligently evaluating) the risk of launching new products; changing management environments; improving productivity; prioritizing design criteria; marshalling resources to get the most benefit and managing effectively through focusing on the customer.

"How" to read this book

This book is written for varying levels of familiarity of QFD and other "quality" methodologies. Often, in a book that may be written for a variety of audiences and disciplines, authors sometimes recommend a "selective" approach to reading the book, outlining which chapters specialized audiences should read. It is the authors viewpoint (as expressed in the very first chapter) that part of the problem in most organizations is a reluctance to expose oneself to viewpoints and information typically aimed at another specialized group — i.e., engineers often ignore management-related information, and management often ignores technically-oriented information. This book is written to flow from Chapter 1 through the Appendix in a coherent integration of QFD from how it fits in a larger scheme, to understanding its functionality, to seeing it in action. For this reason the authors are reluctant to recommend "skipping" chapters that may not be as essential to engineers, for example, as they are to the executive management. But the authors also understand that not all individuals have the time or luxury of a "slow read" and will, therefore, make some suggestions for those how want to get the most from this book in the shortest amount of time, depending on viewpoint. It is our hope that readers will use this book as a reference guide and find new application for the information it contains as the viewpoint of its owner may change from job to job.

From a systems engineering viewpoint, this book offers a first-time look at how QFD fits into a systems approach for design and manufacturing. The authors would recommend that engineers or those in technical positions who are familiar with the concept of QFD, start with Chapters 3 through 8, and Chapter 14, in detail. These chapters will present the details of how to construct the matrix, develop an integrated plan, and offer new insights into the application of QFD in Robust Designs, Systems Engineering and Concurrent Engineering. The engineer can then return to the first two chapters to gain an overview perspective, and read Chapters 9 through 13 at a later date to understand its application in a business environment, and how to facilitate a QFD pilot program.

For those readers who are more interested in how QFD fits into a management scheme, the authors would recommend reading Chapters 1 through 5 in detail, and then turning to Chapters 9 and 10 which deal with business re-engineering and management perspectives. These chapters will give an excellent overview of the power of the technique in business reengineering and how to implement it in an organization-wide program. The remaining chapters dealing with the building of the QFD matrix and use in various situations can be read when the executive has a little more time to concentrate on the details.

For readers who are uninitiated to the concept of QFD, this book provides a step-by-step detailed description of the use of the technique with examples and application recommendations. QFD is not a simple cookbook tool but a complex analytical technique, especially when combined with other statistical and management concepts. A reader who is unfamiliar with QFD will not be expected to close the book and facilitate a QFD matrix analysis on a complex design. There is too much detail to assimilate in one reading, and QFD facilitation training is a whole process in itself. But the first-time reader can understand the concept, how it fits into a customer-focused organization and how it is used to create better processes, products and services. If the reader is a student, this book provides an excellent overview of how QFD can be used in a business, and will provide the engineer, or the business major, with a jump start on techniques in use in business today.

For those who have used QFD or at least been exposed to one of the schools of thought on its application, this book offers an in-depth analysis of the technique, application nuances and recommendations for pilot programs and facilitation guidance. This book also offers some new applications of QFD in management, technical and support roles that may open up new vistas into the power of QFD as a decision-making analysis tool.

Who should read this book?

Everyone involved in the organization should read this book.

As pretentious as that statement sounds, the authors firmly believe that the dissemination of information in an organization, especially one undergoing change — and which organization isn't? — vastly increases the odds for success. This book has information for those at all levels of an organization. The authors have crafted the subject matter for both those who are technically-oriented and those in management positions who are integrating the varied disciplines found within any organization. Executive management will benefit by reading this book because it discusses integrating company goals and missions into daily activities and operations. Engineers will benefit from a detailed discussion of how to implement QFD in a design process and carry it through to product completion. Marketing people can focus on how to really meet the customer's needs. Those in manufacturing, production control and procurement see how their activities contribute to the overall success of a new product or service, and how to work with the design team in the early stages of the product.

The authors hope that no matter from what area of emphasis or skill specialty the reader approaches this book, he or she will endeavor to expand his or her knowledge base and understand the system within which QFD reaches its maximum effectiveness. As in playing craps, it's a whole lot more fun betting with the roller. The more each person of a team understands about QFD, the better the chances of betting on the right odds for success.

Finally, there is an assumption the authors make about the readers of this book. The second maxim at the beginning of this Preface suggests that throwing the dice away certainly prevents losing on the craps table. However, in business, not taking any risks means not playing the game at all; and if you don't bet, you don't win either. Whatever your comfort level is in accepting risk as a part of business, the authors assume you've come to play!

The authors

Barbara A. Bicknell is the President of Bicknell Consulting, Inc. She is a nationally recognized speaker and consultant on Quality Function Deployment, Concurrent Engineering, Systems Engineering, Integrated Product Design and Re-engineering. Ms. Bicknell has managed leading aerospace operations achieving breakthrough results on issues involved in simultaneous product and process developement using QFD. She developed and automated a concurrent engineering capability for several large spacecraft programs while manager of the Systems Engineering Implementation of Concurrent Engineering Program at Martin Marietta Defense Space Systems Company. She also facilitated numerous QFD projects to support, air traffic control system definition, technical training assessments and strategic planning. These projects included multifunctional design teams. Ms. Bicknell has a strong commercial clientele including companies such as 3M, Martin Marietta Astro Space, General Motors, Ford, Valeo, Intermedics Orthopedics and McData, and government entities including the University of Denver and the University of Colorado.

Ms. Bicknell had an extensive technical background of 11 years experience at Martin Marietta including assignments as technical lead on a space station contract, lead propulsion analyst for the Titan Reassessment Program and principal investigator of a low-gravity resupply experiment on board a NASA KC-135 aircraft. She has authored numerous technical papers and reports on topics including spacecraft and launch vehicle advancements in liquid propulsion, low-g venting, Total Quality Management, Systems Engineering and Cryogenic propulsion. Ms. Bicknell has received Technical Achievement Awards from both Martin Marietta and NASA. Among her most recent publications are contributions to the books, *Total Quality: A Textbook of Strategic Quality, Leadership, and Planning*, and *Basic Statistics: Tools for Continuous Improvement*. Ms. Bicknell has a Master's in Engineering from the University of Colorado, Boulder (1993) and received her BSME from the University of Notre Dame (1980).

Kris D. Bicknell is Vice-President of Bicknell Consulting, Inc. and is a consultant in QFD, Quality Decision Making Processes, Business Re-engineering and Process Improvement, primarily in the commercial and environmental industries. Mr. Bicknell has presented at the Annual TQM Conference in 1993 and 1994, sponsored by the National Business Quality Foundation, Inc. He has a broad background including corporate consulting, environmental and legal experience. He has experience as an environmental and trial attorney who has represented both large and small industrial clients at Superfund sites in Colorado, Utah and Wyoming. He is currently working with companies to implement quality initiatives for Re-engineering, Process Improvement, Strategic Planning, QFD and waste minimization. Prior to practicing law, Mr. Bicknell worked at Martin Marietta as a program representative and proposal writer.

Mr. Bicknell is currently a Board Director for the Rocky Mountain Association of Environmental Professionals, a member of the Colorado Hazardous Waste Management Society (CHWMS) and the American Bar Assocation, Colorado Bar Association (Environmental Litigation Section) and the Denver Bar Association. Mr. Bicknell has authored an article titled "Harmony in the Hazardous Waste Universe: An Integrated Approach to Waste Elimination, Reduction and Cleanup Utilizing Systems and Statistical Analyses, and Design of Experiments," which he presented at the 1992 CHWMS Conference in Denver.

Mr. Bicknell received his JD from the University of Denver in 1986, and has a Masters Degree in Communications from the University of Texas (Austin) (1980). He received a BA in English (with Distinction) and a BS (with Honors) from the University of Colorado (Boulder) (1978).

Table of Contents

11 Facilitation Guidance

12 Developing a QFD Pilot Program

13 Case Studies and Examples

chapter one

Working toward an integrated systems quality paradigm

Nothing succeeds like success
Alexandre Dumas the Elder, 1854

1.0 Introduction

Alexandre Dumas probably was not referring to business methodology when he penned "Nothing succeeds like success" 140 years ago, but his words of wisdom may have more to them than first imagined when considering today's business climate. The issue of success has always been paramount in business, but the methodology in achieving success has always been subject to debate. Much of what defines success in business — profit, growth, market share, and longevity — has remained constant while the management and technical engines that drive successful businesses constantly change. Organizations are constantly searching for strategies to gain a competitive edge and strategies that worked in the past are constantly being called into question by the changing markets. For example, being large or controlling a large share of a market — while sound business strategy for the past decades — does not guarantee success in today's business climate as evidenced by the declining fortunes of General Motors, IBM, and American Airlines. Bigger is not always better when trying to be a lithe contender and react quickly to changes in market forces. Better is better. Given the growing volatility of markets and the increased competition for those markets during the past two decades, management and businesses are more willing to subscribe to new and sometimes unconventional strategies to cut costs, increase production, and get products to the market faster. In addition, there is really no such single entity as "American Business." Businesses in this country are as diverse as the people who manage and work in them. What may work for one business in achieving success may not work in another for a variety of reasons: size, management style, customer base, etc.

At the same time that conventional management styles have been called into question, businesses are beginning to question the "Yankee know-how" that has distinguished and nourished the American competitive spirit for most of the last 200 years. There is a growing feeling that American business independence may also be part of the recent difficulty the U.S. has had in competing globally. Reliance on finding ways to do things better and the independent nature of our businesses have not created an environment for cooperation among suppliers and manufacturers or functionally linked products. In addition, the general paradigm for a successful business often relies heavily on the personal vision of a founder or the "reactive" ability of the management to attack problems as they arise. The entrepreneurial spirit embodied by Steve Jobs (Apple Computers) and Bill Gates (Microsoft) are examples of how we view the successful American business, and increasing numbers of articles decrying an end to the American business dynasty point to our perceived inability to create more of these wunderkind.

But the reported "death" of American competition in business has been greatly exaggerated (our apologies to Mark Twain!). The entrepreneurial spirit that has sustained us in the past is the same spirit that is compelling companies to reach out beyond "business-as-usual" approaches to compete. Times of change are also times of chaos as a new formula is sought to replace an outdated one. For example, one trend that is emerging as the new benchmark for business operation is the shift from reactive to proactive management, and the emphasis on creating customer-focused organizations. This trend accounts for much of the philosophy of "new" management styles such as Total Quality Management (TQM) and has given rise to a new business terminology such as "change management" and "worker empowerment." "Quality" is another term that has become the new beacon by which businesses are to guide their ships through treacherous national and global marketing waters to reach their destination — success. However, let the buyer beware because, as Daniel M. Tellep, Chairman of the Board and CEO of Lockheed Corp. warns, "Not all quality efforts are created equal."[1]

Throughout all of this talk about quality, there is one common denominator that businesses use to measure any new approach or method: businesses will not use ideas or concepts that do not achieve results; and, more importantly, not just results but immediate results. American businesses, especially those with more entrepreneurial management styles, are notoriously impatient when it comes to implementing changes and subscribe to the "Microwave Management" philosophy — pop the technique into the organization and wait for a quick result. As one president of a company told the authors, "Don't tell me about *how* to do this, just tell me *what to do*."

This book is about a methodology that achieves results — quickly — and discusses both the how and the what; it is not a book on microwave management. Most things in life that are quick and easy are hardly worth the time to try; Quality Function Deployment (QFD) is not quick and easy but it is effective. However, in order to understand how this methodology functions, it is first important to understand what is so different about the marketplace now that requires new thinking.

1.1 Competitive strategies to enhance productivity

Every manufacturer or business knows that there are certain key concepts for being competitive. They include, but are not limited to, competitive pricing, product features or services offered, identified customer or market niche, and manufacturability of the product. Most businesses also have an intrinsic understanding of the market and the requirements to be competitive and to remain in business. However, the difficulty arises in how to accomplish the broad goal of being "competitive" in a methodical and repeatable fashion. This becomes more difficult today in a business climate where the time between product concept and product sale times is shortening. At the same time, the impact on a company of failed or less-than-successful products is increasing. Now, more than ever, a company needs to know and understand how to focus on the customer using an integrated systems approach to best assure continued success of existing product or services lines, to best evaluate potential success of new product concepts and to deliver those products in the most competitive fashion. *Fortune Magazine* puts it this way:

> Whether you sell $100 million planes or 79-cent pens, your buyers have changed enormously in the past few years. Their demands are lengthening; their patience is shrinking. Epochal shifts in the global economy have given them a sultan's power to command exactly what they want, the way they want it, when they want it, at a price that will make you weep. You'll either provide it or vaporize.
>
> Sound frightening? It is. But times of tumult are when fortunes are made. Think of it this way: The opportunity to understand the new consumer before your competitors do is probably one of the richest you will ever meet.

Business in the 1990s will prove to be more challenging than ever before. Competition will continue to become more global every day; trade barriers will continue to erode; and quality standards will become more stringent. Given the increasingly complex nature of the business environment, and the necessity for any successful business to be able to adapt quickly and efficiently to changes in market conditions, how will businesses be able to compete in the 1990s?

In general, a company in the 1990s must be *focused*. It must implement a long-term, systematic approach using integrated quality techniques. Too many companies have not combined integrated quality techniques with corporate plans or strategies, resulting in incomplete results and abandonment of quality techniques. Simple automation will not singularly answer this question. A 200-mph car can get you to your destination faster, but you do have to know where you are going first!

In a time of tighter budgets and pressures to perform, management must be able to use high performance where it counts. Resources must be concentrated to ensure superiority in areas that customers value the most. This means focusing on critical processes and using analytical techniques to validate performance. Concurrently, in order to focus on these resources, business must use methodology that enables them to identify critical processes and *accurately* measure a customer's pulse. Only after processes are under control and areas for improvement are identified can a business employ technology to automate processes and techniques to gain further advantage.

Training employees and management will also be a critical element in the successfully competing company of the 1990s. Most companies recognize the need for employees who are effective in performing their daily tasks and recognize the need for additional employee training to upgrade skills when new technology is acquired. However, many companies fail to truly analyze the current skill levels of their employees — whether they are management or nonmanagement — and to implement training programs to address any gaps in skills. In 1992, *Business Week* reported that professionals and executive managers are those who most often receive training (Figure 1-1). In contrast with those statistics, *Business Week* reported that in terms of who needs training, approximately 16 million workers need "new technology" training while 17 million workers needed basic skills (three "R's") training.[2] Figure 1-1 reveals that management or professional people receive the most job training, but the machine operator, transportation worker, laborer, or production/craft worker receives the least amount of training in terms of percentages of total numbers, despite their greater need for training. Why is this important? American businesses, overlooking the impact of undertrained workers on productivity and the competitive edge, are not putting the training dollars where they will achieve the greatest results. The article highlighted the potentially alarming statistic that there are 11 million employees in need of customer-service training. In a country where service industries are currently growing, are projected to continue to grow, and emphasis on the customer is paramount, customer-service relations can only become more critical and a key element for the customer-focused business.

The need for training is often an overlooked element in meeting quality targets and doing process improvement. An example of the need for proper evaluation of employee skills, and subsequent training to fill in the gaps, is given later in this book in Chapter 13. Training can also be a relatively low-cost investment that can yield astonishing returns (see Chapter 13, Section 13.1).

Despite the compartmentalization of many corporations and industries themselves, the business environment of the 1990s will require the establishment of alliances between contractors, suppliers, and government agencies. "Teaming" is taking the place of competition or self-centered negotiation in the vertical product chain. Finally, the efficient blending of technology with

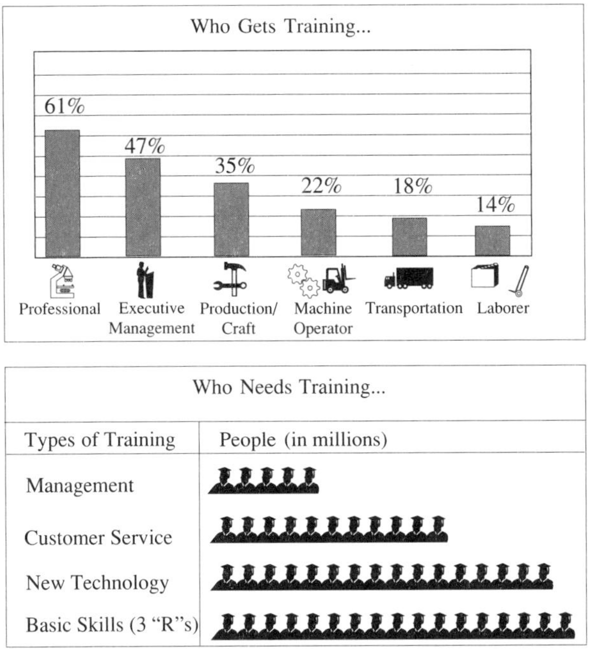

Figure 1-1 A critical competitive strategy is maintaining a highly skilled workforce. (From *Business Week*, p. 86, Special Bonus Issue, 1992.)

marketing knowledge will be a benchmark of a successful company of the 1990s. As products become more interdependent, especially in the computer and multimedia markets (e.g., the attempted Atlantic Bell/TCI merger),[3] being able to effectively tap existing resources and market to a system of interconnected products and services will become more critical. In many ways this is what "quality" is all about, and a key technique for implementing and obtaining quality in any organization is a method called Quality Function Deployment (QFD). QFD is a method for mapping and prioritizing customer needs and wants into product, service, and business operations to optimize organization performance and meet or exceed customer expectations.

1.2 *What is "quality" about Quality Function Deployment?*

Reams of paper have been consumed in recent years by various pundits, scholars, and watchers of American business lamenting the "quality crisis," which is the alleged lack of quality in American products and services when compared to foreign competitors. American quality-bashing has become so prevalent it is almost a cliche. The harsh reality is that the U.S. is losing ground to foreign competition. However, what some experts have failed to recognize is that in this ever-more-complicated world, any attempt to explain this reality by saying that Americans are simply not quality-conscious fails to adequately explain the forces at work in today's international and domestic marketplaces. While some Americans have been concerned with producing quality products, the complexity of modern technology has dulled the instinct to develop the methods to ensure continuing quality in existing and new products and services. We are trying to attack new problems with outdated quality skills. The term "quality" has evolved and American businesses need to use new approaches to meet the expanding "quality" problem.

As systems have become more complex, individuals have had to acquire specialized knowledge to support the new technologies. This increasing need for specialized knowledge results in problems translating technical and program knowledge across functional organizational boundaries. People are less able to communicate because of difficult, complex terminology. In addition, specialists tend to remain secluded in their areas of expertise causing significant differences in individual focus and emphasis during the development of a product or system. The clouded views of the system characteristics and requirements ultimately lead to successive reworks and program cost overruns. Often this is identified as a "lack of quality."

The most compelling evidence of the need to integrate design knowledge can be found in analyzing cost drivers in major manufacturing firms. In 1987 the typical American factory spent 20 to 50% of its operating budget to find and fix mistakes. In addition, as many as one quarter of factory personnel did not produce end items but reworked items that were not done correctly the first time.[4] Finally, statistics indicated that as much as 80% of all quality defects occurred either during the design phase of product development or as a result of purchasing policies that valued low price over quality. Even as late as 1991, industry executives were still quoting the cost of repair and rework at an unacceptable and expensive 15% of the cost of any product or service.[5]

Such statistics drove companies to examine the way they created new products and did business. Since the late 1980s companies have sought to reduce quality problems by eliminating them in the design phase. This has been a key competitive strategy that American businesses used to compete in the global marketplace. These "quality" failures were not unique to American manufacturing. The Japanese were experiencing similar problems in integrating design knowledge in the 1970s. However, our overseas competitors moved more quickly to analyze the source of the problem. Thus, to prevent costly rework, a disciplined method was developed through Japan's Total Quality Control Program (Japan) as a vehicle to force horizontal integration of design knowledge. The technique — now known as Quality Function Deployment (QFD) — reflects an integrated product design process but offers advantages by using a format and language understood by customers who become part of the system design process. QFD (Japan) provides a mechanism for multifunctional teams to capture the knowledge of all of the stakeholders in the organization to develop a full set of accurate requirements and fully integrate those requirements into system designs.

QFD is one of the tools that enables a company to focus on its customers and incorporate competitive principles. It is the keystone when building a program to achieve success and a tested approach to managing products, processes, and operations.

1.3 QFD — The road map to repeatable success

One of the problems in discussing any single method used to improve design, production, manufacturing, or processes is that it is usually discussed from a myopic viewpoint excluding other methods or elements that interact. By doing so the reader often is not given the whole picture first as a panorama within which the subject can be discussed. This book attempts to remedy that deficiency by giving the reader a road map for seeing how QFD applies in the context of an integrated systems and management viewpoint. The main focus of this book will be to present QFD with an eye toward implementation and facilitation of the methodology. How it fits into the overall scheme of product and process development and how those elements fit into the operational structure of the organization are important in understanding QFD's total affect. This is done in a step-by-step approach. Chapter 2 will give the reader the panoramic view to introduce the context for application of QFD. It builds the framework within which QFD functions the best. The goal is to show the reader how QFD is part of a larger management system and how it applies to both technical and business operations. Chapter 3 addresses QFD directly, giving a brief history of its development, its benefits, and the various methods of QFD itself. Chapter 4 covers the application of QFD and how to build the Core QFD Matrix central to its application in any scenario. Chapter 4 is not only written to give the philosophical construction of the Core QFD Matrix but it is also presented in a building-block fashion to help the reader see how it is used.

Chapter 5 discusses the interpretation and analysis of the Core QFD Matrix, and how to spot inconsistencies in designs and gaps in information. Chapter 6 discusses how to create a strategic plan from the matrix to implement the prioritization of elements. The authors believe this is often a missing step in using QFD. Once a matrix is constructed the next step, an integrated plan for implementation, is crucial for applying the knowledge gained in using QFD.

Chapter 7 looks at QFD and the robust design, focusing on specific applications to the development of products, and discusses what is beyond the Core QFD Matrix. It is important to understand how QFD relates to other technical and management tools to understand its role in the systems approach. In Chapter 7 the reader is shown the power of QFD in its ability to not only interact with other technical tools such as experimental design, functional analysis and statistical process control, but also how QFD is critical to the success and optimization of these tools. Chapter 8 takes a closer look at systems engineering and concurrent engineering in relation to QFD, still within the framework introduced in earlier chapters. Chapter 8 shows how systems or concurrent engineering and QFD are inseparable. Chapter 9 looks at the application of QFD to process improvement and business operations, key elements to the full implementation of a QFD agenda.

Chapter 10 goes beyond the paper application of QFD and demonstrates facilitation guidance. The reader will not only see how QFD should be implemented, but will also come away with an understanding of potential pitfalls and successful approaches. The goal of Chapter 10 is to give a practical background in the methodology. Chapter 11 gives examples of the perspectives involved in a QFD — executive, management, employee, and facilitator — because the authors believe that the key to successful application of QFD is understanding the organizational roles involved. Chapter 12 is a look at how a pilot program can be created based upon the authors' experience with QFD implementation in a variety of organizations.

Finally, Chapter 13 looks at some actual case studies of QFD implementation in product, manufacturing, and business operations examples. The case studies are presented after both the mechanics of QFD and the practical application of the methodology to help the reader better understand the individual applications in the case studies. The authors present some conclusions and reflections in Chapter 14 for the reader to consider.

Having given the reader the road map, the next step is to understand the total quality paradigm in which QFD attains its optimum efficiency.

1.4 Chapter summary

1. To stay competitive in the global marketplace, management methods are becoming more proactive employing a long-term systematic approach using quality techniques to:

   a. Focus the entire organization on the customer;
   b. Optimize resources to get high performance where it counts by:

      - Concentrating resources to ensure superiority in areas customers value most
      - Improving critical processes
      - Using analytical techniques to validate performance
      - Employing technology to automate processes (after processes are under control) to take advantage of productivity gains in the streamlined operations;

   c. Provide training to maintain a highly skilled work force;
   d. Establish alliances between contractors, suppliers, and government agencies; and
   e. Blend technology with marketing knowledge efficiently.

2. QFD is one of the tools that enables a company to focus on its customers and incorporate competitive strategies.

3. QFD is a method for mapping and prioritizing customer needs and wants into product, service, and business operations to optimize organization performance and meet or exceed customer expectations. It reflects an integrated product development process but offers advantages by using a format and language understood by customers who become part of the system design process.
4. QFD provides a mechanism for multifunctional teams to capture the knowledge of all of the stakeholders in the organization to develop a full set of accurate requirements and integrate those requirements into systems designs.
5. The focus of this book is on the implementation and facilitation of QFD into your organization to make sure you are on the road to repeatable success.

chapter two

Repeatable Success Method™: A systems integration quality paradigm

Everyone can enjoy random success, but one is advised not to build a career on it!
The authors, 1993

2.0 What is quality?

The above quotation is not meant as a Machiavellian commentary on today's business climate, but rather as a corollary to the Dumas quotation from Chapter 1. It points out that the process of achieving success can be as mysterious as determining the cause or causes of a failure. We often do not analyze the course of success as minutely as the root causes of failures, both in our personal and business lives. As a result, it is an ill-advised course of action to pursue "success" as merely a series of uncontrolled random events — a laissez-faire attitude that "if it worked once, it will work again!" An often-overlooked function of Total Quality Management, for example, is to recognize what a company is doing right, as well as what an organization can improve. Thus, not just success, but *repeatable* success is the key to progress and long-term benefits. But how do we achieve "repeatable success" and where do we fit in QFD?

Over the last 10 years the business of doing business has changed immensely. One of the key ingredients of a successful business formula is a new focus on quality systems: quality in the products we produce; quality in the way we manage our businesses; quality in our everyday activities. But as single-minded as we are about the need for quality, we are just as divided on how to obtain that quality, or even concisely defining quality.

What is quality?

Quality has been described in many ways. Joseph Juran describes quality as consisting of two concepts:

- One form of quality is income oriented, and consists of those features of the product which meet customer needs and thereby produce income. In this sense higher quality usually costs more.
- A second form of quality is cost-oriented and consists of freedom from failure and deficiencies. This higher quality usually costs less.[6]

Juran's definition is focused on the function of the product and its application or intended application. Taguchi similarly measures the quality of a product in terms of the total loss to society due to functional variation and harmful side effects. The *American Heritage Dictionary*, Second College Edition, has among its definitions of quality the following:

> 1. The essential character of something; nature. 2. a. An inherent or distinguishing characteristic; property. b. A personal trait, esp. a character trait . . . 3. a. Superiority of kind . . . b. degree or grade of excellence . . .[7]

The first three definitions have some relevance to what most businesses may consider in a definition of quality for their goods and services: a unique characteristic, a product's "nature," and a degree of excellence.

Another definition proposed by the Ernst and Young Quality Improvement Consulting Group is

> 1. Conformance to specifications — quality is defined by the relative absence of defects.
> 2. Meeting customer requirements — quality is measured by the degree of satisfaction with a product's characteristics and features.[8]

Here we see the definition of quality including specifications and the importance of what the customer wants. Quality is defined by the degree of customer satisfaction. Almost every writer who has talked about quality has his or her own definitions of exactly what it means. Definitions, though, have the almost unavoidable tendency to define by exclusion or limitation. The authors believe a better "definition" of quality is to describe it as an outcome or goal: "quality" is staying competitive through continuous improvement, being profitable (or reducing costs for nonprofit organizations and governmental entities), and being able to repeat success by having products, processes, and operational systems under control. The importance of "quality" is not in its meaning as a term, but its value as a system. As Thomas J. Murrin, former Deputy Secretary, U.S. Department of Commerce, said, "After 36 years in American Industry, I've concluded that quality improvement is the *only* business activity that simultaneously satisfies customers, motivates employees, comforts investors, teams suppliers and wins media and public approval . . . What other effort might simultaneously gain the same results?"[9]

Quality is being seen more and more by business as a system involving every aspect of the business, whether defined as a trait or as an outcome.

However, the tough part of quality is not **recognizing** its power or effect, but how **to achieve** it on a daily basis in every aspect of a business.

2.1 *The Repeatable Success Method™*

Much has been written on quality systems and the utilization of such systems in business. The most common umbrella for describing these systems is Total Quality Management (TQM). TQM has been used to describe methods of such well-known quality experts as Deming, Juran, Crosby, and Taguchi and also has been used to include quality techniques or tools such as QFD, Functional Analysis (which is a subset of Value Engineering), Experimental Design, and Statistical Process Control (SPC). Other umbrella terms that employ some of these techniques are

- Concurrent Engineering (which focuses more on the design and manufacturing element of quality integration)
- "Re-engineering" (which is a phrase of more recent development coined by Michael Hammer and often associated with company downsizing and business operations).[10]
- Process Improvement (which deals with the analysis and improvement of processes within a business environment).

The point here is not so much to define what the methodology is called but to discover what collection of management and analytical tools produces repeatable success, and how to institute those methods efficiently in business.

Recently, some methodologies have suffered a reversal of fortune, such as TQM. TQM has fallen from favor with some members of the business community because of a perceived failure to deliver as promised.[11] Some business journals have printed articles about the demise of TQM and its progeny; but the authors believe this more accurately reflects the inability of some companies to make the fundamental changes required, rather than the invalidity of the processes. In addition, the authors have seen first-hand in their consulting the impact of incomplete adoption of quality techniques without a strategic plan incorporation. For example, buying into automation without process analysis only results in making bad parts faster.

The failures or dissatisfaction with quality programs can usually be traced to one of several problems that may or may not have any connection with the program itself. Failure of a quality program is often blamed on the method, or alleged failure of the employees and workers to accept change. These are red herrings and usually conclusions reached by management. There is certainly a strong interest by American workers to want to perform well and embrace quality, even after programs have failed. A case in point arose at the 1992 TQM Conference in Colorado Springs, Colorado. The

Annual TQM Conference attracts the military, government agencies, industrial companies, high-technology companies, and aerospace companies from all over the U.S. The authors were presenting a workshop on basic QFD and requested topics from the audience of about 80 people. An engineer from a large aerospace company wanted to know how to restart a quality program that had been tried and abandoned several years earlier at the company. The authors thought this type of problem for a QFD analysis was not particularly interesting to the rest of the participants, so alternative problems were solicited from the group and a vote was taken to select the problem topic. To the surprise of the authors the "restarting a quality program" (referred to by most of the crowd as restarting TQM) won by a clear majority. This was only the first indication to the authors that interest in TQM and quality programs was strong, but successful implementation was lacking. The second indication came on the second day of the conference when a new audience also was heavily in favor of working on restarting a quality program. Debriefing some of the conference participants after the QFD sessions yielded some interesting insights:

- Quality programs were, in general, favored by many employees and management.
- Excitement and expectation was generally high at the beginning of each program.
- Quality programs, usually under the guise of TQM, were initiated by working on developing "teams," including cross-functional teams. However, repeatedly, team members expressed frustration at the lack of guidance, direction and structure for interaction. Consequently, interest waned and it was soon business as usual.
- With some companies, quality actually worsened because teams became disorganized, schedules were missed, and chaos ensued.
- "Empowerment" was often preached by management; but workers felt that if they made decisions not blessed by management, they risked a negative impact on their careers.

One of the most insightful, and disturbing, results of this experience was discovering that several of the individuals who wanted to restart quality programs came from companies heavily committed to Baldrige criteria with active TQM programs. Either communication from employees to management was being blocked or the efforts of the companies in quality were not perceived by employees as quality management. Actually, this should not come as a surprise to anyone who has attempted to implement any new program at a company; but how can a company choose the right path to implementation and avoid the pitfalls of an improperly or poorly structured quality initiative?

Based upon existing literature, experience, and facilitation of integrated quality techniques, the authors have developed a common denominator for approaching quality methodology. That denominator is knowing what drives quality and knowing how to repeat and improve on that knowledge. It also involves a methodical approach to implementing a quality paradigm from a small pilot program to organization-wide implementation. The authors call that common denominator the Repeatable Success Method™. The Repeatable Success Method™ or RSM™ describes the group of integrated quality tools and methods that incorporates QFD as a central feature. The authors have selected this term because, like the goal-oriented definition of "quality" given above, it is more descriptive of the goal and the process of instituting quality practices, and looks at quality methods from a hierarchical and integrated perspective.

As mentioned in Chapter 1, the power of QFD is in its relationship to a whole. Because most published examinations of QFD focus on the use of the tool in a specific product or service problem-solving context, it is the purpose of this book to analyze QFD as part of an integrated quality approach so that the reader understands the interplay of quality methods and how QFD is the key element of the methodology. Without understanding how QFD relates in an overall quality paradigm, it becomes more difficult to implement QFD at either the program or company level.

The authors offer the following definition of RSM™:

> A systematic approach that combines integrated quality techniques to enable people, processes, analytical methods, and information systems to efficiently and effectively produce products and provide services with repeatable and improvable quality attributes. RSM™ also provides traceability of daily operations back to the corporate vision and customer expectations with the means for monitoring and enhancing productivity.

The above definition not only incorporates the idea that quality is a key consideration, but also that the overriding process must be repeatable. RSM™ represents an integration of four key elements: people, processes, analytical methods, and information systems (or automation).

2.2 The elements of the Repeatable Success Method™

RSM™ places quantitative and qualitative quality elements in a hierarchical framework for a better understanding of the interplay among the various elements. RSM™ can be viewed as a "pure" process methodology in that it can apply across the board in various industries, whether the product is a complex spacecraft, a simple consumer good, or a service. Figure 2-1 illustrates how RSM™ puts all the elements of quality methods together. As mentioned above, the four elements of RSM™ are (1) People; (2) Processes; (3) Analytical Methods; and (4) Information Systems. The four elements are viewed as building blocks because each "builds" on the previous elements. The functions of these elements are described below.

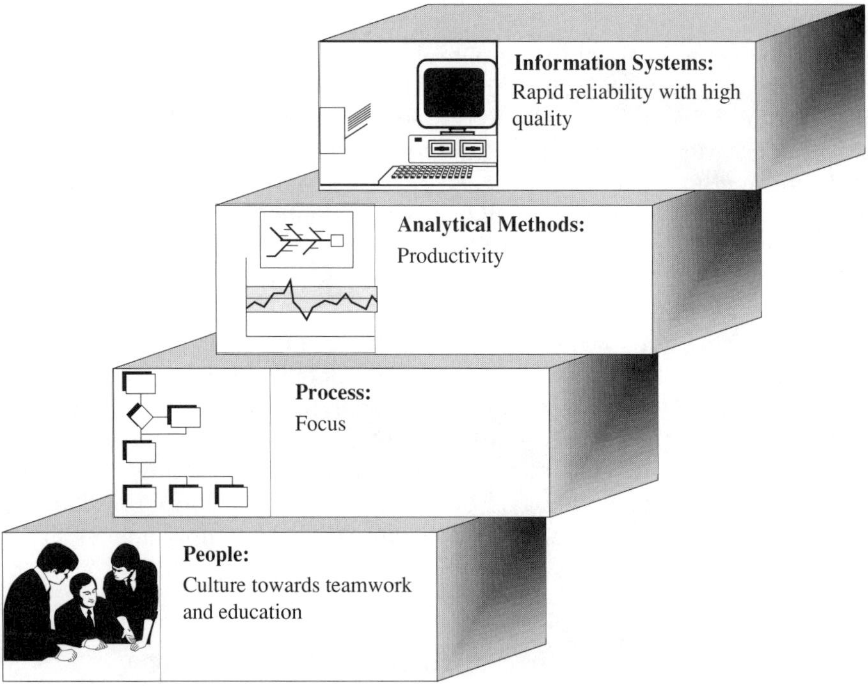

Figure 2-1 Elements of the Repeatable Success Method™.

Figure 2-2 expands upon Figure 2-1 by detailing the elements that can be found in each of the four building blocks. The list of elements within each building block is not meant to be all-inclusive, but instead to be representative of the type of tools, training or philosophies that would apply in each case.

2.2.1 People

In order to institute RSM™, or for that matter any quality-related improvement program, the first element always involves people. Without creating a culture educated in the techniques of teamwork and the philosophy of empowerment, all of the other elements become infinitely more difficult to institute. This step is referred to often as traditional TQM, because it involves the building of a teamwork structure, such as the Quality Circles used by the Japanese. However, some companies have confused unguided teamwork and company boosterism with the completion of a TQM program. This usually results in a lot of motivated people with nowhere to go and ultimate disillusion with the process. Therefore, a shared vision or destination must go hand-in-hand with team training, productivity improvements, and development of new products and services. Throughout the book the authors focus on not only the mechanics or theory of QFD, but also on identifying the people elements involved at that particular stage. Chapters 10 and 11, dealing

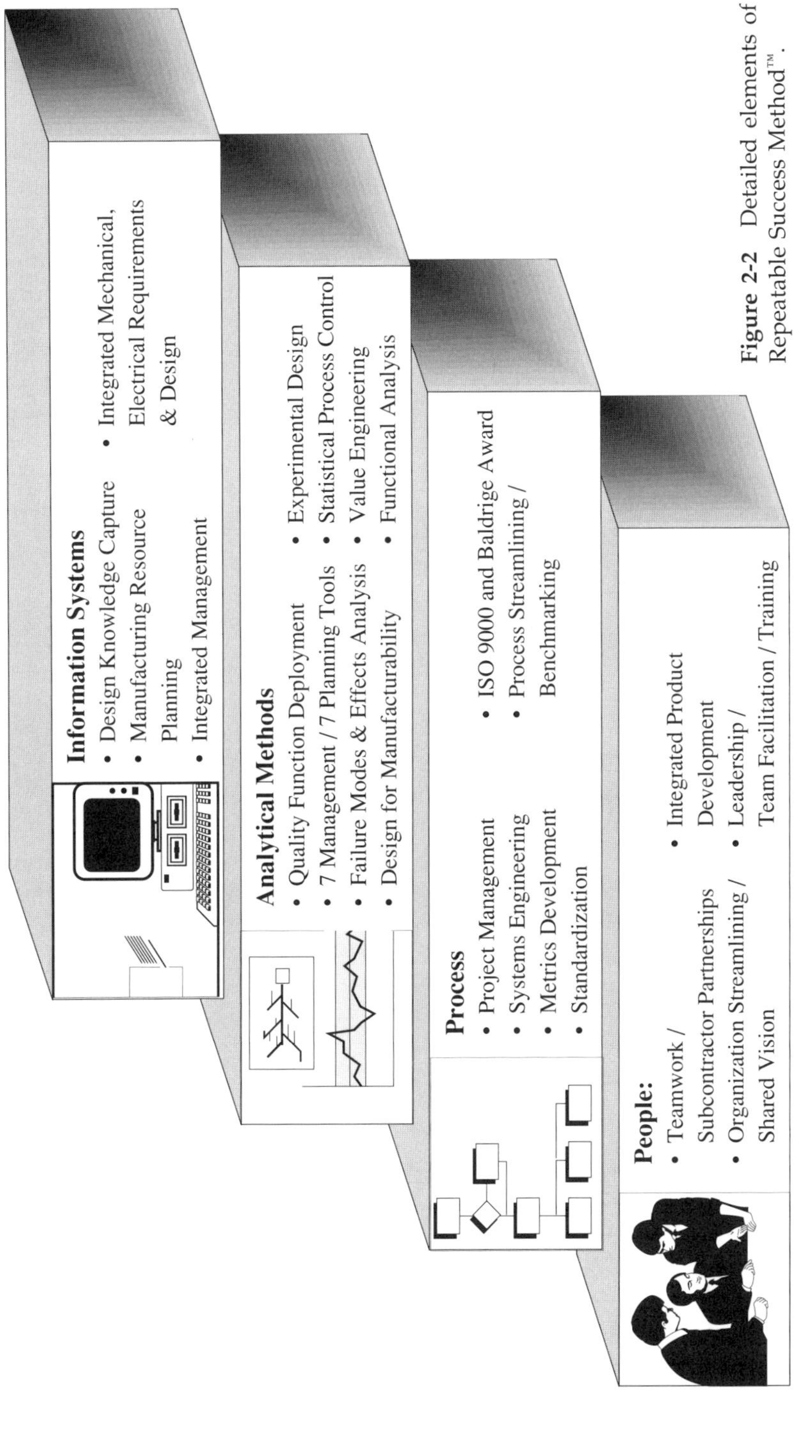

Figure 2-2 Detailed elements of the Repeatable Success Method™.

with facilitation and participant perspectives, focus heavily on the management and involvement of people in the process. One key point that should never be forgotten when doing QFD, or implementing any quality-related program, is that these methodologies are there to assist *people* in making their companies more effective and competitive. Quality initiatives are not an end in themselves.

2.2.2 *Process*

Once the concept of cooperation and teamwork has been put in place, the next logical step is to concentrate and focus on processes. This element is typically characterized by standardization and documentation of processes using guidelines such as the Baldrige Award criteria, ISO 9000, and process re-engineering. Benchmarking and process streamlining would also be utilized in process focusing. This element builds on the basic element of people and culture because once a team has been constructed — or a culture oriented towards change has at least been presented to the organization — benchmarking current processes become easier and more effective. This is a natural result of the team working toward a common goal and having focus.

2.2.3 *Analytical methods*

Analytical methods are then used to perform such functions such as process improvement and productivity enhancement. This is where QFD plays its major role. Not only is QFD the building block of all the analytical processes, but it can also be the starting point for an RSM™ program. RSM™ application cuts vertically through the four elements, making it an ideal catalyst for a pilot plan implementation, as is discussed later in this book. In addition to QFD, other tools — Functional Analysis, Experimental Design, and SPC — are used to focus processes and to assist the people in moving toward a shared vision. These are some of the analytical tools that allow measurement, benchmarking, and control of a quality program from vision or concept to process or parts characteristics. The combination of using QFD in conjunction with these other techniques yields a streamlined, integrated approach that is much more difficult to attain without the use of the QFD analysis matrices for prioritization.

2.2.4 *Information systems*

Finally, once the analytical methods have been used to best improve processes and prioritize improvements, technology advancements and information systems can be selected to automate the processes and yield additional productivity enhancements. Too often, quality-improvement programs start with automation prior to a careful analysis of an organization's current structure and process strengths and weaknesses, which results in the selection of a technological solution that may not be appropriate for the organization and does not obtain the "biggest bang for the buck."

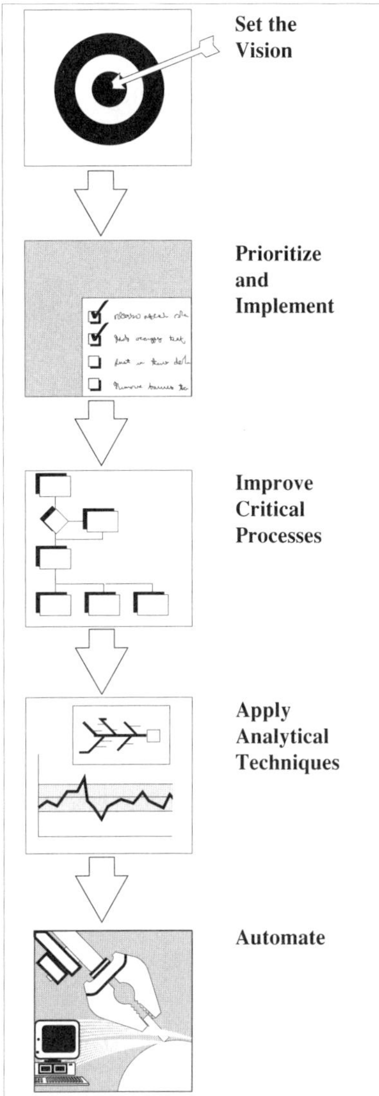

Figure 2-3 The implementation order of RSM™ elements is critical because it methodically establishes a traceable road map for success.

2.2.5 Order makes a difference

The order in which quality methods are implemented plays a key role in the success of any program. Figure 2-3 illustrates the proper order for implementation of RSM™ elements. Just as Figure 2-2 illustrates how people, process, analytical methods, and information systems combine to "build" an approach, Figure 2-3 demonstrates the methodology ensuring proper implementation to achieve the best results.

Critical to the success of any quality plan is first **setting the vision**. What is the goal of the implementation of quality? Is it company-wide or only program-wide? These questions seem obvious; but in the authors' experience, companies sometimes start a program of TQM or quality implementation with no clear goal or expected outcome. The inclusion of the first RSM™ element — people — is central to brainstorming ideas to determine the vision that will carry the quality program implementation through to its conclusion. In essence, this first step creates the destination as well as the map.

After a vision has been created, the next operation is to **prioritize and implement** that vision. This is done in a variety of ways such as formally brainstorming ideas (which is discussed in more detail later), and using methods such as QFD to prioritize the steps in the quality program. People, data, and processes are used to give direction to the vision, and provide a plan for integrating activities.

Improving critical processes is the third step to implementing the RSM™ elements. In this step processes are mapped and improved. Critical processes are identified and steps that do not contribute to the end result are eliminated. Often, this is the stage where re-engineering is applied; but without the first two steps — vision and prioritization — process improvement can miss the mark. Any process can be broken down and analyzed for improvement; but without knowing the goal of the improvement and which critical processes to analyze (as they relate to customer expectations and corporate objectives), a great deal of money can be spent without significant benefit. A major consulting firm specializing in re-engineering found that some clients who had engaged in process re-engineering had huge improvements in processes — 60 to 80% reductions in cost and cycle time — but only modest improvements at the business-unit level. The reason: the changes did not matter to the customer.[12] Improvement without analyzing goals, prioritizing costs, and examining potential impact can be misleading or ineffective at best, and potentially debilitating depending on the cost of the "improvement." A computer company that sought to improve the expertise of its sales staff based on the assumption that such training would be wanted by the customer spent tens of millions in a re-engineering effort. The customers did not care about the expertise of the sales staff — they wanted lower prices.[13]

After critical processes have been improved or targeted for improvement, the next step is to **apply analytical techniques**. Knowing what to improve and taking steps to conduct the improvement positions an organization to apply data gathering and measurement techniques such as SPC to see if improvement is occurring and to benchmark the improved process. Applying analytical techniques produces the greatest benefit at this stage of a quality effort because the data being gathered is tied directly to the goals defined by the effort and is focused on processes that have been prioritized and identified as critical to the product, service or company operation.

The last step is **automation**. The quality program can now select new methods of automating processes, production, or service because the organization knows that it has directed its efforts towards a specific goal and worked on those elements that will be capable of the "biggest bang for the buck."

Critical to Figure 2-3 is understanding that the four elements of quality methods play a role in each of the five steps for implementation. The authors have found that many "failed" TQM programs have not been accepted or have not been successful because the order of implementation was not followed, or elements were left wholly unaddressed. Figure 2-3 is a road map for successful implementation of a quality program, whether it is at a program level, or at a company-wide level.

2.3 *QFD is the heart of RSM™*

Having introduced the principles of RSM™ and how it operates as an integrated quality paradigm, the next step is to show how it can be incorporated into an organization, and where QFD fits into a management model.

Strategy provides the engine behind any company or entity, and successful implementation of strategy yields success. But strategy does not exist in a vacuum. The first step the authors use in implementing RSM™ is **setting the vision**, and one of the best ways to do that is to create a strategic management model that incorporates the principles of RSM™. The importance of a company knowing where it is going and how it is going to get there cannot be overlooked or underestimated. An organization reflects the attitude of its management and if a clear plan is not apparent or communicated to the organization, the chances of it being implemented are minimal. In 1992, Charles A. Aubrey, II, then-president of the American Society for Quality Control (ASQC), said:

> Genuine progress requires that senior management take a strategic view of quality: setting priorities, identifying what's most critical to the success of the enterprise, and focusing improvement efforts on the customer. These are decisions that only top management can make.[14]

Although Aubrey's words ring true, how does a company incorporate a "strategic view of quality?" The key lies in how a company incorporates its vision into its daily operations. Figure 2-4 illustrates the RSM™ strategic management model showing where QFD is used to ensure that the mission (vision) of a company is translated in action via products and operations, and how feedback makes the model a dynamic one. The authors have found time and time again that although companies spend a great deal of effort on corporate strategy and planning, much of that effort is wasted or ineffectual without methodology for implementation. QFD is the heart of RSM™ because it provides that link between strategy and implementation. The elements of the model use QFD to track the company mission into the daily operating strategy using measurable targets. How QFD does this will become clearer in later chapters.

The elements of the RSM™ strategic management model are fairly straightforward:

- Mission — The mission of a company is usually described as a unique purpose. It also describes the products, services, markets, and technological areas of emphasis.
- Company Profile — The Company Profile lists the company assets, both tangible and intangible, including resources and strengths. The Company Profile usually mirrors the Mission in some way. Company weaknesses or areas of challenge are also listed here.
- Operating Environment — The Operating Environment contains a description of the limiting factors on the company, such as geographical region or physical size. It can also include conditions and choices that affect strategic operations, both controllable and uncontrollable.
- Long-Term and Short-Term Objectives — Company objectives related to time are usually a feature all organizations have in writing somewhere. However, they should be linked directly to the Mission, Company Profile, and Operating Environment. If a company's objectives are outside the Mission or beyond the Operating Environment of the organization, then these objectives should be re-examined.
- QFD — QFD offers the bridge between the above elements and the daily operations of the company by incorporating them into development of new products and services and operational streamlining or "business re-engineering." It does this by mapping the customer requirements reflected in the Mission and Objectives into the products and operations.
- Institutionalize — By this the authors mean that the successful methods that define doing quality become part of the daily operation of the organization.
- Feedback and Control Mechanism — For all management systems there should be a feedback and control loop that gives input back to the Mission, Profile, and Objectives for re-evaluation. This keeps the model a dynamic one and creates the mechanism for additional change and improvement.

To help in understanding the elements of the model, and to provide an example that will be further developed throughout the book, the authors have selected the development of a video game. Most of us, either through personal or vicarious experience through our children, have had some exposure to the video game craze that has swept the country. Without debating the merits of the pastime, video games offer an excellent example of a product that is fairly technical in nature, but also reaches a mass audience in an industry that is highly competitive and constantly changing. The product or video game that we will be developing will be for the mythical company, Compulsive Video Mania (CVM), whose motto is "Play our games once and

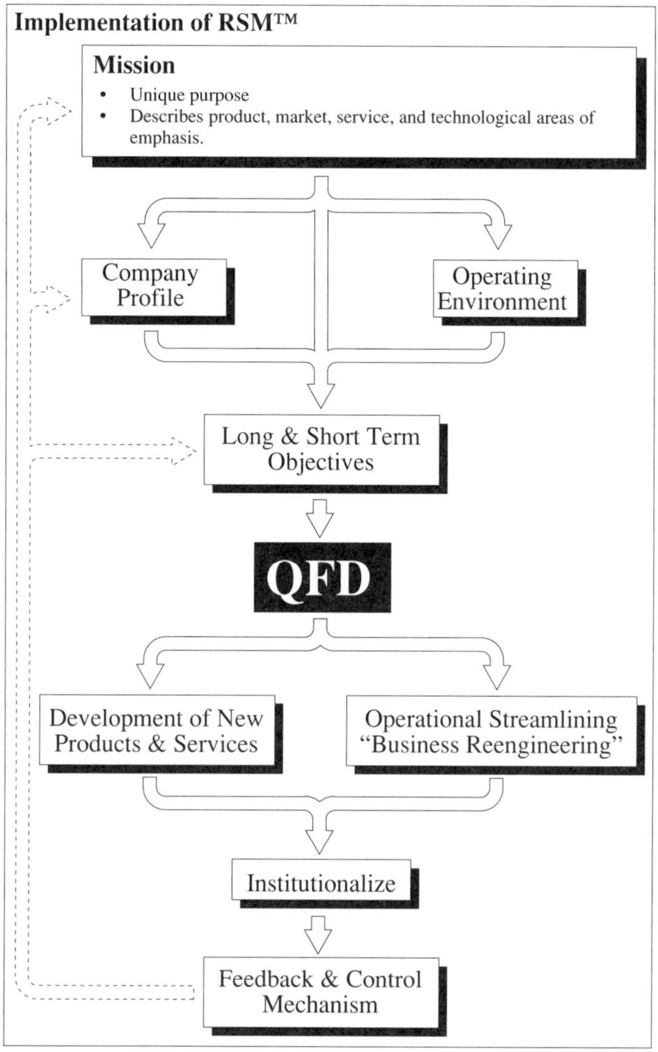

Figure 2-4 QFD acts as a bridge between the vision of a new product and service and the streamlining of critical product and service features and the associated business processes.

you'll be hooked!" To show how a company such as CVM could utilize the model, Figure 2-5 lists elements pertaining to each of the sections of the RSM™ strategic management model. The CVM example strategic plan does not contain the detail that would go into such a plan, but offers specific types of data that could be found in each of the sections. The reader should take a few moments to examine each of the sections of Figure 2-5 and how the mission of CVM tracks into each of the other plan areas, even down to a mechanism for feedback — prototype assessments.

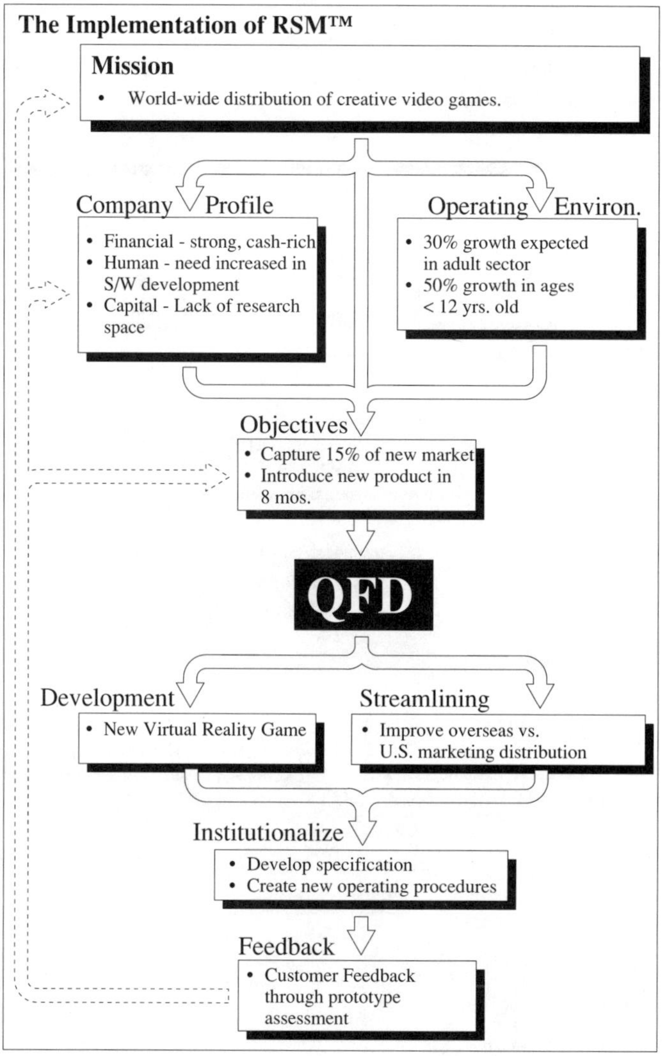

Figure 2-5 An example of bridging the vision of a new product and the streamlining of critical product and service features and the associated business processes for a video game.

The authors have used the RSM™ plan to begin the implementation of a pilot process improvement plan with QFD as a central feature. It can also be used at a program level to focus the team on its immediate mission and give the vision necessary to act as the glue to hold the process together.

Now that the reader has seen how QFD fits into the overall philosophy of an integrated quality paradigm, the next step is to introduce the analysis method itself in Chapter 3.

2.4 Chapter summary

1. The authors have described "quality" as staying competitive through continuous improvement, being profitable (or reducing costs for non-profit organizations and governmental entities), and being able to repeat success by having products, processes and operational systems under control. The importance of "quality" is not in its meaning as a term, but in its value as a system, and how to achieve it on a daily basis in every aspect of business.
2. The authors have developed a systematic approach, RSM™, to efficiently and effectively produce products and provide services with repeatable and improvable quality attributes. Critical elements used in the systematic implementation include people, process, analytical methods, information systems, and automation.
3. Order makes a difference in implementing RSM™. At a high level the steps are simply:
    - Set the vision;
    - Prioritize and implement;
    - Improve critical processes;
    - Apply analytical techniques; and
    - Implement information systems and/or automate.
4. QFD is at the heart of RSM™ and the strategic quality management of an organization, bridging the gap between corporate strategic goals and the development of new products, new services, and business operations. QFD provides strategic mapping, prioritization, and development of performance measures based on customer needs and corporate objectives.

chapter three

What is QFD?

I think it is impossible to really innovate unless you can deal with all aspects of a problem. If you can only deal with yokes, or whites, it is pretty hard to make an omelette.

Gene Amdahl, founder of Amdahl Corp., from *The Change Masters.*[15]

3.0 The genesis of Quality Function Deployment

Quality Function Deployment is not a quality initiative in itself. It is a systematic means of analyzing customer requirements and deploying them into product, service, and business operations. The genesis of the approach goes back to the 1960s.

The term QFD is derived from six Chinese/Japanese characters that translate as follows: "hin shitsu" (qualities); "ki no" (function); "ten kai" (development, deployment, or diffusion).[16] The translation does not correspond exactly, as is the case with many translations from Asian languages into English. However, the concept is to deploy or diffuse all the necessary product functions through the use of design qualities or performance requirements. It is through this process that "quality" is maintained through a system.

The genesis of QFD is traced to Dr. Yoji Akao, who was chairman of the QFD Research Committee of the Japan Society for Quality Control (JSQC) from the early 1970s until the committee's disbanding in 1987 upon the successful completion of the promotion of QFD in Japan.[17]Akao has written numerous QFD articles and is credited with being the first to bring the concept of QFD to America through an article published in October of 1983 in *Quality Progress*, the monthly journal of the American Society for Quality Control (ASQC) . Akao was involved with the first large-scale applications of QFD at the Kobe Shipyard of Mitsubishi Heavy Industries in the early 1970s. After the introduction of QFD in America by Dr. Akao and others, the first QFD case studies began to appear in the mid-1980s. Since that time, QFD has been used by a wide range of companies in an equally varied array of industries including 3M, Eastman Kodak, Martin Marietta Aerospace Group,

Westinghouse Corporation, AT&T, Bell Labs, Digital Equipment Company, Ford Motor Company, and Proctor and Gamble to name but a few.

QFD is a systematic approach mapping the customer's needs into definable and measurable product and process parameters, using matrices and other quantitative and qualitative techniques. There are several approaches to QFD and although the basic elements of QFD are not difficult to comprehend, application of QFD can be an exhaustive and tedious process. When used correctly, QFD provides a powerful tool for consistently communicating and transferring information through highly technical and adversarial or competing processes. This leads to focused technology development, improved services and processes, and improved business operations. QFD prioritizes conflicting system requirements and helps to resolve the divergent needs of the parties involved, which include the customer or customers and the various company operation and function representatives.

3.1 QFD — The fast lane to competitiveness

QFD is often cited as the key to getting high performance where it counts by focusing an organization on the customer's needs and prioritizing those needs to produce competitive products or services. The benefits of QFD in removing quality problems that would appear in the process or manufacturing stage of a product are undisputed. This is accomplished by prioritization and analysis in the design phase. This is where QFD produces its most dramatic results. However, QFD is also a powerful method of analysis for process improvement, service improvement, and operational streamlining. Companies as diverse as Goodyear Tire & Rubber Co. and Haworth, Inc. (a furniture manufacturer) have seen the benefits of using QFD in both meeting the customer's needs and anticipating those needs in new products.

At Goodyear, QFD was used in designing a new tire series for commercial trucks called the 300 Series Unisteel radial tire. Goodyear enlisted a multidisciplinary and cross-functional team of engineering, design, materials, compounding, retreading programs, testing, field engineering, sales, marketing, and manufacturing to use QFD in the computer design and testing of the product. The result was a tire that provided as much as 15 to 30% improvement in tread life incorporating more than 50 changes made in Goodyear truck tires since the mid-1980s.[18]

For Haworth, a cross-functional work team employed QFD to develop a new line of furniture called PREMISE.[19] Through the QFD matrices, the team identified conflicts and addressed them before they became costly problems. In the process of developing a detailed list of customer needs, Haworth employees targeted a key customer concern: compliance with the Americans with Disabilities Act (ADA).[20] This gave them an advantage over their competition by designing office furniture to help employers accommodate disabled workers.

The results at Goodyear and Haworth support a growing realization that although using QFD may require additional time and effort in the up-front

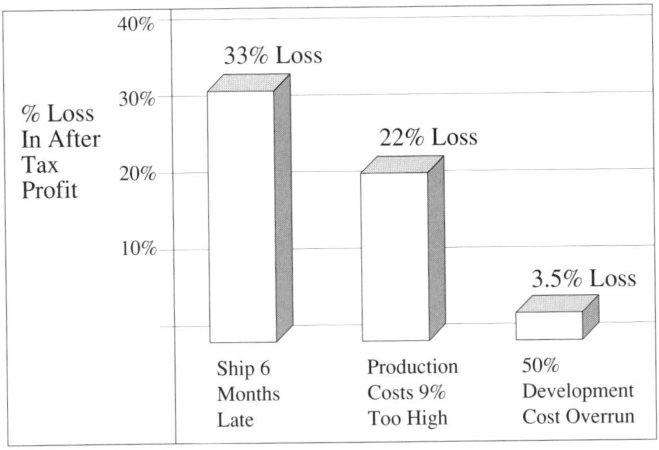

Figure 3-1 Sensitivity of profits over product life (assumes 20% growth rate, 12% annual price erosion, 5 year life). (From Shina, Sammy G., *Concurrent Engineering and Design for Manufacture of Electronics Products,* Van Nostrand Reinhold, New York, 1991.)

design of a product or service, the payoffs can more than compensate. For example, statistics have shown what most companies know to be an axiom: shipping a product 6 months late dramatically affects profits reflected as a profit loss in after-tax dollars (see Figure 3-1). However, what Figure 3-1 also demonstrates is the relatively minor impact of a 50% development cost overrun for a new product in after-tax profits. If a company ships a product 6 months late, there can be a 33% after-tax profit loss due to market loss, whereas a 50% *development* cost overrun yields a net loss of only 3.5% in profits. This fact emphasizes that it is better to pour more resources into the front end of a product design in order to meet or beat a ship date than to invest in correcting errors or quality problems in the product during the production stage that could result in shipping delays. This is especially evident in industries where product life cycles are having a shorter and shorter span, such as in the electronic or computer industries. Figure 3-2 illustrates that engineering change costs increase logarithmically during a product life cycle, once again demonstrating that changes late in the development cycle obviously have the most impact on the cost of product development. Figures 3-3 and 3-4 demonstrate the leveraged affect of the design phase in a specific industry in terms of total product cost. These figures underline the necessity of utilizing all the resources available to a product design team to come up with not only a marketable product, but also one that avoids costly changes in mid-production or in attempting to eliminate after-shipping quality problems. In a marketplace that is increasingly competitive and has consumers who are becoming more demanding of not only quality but also value, companies must be on the mark the first time. What is disconcerting is the fact that statistics are showing companies to be missing that mark in a major way:

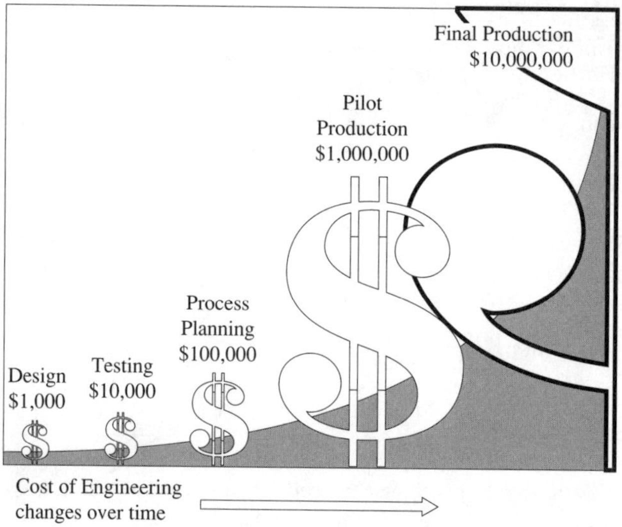

Figure 3-2 Engineering change costs increase logarithmically during the production life cycle. (From *Business Week,* April 30, 1990.)

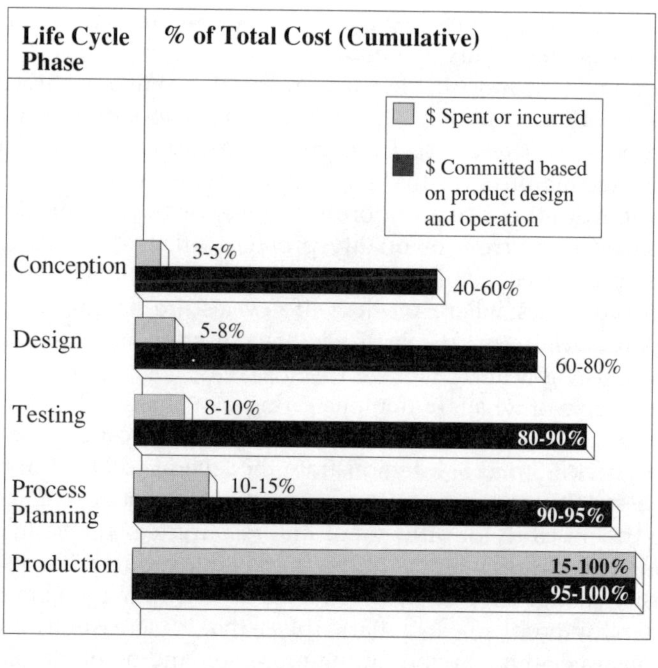

Figure 3-3 Leveraged effect of design phase in the electronics industry. (From *Business Week,* April 30, 1990.)

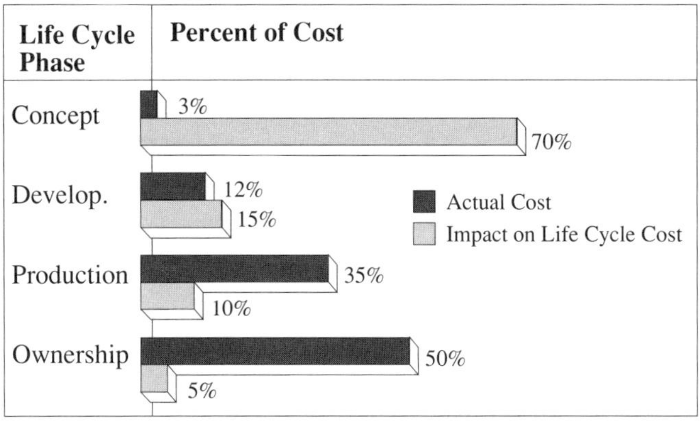

Figure 3-4 Leveraged effect of the design phase on life cycle costs. (From *Military Electronics,* August, 1980.)

> Group EFO, a research outfit in Weston, Connecticut, figures that in 1992, 83% of new products failed to clear the performance hurdles set by their developers.[21]

If companies are focusing on the customer and designing requirements to meet those needs, than either the companies have misjudged the consumer or are flat-out not producing new products to satisfy those needs. This is in spite of companies spending $3.5 billion with the 50 largest research firms to learn about their customers.[22] In a poll taken by the Center for Strategy Research in Cambridge, MA, one company executive made a comment that may be right on the mark: "Market research can be extremely valuable, but most of what is done right now is so poor as probably to be fraudulent."[23] Assessing customer needs is critical to a product's success, and accurate assessment of customer needs is critical to the success of a QFD analysis.

If customer needs are correctly assessed QFD provides a method to analyze both product and process development simultaneously, which decreases or even eliminates the chances of costly design changes and missed schedules. Figure 3-5 demonstrates the concept of concurrent engineering, the basic principle of which is to iterate and integrate the design of a product and the design of its manufacturing and logistics processes with specific focus on achieving a lower production cost, shortest schedule, and robust quality. The old development model of first doing the product development, then doing process development to get a verified design, and then going into production is no longer viable in today's competitive global market. Unfortunately, that model soon slipped into the "concurrency" model in which product development, process development, and production overlapped each other. This method yielded the most "in-production" changes because the design and the process definition were not solidified in a verified design until well into the production cycle, causing expensive changes. Concurrent

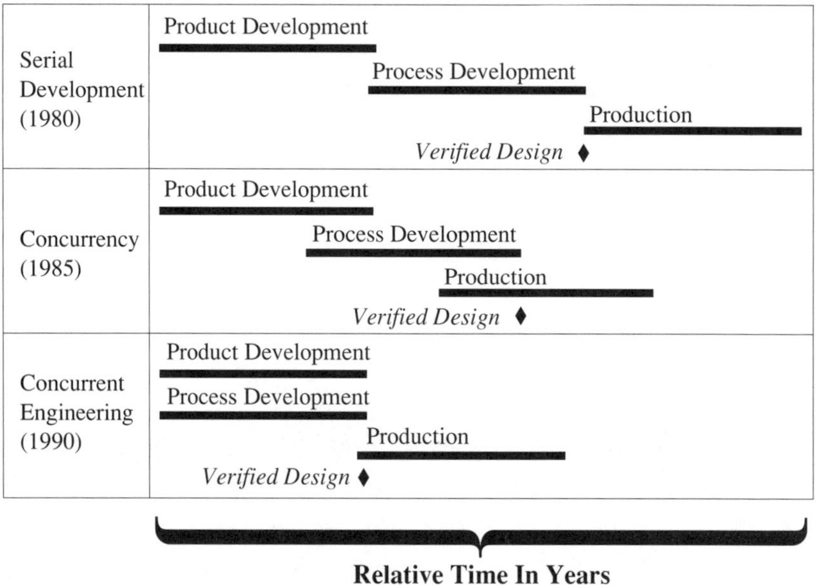

Figure 3-5 Efficiency achieved through simultaneous product and process definition. The basic principle is to iterate and integrate the design of a product and the design of its manufacturing and logistics processes with specific focus on achieving a lower production cost, shortest schedule, and robust quality.

engineering develops both products and processes simultaneously, requiring heavy interface of design, manufacturing, and procurement elements in a company. QFD is the best tool to accomplish this interface and reduce design and manufacturing time making a company capable of rapid development of new products that satisfy customer needs.

3.2 Benefits of using QFD: "So what and who cares?"

One of the first benefits to be realized by using QFD is improved communication. QFD acts as a vehicle for breaking down barriers between departments as well as with customers, i.e., it opens the lines of communication. The ultimate benefits of QFD are increased market share and larger profits. These benefits are possible because QFD plays a major role in creating products and processes with reduced costs, improved quality, features that satisfy customers, and significantly shorter development times. As a result, products are intrinsically more appealing to potential customers. In essence, as illustrated in the Goodyear and Haworth examples above, QFD brings together all the specialized knowledge of many diverse program elements to analyze products and processes concurrently. Table 3-1 illustrates examples of results and benefits that are common when QFD analysis is properly conducted. The vast amount of improvement in design cycle, production cycle time and lower start-up costs comes as a result of the ability of QFD to

Table 3-1 Proper Implementation of QFD Analysis Yields Substantial Benefits

30% to 50% reduction in engineering changes.
30% to 50% shorter design cycles.
20% to 60% lower start up costs.
20% to 50% fewer warranty claims.
Better systematic documentation of engineering knowledge and business operations procedures.
More competitive pricing due to lower development and start up costs.
More satisfied customers.

anticipate design conflicts and problems before production is started or before prototyping.

The authors have seen dramatic results in companies and organizations of varying sizes, products, and services. Table 3-2 shows the direct benefits that some of these companies have managed when using QFD (Chapter 13 discusses in more detail case studies implementing QFD, and how the results were achieved). These are the tangible and measurable results obtained by the use of QFD in specific situations. There are also somewhat intangible benefits of QFD that are harder to measure, but in many ways as valuable as improved cycle time and increased profits. The authors have listed some of these benefits below:

- Improved interdepartmental and interfunctional communications;
- Improved management/employee relationships fostered by a mutual trust and need to cooperate;
- Breakdown of "fiefdoms" without bloodshed as a result of the need to transfer information absent fear of someone "usurping" power;
- True "empowerment" of employees. True empowerment happens when the employee not only has the power to make decisions but also owns the responsibility and accountability for the major outcomes of tasks, projects, activities, or actions without fear of management retribution;
- Managers feeling more empowered because they have learned to delegate authority and responsibility;
- Improved company morale without slogans, "reward" systems or promises. (This results from a feeling of accomplishment and ownership QFD teams develop. This is not to say that reward systems should be abandoned, but they should be part of the quality program and their role should be defined in the strategic management plan.); and

Table 3-2 Case Study Examples of Benefits Achieved

Project	Results
Martin Marietta Astronautics Group	
Satellite Server Space System	• Reduced proposal time by 20% and increased cost saving by 30%. • Won proposal / used as a benchmark for the industry.
Flight Telerobotic Servicer	• Reduced design time and resources by 75%.
Concurrent Engineering implementation on large space system	• Reduced space system cost by 50%. • Reduced development time 60%.
Die Cut Technologies	
Pilot program for small manufacturer to support small business initiatives.	• 30% - 40% reduction in rework. • 30% shorter production cycle. • 25% lower production costs. • Development of new technically sophisticated markets. • Growth in sales 30%. • Increased employee skill level.
3M	
Technology development of new material through development phase.	• 50% reduction in product development time. • All deadlines met within days of scheduled activity. • Created focused, integrated team environment for tech. transfer.
University of Colorado	
Development of an integrated teaching environment.	• Additional $1.1M appropriated over requested. • Design stage on schedule.
Cobe Laboratory	
Medical device development	• Produced clear project vision using common terminology and customer prioritized requirements. • Allowed cross functional team to participate in development of project goals, thereby developing ownership of those goals aimed at the customer.
Resources Trust Company	
Re-engineering of investment management trade execution services. (Estimated results prior to implementation)	• 66% reduction in customer response time. • 76% reduction in cost. • Reduction in verification steps from 11 to 5. • Increased ability to take on new business.

- The development of a motivating team environment where members have a clear purpose, understand expectations and outcomes, have the right information, are supplied with appropriate support, and have an opportunity to mature.

Another predictable result of QFD is what the authors have now termed the "Disciple Syndrome" (DS). The DS occurs to some individuals who have gone through the full implementation of a QFD analysis and seen the benefits first-hand. As mentioned above, QFD is not an easy process when done correctly. As a result, there is a certain commitment that must be made individually as well as organizationally, and that commitment can turn into beneficial zealotry — sort of a quality jihad — for some individuals. For example, one of the authors was facilitating a QFD session with a cross-functional team in front of a group of evaluators from both the customer (a defense agency) and the product manufacturer. Early into the session, one of the customer/evaluators jumped up and said, "Not this bull — — again!" referring to the use of QFD as a prioritization tool. Seizing the opportunity (sometimes called a "crisis") the author invited the customer to participate in the QFD to resolve the technical problem that had been identified. After several hours, the customer had become fully involved in the process, and the QFD analysis proceeded to its conclusion with positive results. This is not the end of the story. At a major quality conference the next year, the customer sought the author out and said that since that red-letter day of involvement, he had not only sought out opportunities to use QFD but was presenting a paper on how to apply the technique in a new way. In short he had become a convert to QFD.

Even a QFD dilettante can be caught in the alluring power to create order from chaos and produce repeatable results. At a recent TQM conference the authors facilitated a group of 140 participants in an application of QFD within the framework of RSM™, dividing the audience into 10 working teams. One of the team tasks was to complete a brief QFD analysis as part of defining a new service. One team had some difficulty in completing this part of the effort as a result of some internal conflict regarding interpretation of the task elements. The rest of the exercise went well with the group and the authors did not give it a second thought. The next day the authors were sought out by one of the members of the group that was having difficulty and told that the team had met before 7 a.m. that morning to complete the assignment and resolve the conflict. Another case of DS!

The reader now knows how QFD fits into a quality paradigm, its reported benefits, and how it can affect those who try it. The next step is to understand the several accepted approaches to the methodology itself, and the new approach the authors have found the most successful.

3.3 QFD methodology

Although the basic philosophy of QFD analysis is fairly straightforward, it is not a rigid tool like a mathematical formula that can be applied in only one way. The manner in which QFD is applied varies, and it has changed since its first application in Japan by Akao, who developed the approach called the Matrix of Matrices.[24] Since then a second method has been developed and primarily used in American applications. It is generally referred to as the Four-Phased approach and was developed originally by the American Supplier Institute.[25] These two approaches utilize a similar central matrix, but go about application of the analysis in different ways. The two methods are also heavily focused on products and processes.

The authors have developed and used a third approach that uses some of the elements of the other systems but approaches application in a manner that integrates additional tools and analysis. The authors refer to this as the Integrated QFD Approach™. As the title suggests, this approach attempts to integrate QFD with planning, functional analysis, and development of performance targets. It approaches application from a vantage point that allows inclusion of Product Requirements, Service Requirements, and Business Operation Requirements, rather than just primarily focusing on products and processes.

The next sections will briefly discuss the Four-Phased and Matrix of Matrices approaches to give the reader familiarity with them, and will then focus on the Integrated QFD Approach™ in detail. The authors will also discuss the common goals sought by the QFD methodologies.

3.3.1 Four-Phased approach

The four-phased approach to QFD is accomplished by using a series of matrices that guide the product team's activities by providing standard documentation during product and process development. Although the four elements of the approach vary depending upon application, Figure 3-6 illustrates the basic four phases of the approach that structure the QFD matrices into four major activities: (1) Program Planning; (2) Product Design; (3) Process Planning; and (4) Process Control Planning. The ability to trace design features and process needs back to customer requirements is formed by taking the design characteristics from the top of the initial matrix and using them as the left-hand side of the next matrix. This waterfall process continues until specific product and process specifications and manufacturing guidelines (process control methods) result. Traceability is therefore obtained throughout the application. The definition of the activities in each phase is given in Figure 3-6.

3.3.2 Matrix of Matrices approach

The matrix of matrices approach was originally developed in Japan by Akao to create linkages with value engineering and reliability charts such as failure modes and effects analysis. It has been further adapted by Bob King of Goal/QPC and is a much more complicated approach to implementation of QFD, using a series of QFD charts that are to be completed depending

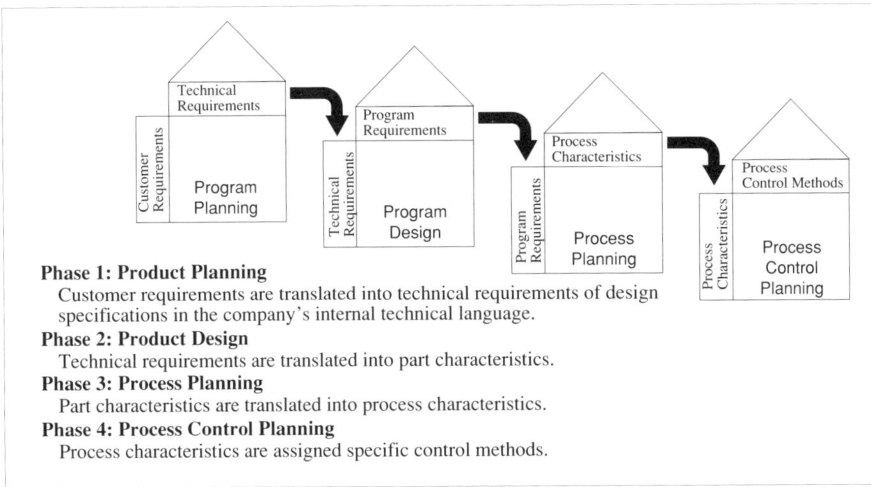

Figure 3-6 American Supplier Institute Four-Phased Approach.

upon the particular analysis of the product. Figure 3-7 illustrates the matrix of matrices diagram. This methodology is probably more attuned to the Japanese approach to manufacturing and product development, and the focus on the detailed construction of a design and manufacturing scheme. This approach — although it maps customer needs against requirements — requires a greater commitment of resources and time to understand and implement, which can be a major drawback when these elements are limited or already stretched too thin. Many of these matrices also map activities currently performed by American and European companies, but not described in a matrix fashion, such as Failure Modes and Effects Analysis, Value Engineering, and Production Operations.

3.3.3 *The Integrated QFD Approach™*

The authors developed the Integrated QFD Approach™ in response to complaints that QFD was too difficult to understand, did not apply to process improvement efforts or service industries, only mapped existing products, did nothing to stimulate the imagination for totally new products, and lacked a plan for implementation. As mentioned above, the Integrated QFD Approach™ takes into account these deficiencies and creates an approach that defines the elements for integration; stimulates the imaginative process for development of new products, services, and operations; includes Business Re-engineering or process improvement efforts; as well as developing an integrated plan for execution. It also represents a departure from the other systems by having as one of its centerpieces the ability to do *continuous product and process improvement*. Figure 3-8 shows how the nine elements of the Integrated QFD Approach™ interact and create a logical flow for the

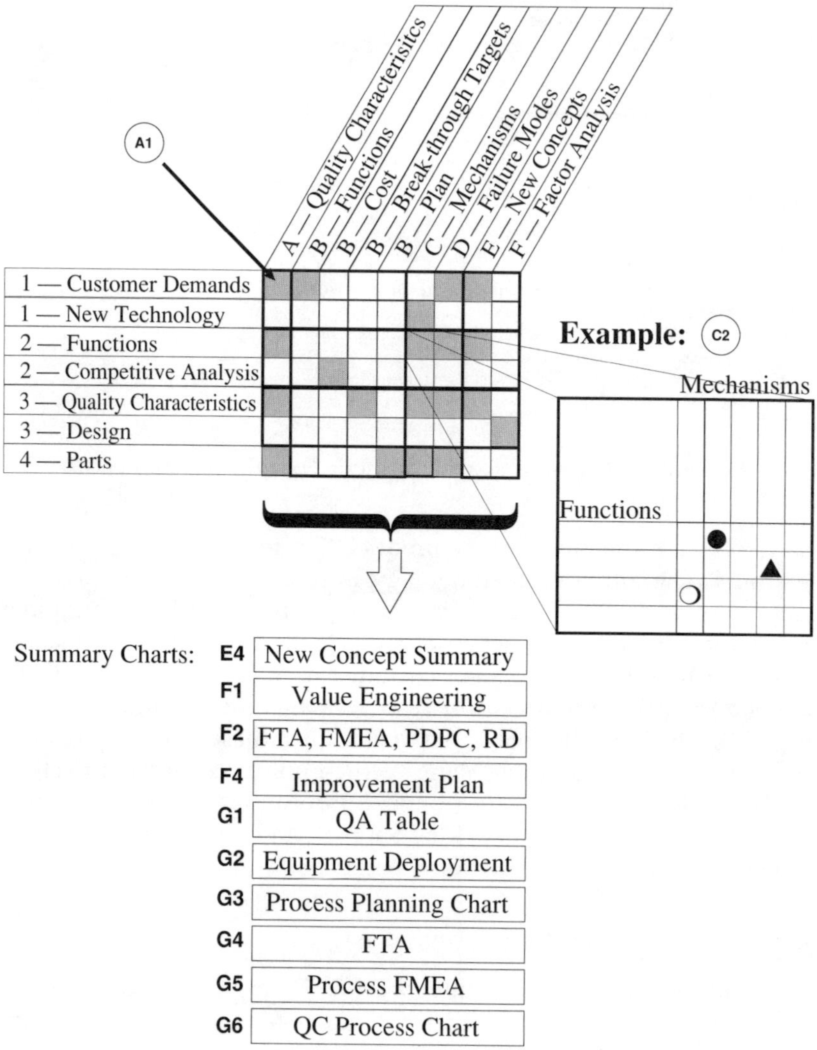

Figure 3-7 Goal/QPC Matrix of Matrices.

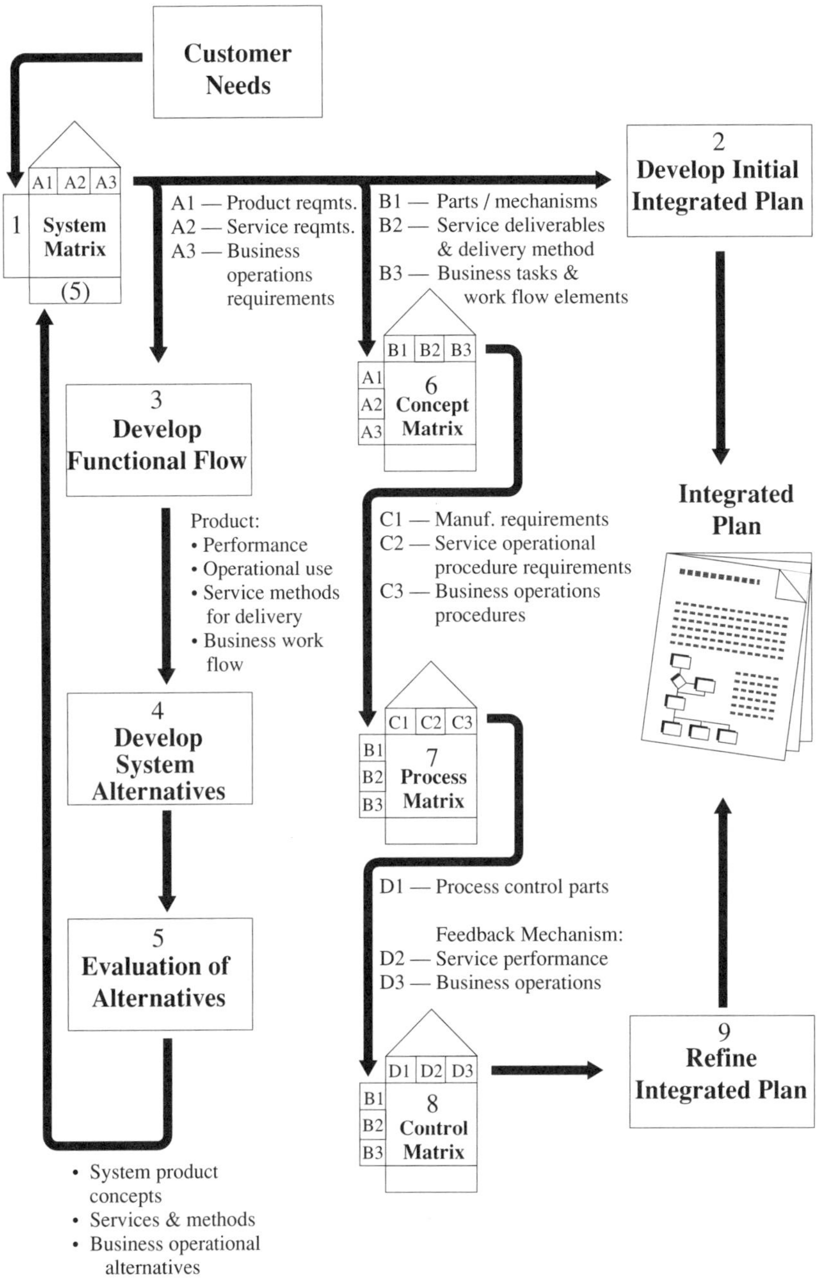

Figure 3-8 The nine steps of the Integrated QFD Approach™.

development or improvement of a product, service or business operation. The nine steps are discussed below:

Step 1. The System Matrix — The first step of the Integrated QFD Approach™ develops the Customer Needs and maps them into three categories of requirements: Product, Service, and Business Operations. By dividing the first matrix into these elements virtually every application of QFD can be accommodated, and every element or requirement developed in response to the Customer Needs will fall under one of these three categories. Companies can either focus on individual elements of the requirements if the scope of the effort concerns only the development of a product or service, or they can involve the entire organization in the integration of defining customer needs against cross-functional goals. Chapter 4 discusses how to develop this initial matrix.

Step 2. Develop an Integrated Plan — One of the most common complaints the authors hear about a QFD application is that after the first matrix is completed, there is a breakdown in implementation of the priorities and data that is developed, which is sometimes a significant cause of abandonment or failure of a QFD analysis. Creating an initial Integrated Plan to implement the results of the initial QFD matrix eliminates this drawback. A cross-functional team that invests in the substantial effort required in assembling information and creating the initial matrix, increases its chances for success dramatically by developing a plan of action. The plan enables the team to focus on accomplishing specific goals that are spelled out. The development of an integrated plan is discussed in Chapter 6.

Step 3. Develop Functional Flows — After the development of an integrated plan, the next step is to develop functional flows for each of the requirements by conducting a Functional Analysis. The purpose of this step is to create a work flow map so the team knows how the requirements are met within the organization and how each function of the company addresses the requirement. This element is addressed in Chapter 7.

Step 4. Develop System Alternatives — There is always more than one way to meet a requirement, and it is critical to develop system alternatives to meet the requirements that satisfy the customer needs. Combinations of different system alternatives allow the QFD team to evaluate the best system based on the priority of customer needs and requirements developed in Step 1 and the Functional Flows developed in Step 3.

Step 5. Evaluation of Alternatives — Now that System Alternatives have been developed, system product concepts, service, methods and business operational alternatives can be compared and a single system (that best meets the customer needs) can be taken to the next level of analysis.

Step 6. Concept Matrix — The product, service, and business operations requirements of the system selected in Step 5 can now be taken to the next level of detail, which is mapping those requirements against Parts/Mechanisms, Service Deliverables and Delivery Methods, and Business Operational Tasks and Work Flow Elements. In Figure 3-8, this is done by conducting another matrix analysis using the same technique as in Step 1. Chapter 7 illustrates the importance of this step.

Step 7. Process Matrix — Now that the top of the matrix in Step 6 (Parts/Mechanisms, Service Deliverables, etc.) has been developed in detail, the next step is to conduct a matrix analysis on those elements by placing them on the left of the matrix and mapping the elements against Manufacturing, Service Operational Procedure, and Business Operations Procedure requirements. This step develops how the procedure requirements will be controlled to meet the requirements on the left hand side of the matrix. In this way, traceability is given all the way back to the Customer Needs in Step 1.

Step 8. Control Matrix — Once the procedures are identified, the last step of matrix development is to find the process control and feedback mechanisms to ensure that procedures are followed (control and measurement). Decomposition and mapping to this level of detail is the most efficient way to ensure that control is maintained to meet the Customer Needs.

Step 9. Refinement of the Integrated Plan — At this stage of the procedure, the previously developed integrated plan can be revised to incorporate tasks and assignments developed since Step 2, and further adjustment to the remainder of the elements can be made accordingly. This is the *continuous improvement* element of the model.

Although each matrix contains elements applying to each of the three focus areas — product, service, and business operations — the team may choose to focus only on the product aspect or business operations, for example. This is appropriate when scoping considerations must be taken into consideration.

The Integrated QFD Approach™ as developed above still has as its central feature the emphasis on Customer Needs. The advantage to this approach is the ability to integrate functional analysis to develop the best way to meet the Customer Needs, and the focus on implementing the prioritized elements of the matrices. The Integrated QFD Approach™ also forms the core of any Concurrent Engineering effort because it allows simultaneous development of design and processes in a methodical and linked manner, as will be discussed in Chapter 8.

3.4 *Common goals sought by the QFD methodologies*

No matter which approach to QFD is used, there are some common goals all methodologies seek to fulfill. Emphasis on the customer is still stressed by

Table 3-3 Common Goals and Differences between QFD Methods

Common goals shared by all QFD approaches

- Drive towards specific means to develop technical requirements.
- Emphasis on setting priorities at each stage of development.
- Consideration of costs, reliability, new concepts or technology, and the company's ability to produce the design or process.
- Use common logic throughout the matrices.
- Use of additional tools and techniques as appropriate.

Different goals shared by all QFD approaches

In addition to the Integrated QFD Approach™, there are generally two accepted QFD techniques: (1) The Four-Phased approach, and (2) The Matrix of Matrices. Although the formats are slightly different they have common goals and elements.

each method as the paramount purpose. No QFD effort can exist without involvement of the customer to the extent necessary to ensure proper identification of customer needs. Mapping of those needs into the technical requirements (however defined by each system) is the next step each method shares in order to prioritize and develop the focused approach to design development. Table 3-3 illustrates additional elements that are common to any application of QFD.

In its original application to new product development, application of any QFD methodology yields a substantial reduction in design-to-production cycle time over "old" linear design methods, in most cases averaging a reduction of 50% (see Figure 3-9). Application of the Integrated QFD Approach™ can achieve even greater benefits in a wider application than just new product development.

The next step in learning how QFD works is to take a detailed look at the central feature: the Core QFD Matrix. This is the engine that drives any QFD effort, and Chapter 4 shows how to construct it.

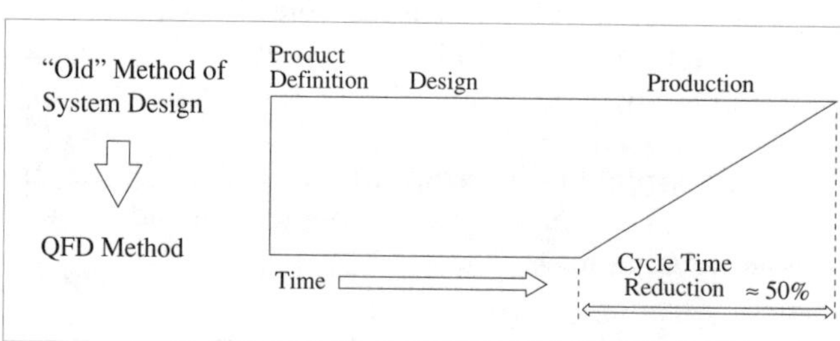

Figure 3-9 Reduction in cycle time due to conversion from "old" systems design methods to QFD method.

3.5 Chapter summary

1. QFD is a systematic means of analyzing customer requirements and deploying them into product, service, and business operations.
2. The genesis of QFD is traced to Dr. Akao, chairman of the QFD Research Committee of the Japan Society for Quality Control, who was involved with the first large-scale application of QFD at the Kobe Shipyard of Mitsubishi Heavy Industries in the early 1970s.
3. The need for QFD is highlighted by the fact that 83% of new products failed to clear the performance hurdles set by their developers in 1992 and the fact that engineering change costs increase logarithmically over a product life cycle.
4. The benefits reported from users of QFD include:
    - Providing 30 to 50% shorter design cycles;
    - Providing 20 to 60% lower start-up costs;
    - Providing 30 to 50% reduction in engineering changes;
    - Improving communications between departments, functional organizations, and customers;
    - Providing a catalyst for creating a motivating team environment where members have a clear purpose, understanding of expectations and outcomes, the right information, appropriate support, and an opportunity to mature; and
    - Providing the framework for employees to become empowered, where not only do they have the power to make decisions but also have responsibility and accountability for the major outcomes of a task, project, activity or action.
5. There are several types of QFD Methods:
    - The Matrix of Matrices, developed by Dr. Yoji Akao, which includes 30 matrices in total;
    - The Four-Phased approach, which includes the Program Planning, Product Design, Process Planning and Process Control Planning matrices; and
    - The Integrated QFD Approach™ developed by the authors.

    All techniques drive toward specific means to develop more complete and accurate requirements with an emphasis of setting priorities at each stage of development.
6. The Integrated QFD Approach™ was developed to: stimulate the imagination for development of totally new products, services, and operations; include Business Re-engineering and process improvement efforts; include service industries and create an integrated planning element for implementing the results. The nine steps to the Integrated QFD Approach™ are as follows:

- Complete the Systems Matrix;
- Develop an Integrated Plan;
- Develop Functional Flows;
- Develop System Alternatives;
- Evaluate Alternatives;
- Complete the Concept Matrix;
- Complete the Process Matrix;
- Complete the Control Matrix; and
- Refine the Integrated Plan.

Each of these steps are discussed in detail in Chapters 4 through 7.

chapter four

Building the Core QFD Matrix

For every complex question there is a simple answer,
and it is wrong.
H. L. Mencken

4.0 The basis of the Core QFD Matrix

H. L. Mencken may be a little too absolute in his conclusion that all simple answers to complex questions are wrong. However, it can be said that when creating new products, satisfying today's more demanding consumer and improving business operations, simple answers may not be totally wrong, but they are more often only *part* of the answer to complex questions posed by these issues. When a company misses the mark on a new product, the answer to the question "why" the product failed or did not sell as well as expected can be simple: "the company failed to satisfy the customer." The answer is not wrong, but it does not acknowledge that the product failed as a result of a combination and succession of decisions in a complex process. The answer above fails to grasp the complexity of the challenge.

The Integrated QFD Approach™ is not a simple answer to the complicated questions of how to meet customer needs — neither is any other approach using QFD. Sometimes we as organizations, as well as individuals, want to believe that solutions are simple and easy, and if a solution appears difficult then it must be the wrong one. We also tend toward single answers to questions: — if sales are down, expand a market; if our competitors sell more, be more like our competitors; if the product fails under heat, make it more heat resistant. These are not wrong answers, they just (probably!) are not the whole answer because we may not be asking the right questions. The Integrated QFD Approach™ encourages asking more questions, and therefore answers more complex questions. It is not easy, but it can be implemented and used effectively and logically to produce answers that *solve* complex questions about interactions between customer needs, interactions between product, service and organizational requirements, and prioritizing those interactions. The central feature of this ability is the Core QFD Matrix.

All of the approaches to QFD use some form of a matrix for analyzing customer needs vs. technical requirements. The various permutations of the matrix are tuned for each specific approach, but the parameters of the matrix involve measuring, comparing, and contrasting two sets of elements. The Integrated QFD Approach™ is no different in this aspect, although the Core QFD Matrix has some elements that are not found in other approaches.

The Core QFD Matrix is the heart of the Integrated QFD Approach™ process. Crucial to the success of any attempt to use QFD in product or process applications is a complete understanding not only of the mechanics of utilizing the Core QFD Matrix, but also the reasoning behind the various sections.

The Core QFD Matrix is based on a matrix that is sometimes referred to as the "house of quality," and many of its conventions are similar. The house of quality is used to initiate a comparison of the customer needs with the technical requirements of a product, and is a major feature of the four-phased approach. The Core QFD Matrix is used in the Integrated QFD Approach™ in several different places, and although it may appear similar in form to the house of quality, it operates somewhat differently in the Integrated QFD Approach™, as will be explained in this and later chapters.

4.1 The Core QFD Matrix overview

The Core QFD Matrix is a product, service, and process planning matrix used to depict customer needs, design/service/business requirements, target design goals, and competitive product or service evaluation. The Core QFD Matrix is more than just the sum of its individual elements. It can be used as the vehicle to conduct the first foray into a total quality environment. As mentioned in Chapter 2, the Core QFD Matrix takes a vertical cut through the elements of RSM™ and can become the focal point for directed team building. This matrix is a fairly flexible approach to implementing QFD that can be configured in several different ways, depending upon the purpose or application of the matrix. Based upon various application environments, the Core QFD Matrix designed by the authors uses eight distinct components. Figure 4-1 illustrates these components, along with the eight steps that should be taken sequentially.

Prior to any construction of the Core QFD Matrix by a design, service, or project team, the effort to be addressed by the QFD process should be scoped by the management and staff. Whenever a multifunctional team is used to construct a QFD, there must be an initial definition of the program objectives, expected outputs, use of data products, and boundary conditions. It is not unusual that a substantial amount of time is spent in this preliminary step in order to establish the ground rules and the scope of the project in terms that everyone understands. A more complete explanation of guidance techniques used in scoping the effort is presented in Chapter 10. The following represents a brief description of the various elements of the Core QFD Matrix, with the paragraph number representing the element depicted in Figure 4-1.

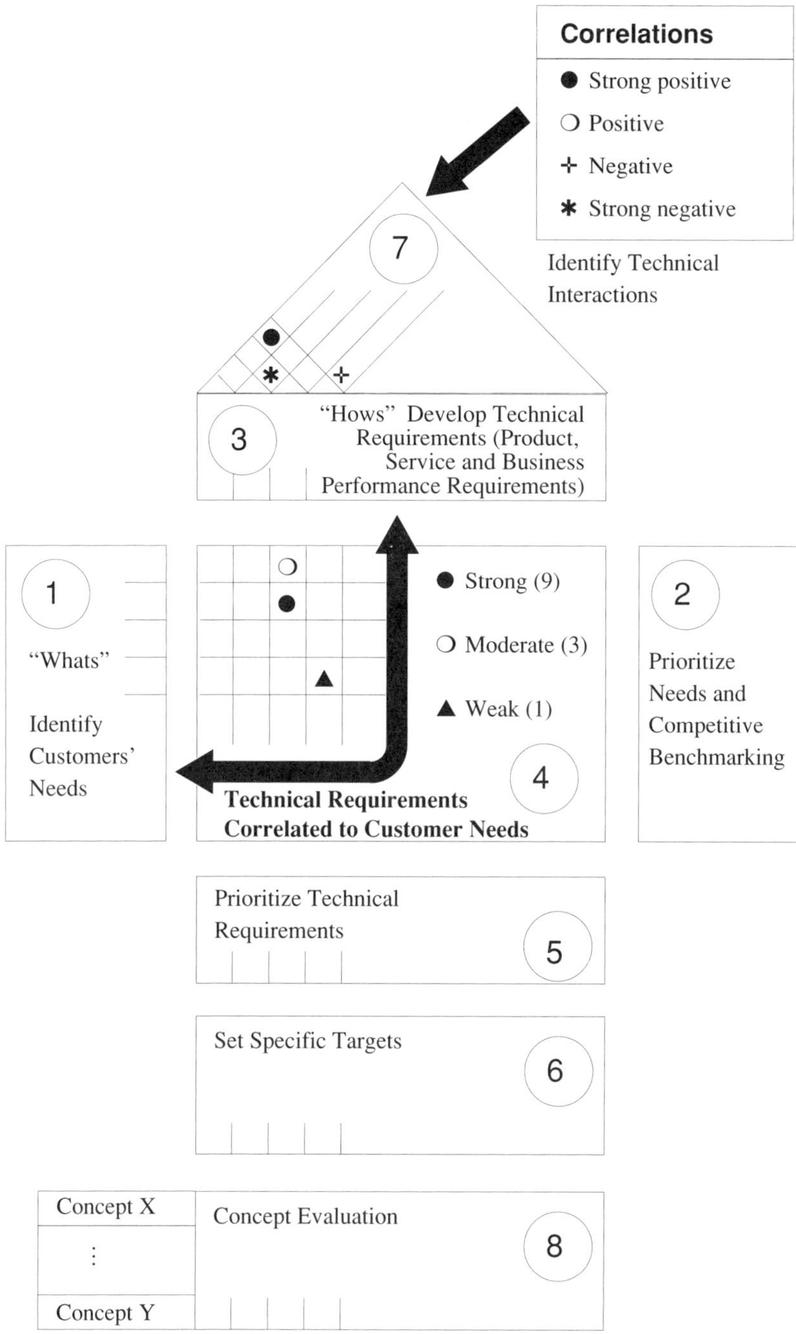

Figure 4-1 The elements of the Core QFD Matrix.

1. **Identify Customer Needs** — The Core QFD Matrix begins by using the customer's physical and functional needs (sometimes termed attributes). These customer needs are often referred to as the "whats" and can be one of the most difficult elements of the matrix to construct. In some circumstances, it is difficult to get the multifunctional team to initially agree upon who is the customer, before even addressing what it is that the customer wants. Sometimes there can be multiple or primary and secondary customers. Customers can be either internal or external to the organization and their needs may be conflicting. Sometimes it is necessary to focus on who is the actual end user of the product or service in order to determine who is the final customer and resolve this conflict in order to bring a team to resolution. There are several categories of customer needs. The first category consists of explicit customer expectations. The second category involves elements a customer expects without telling you, but are soon to be part of the general requirements (e.g., environmentally compatible air conditioning coolants). The last group of customer needs includes expectations for those items not specifically requested because the customer does not know of their possible existence. These are often referred to as "excitement factors." When these features are not present people are not dissatisfied, but when they are present people are extremely pleased. The emergence of air bags and antilock brakes are good examples of excitement factors (prior to their presence because of regulation) that pleased the car-buying customer.
2. **Prioritize Needs and Conduct Competitive Benchmarking** — As mentioned above, the matrix is used to prioritize customer needs and develop competitive benchmarks. The first step in prioritization is to assume that not all customer needs can be equally important. To bring the proper perspective and focus to team activities, the Core QFD Matrix measures the relative importance to the customer of all of the defined needs. Weighted values based on customer's perceptions are derived from the team members' direct experience with the customers or from surveys. Further comparisons are made to determine where the company or project team stands relative to the competition. This is usually accomplished by comparing the customer evaluations of the competition compared to the in-house evaluations of current competitive standing.
3. **Develop Technical Requirements** — The next step involves brainstorming a list of technical characteristics or requirements that will satisfy the customer's needs. These are typically called the "hows" and should be controllable characteristics. The term "technical requirements" is used to group product, service, and operational requirements that will satisfy the customer's needs at the top level of the analysis. This is another key element of the matrix and often requires a substantial amount of time from the multifunctional team. Often one of the most difficult tasks of the team is to distinguish customer needs

from technical requirements. Technically-oriented individuals often immediate jump to a technical solution thinking it is a customer requirement. Therefore, it is necessary to spend substantial time in Steps 1 and 3 in order to create the most beneficial Core QFD Matrix.

4. **Correlate Technical Requirements to Customer Needs** — The "relationship matrix" indicates the impact of each technical requirement on the customer needs. A project team should seek consensus on these evaluations basing them on expert engineering or service knowledge and experience, customer responses, and tabulated data from statistical studies or controlled experiments.
5. **Prioritize Technical Requirements** — The relationship matrix can be translated into design priorities. The relational levels of importance not only define what is most important, but also define what is least important. These results provide the basis for a plan of action to implement the identified technical requirements.
6. **Set Specific Targets** — At the bottom of the Core QFD Matrix, targets are set for each corresponding characteristic to measure the process, service, or product design. If the characteristic is not measurable, then further definition is required. If a characteristic cannot be measured then there is no way to verify that it has been adequately implemented into the design, integrated into the manufacturing process, or included in the service.
7. **Identify Technical Interactions** — Measurable does not necessarily mean conflict-free. Very rarely does a system not have conflicting design elements. In a QFD matrix these correlations between the system characteristics are captured at the top of the matrix in what is traditionally referred to as the roof (hence the designation of the matrix as a "house" of quality). The distinctive roof on the Core QFD Matrix assists engineers or members of the multifunctional team in specifying various characteristic features that need to be improved collaterally. System characteristics exhibiting negative relationships are traded off to find the best compromise and strong positive relationships are studied to prevent duplication of effort. Sometimes one targeted feature impairs so many others that the team may decide to leave it alone. In complex technical designs or in providing a service based on complex information systems, ignoring this part of the matrix can result in catastrophic conflicts that were not considered in the initial stage of the design or development process.
8. **Concept Evaluation** — Product or service concepts can be evaluated against all technical requirements as related to customer needs. Each independent element of a particular concept can be evaluated for strengths and weaknesses as compared to the performance targets set during the QFD analysis. This analysis also permits the design team or the development team to determine if there are any gaps in the concept's technical solutions that would prevent meeting performance goals. As a result, the concept in total can then be compared to another

concept to determine overall ability to meet the customer's priority needs as determined earlier in the analysis. In addition, the analysis will also yield the ability to determine if there exists a technical solution that incorporates the best of several concepts.

The above represents a brief description of the concepts involved in the eight-part Core QFD Matrix. Having been presented with the philosophy of these eight elements, the next step is to examine each individual element in more detail.

4.2 Capturing the "voice of the customer"

The principles found in TQM and the teachings of W. Edwards Deming (one of the premier American authorities on quality) have long advocated focusing on the customer as the key to better selling and, ultimately, increased profits. Also central to the concept of QFD is being customer-focused — but what is being customer-focused?

QFD relies upon the principle of listening to the "voice of the customer" which is a term that is often used when describing what the customer desires from a product or service. Finding out what the customer wants is an industry in itself, and QFD relies on accurate information on the customer. However, before defining what it is the customer wants, one must first define who is the customer, and that depends on the scope or goals of the QFD analysis itself.

Figure 4-2 illustrates a stepped approach for organizing the effort to determine the "voice of the customer." Primary to the stepped approach is organizing the efforts of the QFD study and establishing the team. Team selection is always difficult in any situation, and often the success of a program relies heavily on the composition of the team. In general, a multifunctional team that can address all of the functions and departments that may be impacted by the QFD analysis needs to be assembled (Chapter 10 discusses how to assemble the skills necessary to address a QFD analysis). This does not mean that a team will have a representative from each department. Individuals who can interface with several departments or skill areas may satisfy the requirement.

The success of a QFD study also relies upon a clearly focused goal. For example, a QFD can be a laborious and tedious process, and if the team members are new to the QFD philosophy it may be prudent to design a QFD initial effort. This effort should last 3 to 4 days and be devoted to team building and establishing the top-level Core QFD Matrix for the new product or service. A team leader or program manager should have in mind what is expected out of the QFD prior to the start of the study. If the QFD is only to establish top-level systems requirements in response to customer needs, that effort should be explained clearly and succinctly to the team members so they do not misunderstand the goals of the effort. The authors have worked

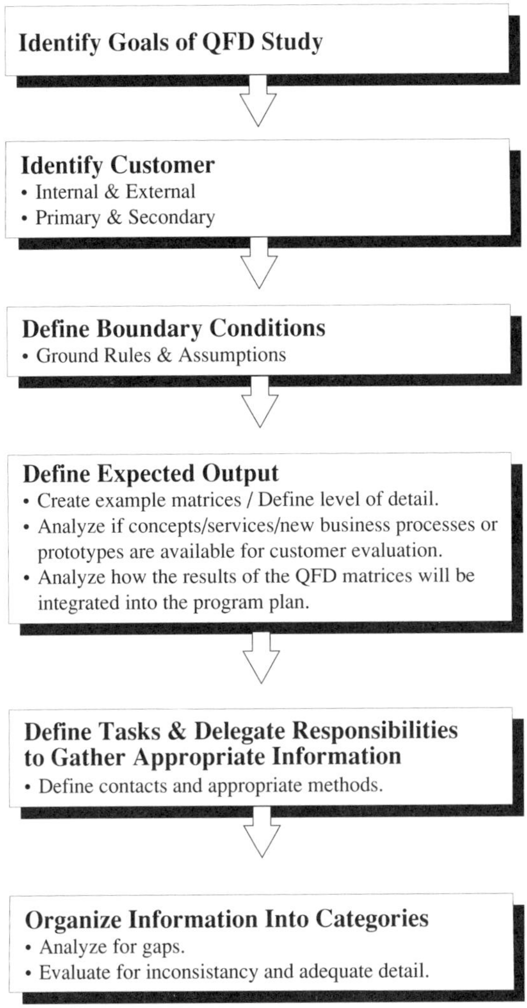

Figure 4-2 Steps to the "voice of the customer."

with companies where QFD had been attempted earlier but failed because the effort bogged down in details and technical minutiae that prevented the creation of a systems or overall Core QFD Matrix (Chapter 10 discusses methodologies for defining the scope of the QFD and conducting team selection).

Once the goals of the QFD analysis have been established, the next step is to select the customer or customers. Customers can be internal and external. Internal customers can be other departments, other divisions of a company, or other individuals in the organization who make use of information, products or services produced by other internal sources. Prior to engaging in the QFD process, the team must first determine who the customers are, both internal and external. This is a very critical step in the process because it will drive the gathering of information to determine the needs of the customer or customers. In the process of examining this question, most teams soon realize that there is more than a single customer. For example, in developing a new computer component piece, the component manufacturer may realize there are several customers: the ultimate customer will be the consumer who will use the assembled computer; another key customer will be the computer integrator who is building the entire product, who may have certain technical needs of the component that are irrelevant to the ultimate user (the consumer); there can also be a third-tier customer, who is internal to the component manufacturer, e.g., a manufacturing department may need certain processes identified in advance in order to prepare the manufacturing process and meet the scheduled delivery date. Based on considerations such as these, it may be important to group or segment customers according to certain similarities. Once this has been done, a primary Core QFD Matrix can be designed to determine the needs of the primary customer and then secondary matrices can be constructed to determine whether or not there are any additional customer needs to be met, or if there are conflicting requirements for any two customers.

Once the customer, or customers, have been identified it becomes necessary to create a structure for determining the customer needs. This structure identifies the boundary conditions for gathering information on customer needs. Part of defining the boundary conditions includes budget available for developing customer information, time limitations, assumptions that need to be made about the customer, and resources that are available to all members of the team. In addition, the type and extent of customer information to be gathered must be defined. For example, there are many techniques for gathering information on customer needs and desires in a product. Some of those techniques are listed in Figure 4-3. Typically, most organizations employ their marketing departments, or an outside marketing and information gathering agency. These groups define how to gather information on both customer needs or desires for a new or existing product and potential competitive benchmarking for any other products or services that may already exist. Figure 4-3 is not meant to represent an exhaustive list of potential marketing and information gathering techniques, but instead to give some guidance in scoping the gathering of information for a new product or service. The resources available to conduct marketing research and gather information must always be considered and bear a relationship to the risk for the new product or service and its potential for profit. Often the first task for a product or service team will be to determine if enough data already exists to do a preliminary QFD analysis mapping the customer needs against technical requirements. This is a difficult decision to make because often it is not known that certain data or information regarding customer needs is missing until the initial top-level QFD is completed, or at least attempted. There is no hard-and-fast rule that states once a QFD analysis is started the team cannot

 **Survey** • Mail / telephone • Comment cards • Questionaires	 **Focus Groups** • Facilitator guides small group through questions
Interviews • Face to face • Telephone	**Product In Use** • Displays • Clinics
 **Mystery Shoppers** • Role play customer and record observations	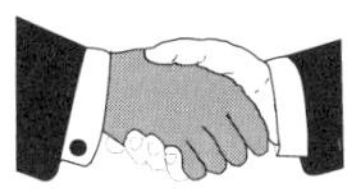 **Natural Field Contact** • Sales meetings • Service calls • Trade shows
 **Feedback** • Employees and users discuss competitor's products.	 **Complaints** • Letters / cards
Direct Visits • Watch and record	**Warranty Data** • Service Records • Cards / letters
 **Hot Lines** • Toll free	 **Sales Records** • Monthly • Reports/parts sales
Publications • Government • Independent labs • Trade journals • Consumer magazines	

Figure 4-3 Techniques for obtaining new information.

stop to conduct additional data gathering in order to ensure that the customer needs are truly identified and are being incorporated into the design or service development process. It is incumbent upon the team to determine how the data is going to be gathered, what techniques will be used, who will gather the data, and finally how the data will be analyzed.

In facilitating a QFD effort, the initial brainstorming to determine who the customers are and what data or information needs to be gathered can serve a dual purpose. First, it gives direction to the team. Second, this process can easily lend itself to the initial team-building effort and break down some of the barriers that may exist between disparate functions or departments in a company. Sometimes these conflicts develop because of a lack of communication and contact, or because the individuals from the departments may never have had the need to work together in a team environment (Chapter 10 discusses additional ways to not only select the team, but also to develop team cohesiveness in the initial phases of the QFD process).

Although there is a substantial amount of information available in existing literature on gathering customer-related data, the authors have included Figures 4-4 and 4-5 as examples of how to document market research information, and guidelines for survey and interview questions. Figure 4-4 is a blank sample marketing research document and Figure 4-5 is a completed form using some data of a potential customer for our imaginary video game company, Compulsive Video Mania.

An organization may wish to enlist the services of a professional market research firm to conduct focus groups, detailed and focused customer surveys, or other tools either before or during the QFD effort. In the process of analyzing the customer information the team should look for gaps or inconsistencies in the information.

The authors have found that this stage of the QFD process is often overlooked because some organizations think they are in tune with the customers' needs if they currently are in a profitable position or they have conducted past marketing research. This is where a multifunctional team conducting its own analysis of existing data or looking at a new product design may determine that there are either information gaps or outdated assumptions on what customers really want or need. Marketing and business history is full of examples of misdiagnosing the pulse of the customer. The Edsel is a classic example, but recent times have also seen failures of greater magnitude such as Coca-Cola introducing "New Coke" in 1985 — the return of "Classic Coke" did not take very long!

The important point in addressing the "voice of the customer" is to fully understand the needs of the customer, and that is often best done by asking the customer directly. This is why more and more companies are bringing the customer into the design and development process and making the customer part of the multifunctional and multidisciplined team.

Once the above groundwork has been done, the next step is to begin constructing the Core QFD Matrix.

Market Research Documentation

Identification No.

Customer Profile ❑ Internal ❑ External

Characteristics:

Voice of the Customer (recorded data):

Product Use

What:

Where:

When:

Why:

How:

Revised

Performance Features

Reliability

Additional Comments

Figure 4-4 The Market Research Document.

4.3 Constructing a Core QFD Matrix

The Core QFD Matrix represents the information gathered and the decisions determined by the product or service team in the QFD process. The Core QFD Matrix is a very compressed and compact mosaic of information compiled, discussed, argued over, fought over, and finally compromised on by the team. In implementing a QFD effort, many questions arise as to how the information gathered and developed by the team gets translated into the

Market Research Documentation

Identification No. 888

Customer Profile ☐ Internal ☒ External

Characteristics:

young woman 24yrs old Assistant Financial Manager

Voice of the Customer (recorded data):

"Easy enough to figure out without reading the instructions"
"Great graphics"

Product Use

What: video games

Where: At home

When: At night, during holidays

Why: Stress reliever

How: connected to a big screen TV

Revised

- Easy instructions
- Good visual effects

Performance Features

- Optional self guided tutorials
- Good graphics

Reliability

- Mean Time Between Failure > 900,000 cycles

Additional Comments

- Would like modular upgrade capability from 2 to 8 players

Figure 4-5 Video game example of the Market Research Document.

format represented by the matrix. Where do the customer needs come from? How does the team come up with the technical requirements?

One of the most effective ways to understand the operation of the Core QFD Matrix is to walk through a specific example for each of the various elements. We will use the development of a video game by our imaginary company Compulsive Video Mania (CVM) to illustrate not only the mechanical functions of the Core QFD Matrix, but also the conceptual relationship of the elements. Most of us, either through personal or vicarious experience through our children, have had some exposure to the video game craze that has swept the country. Without debating the merits of the pastime, video games offer an excellent example of a product that is fairly technical in nature, but also reaches a mass audience in an industry that is highly competitive and constantly changing. The president of the company has selected key team members from the following departments: marketing, design, software, hardware, procurement, human factors, test, manufacturing, subcontracts and finance to construct the next generation of video games. The team has decided to use the QFD process in an effort to get the jump on the other video game manufacturers. The QFD team first determined through its marketing that its customer would be 8-year-olds to adults who currently owned some variety of video game equipment. Although this was a fairly wide demographic group, the team decided that it may later conduct additional QFDs to target certain age groups or demographic groups within that bracket. The team also determined that its secondary customer might be the manufacturers of the QFD game players. These were secondary customers because CVM had to be able manufacture games that would play on the various video game equipment currently on the market.

4.3.1 Techniques to support the development of a Core QFD Matrix

There are three analytical tools that can be utilized effectively both in the initial stages of the QFD and in later stages of full integrated product development. These three tools are (1) brainstorming, (2) affinity diagramming, and (3) tree diagramming.

In determining the initial customer needs it is helpful to use a technique called "brainstorming" that creates an open environment for creativity and taps the individual resources of all the team members. When determining the initial customer needs, the group should list all attributes including those determined by any form of market research and perceived or suspected attributes demanded by the customer in the new product or service (tips on the technique of brainstorming are contained in Chapter 10).

Once the customer needs have been brainstormed, it is often extremely helpful — especially given the fact that there are often a large number of customer needs or attributes identified — to group the customer needs through the use of affinity diagramming. Using our imaginary video game company, CVM, Figure 4-6 illustrates how affinity diagramming is used to develop some of the customer needs for the new video game. As the reader can see, certain attributes of the video game can be grouped into elements that have an "affinity" for each other, or objects that are at least similar in their purpose or function. Affinity diagramming also assists the team in eliminating duplicate needs of the customer.

The next tool that can be used is tree diagramming. Figure 4-7 gives an example of the information contained in Figure 4-6, but arranged in a tree

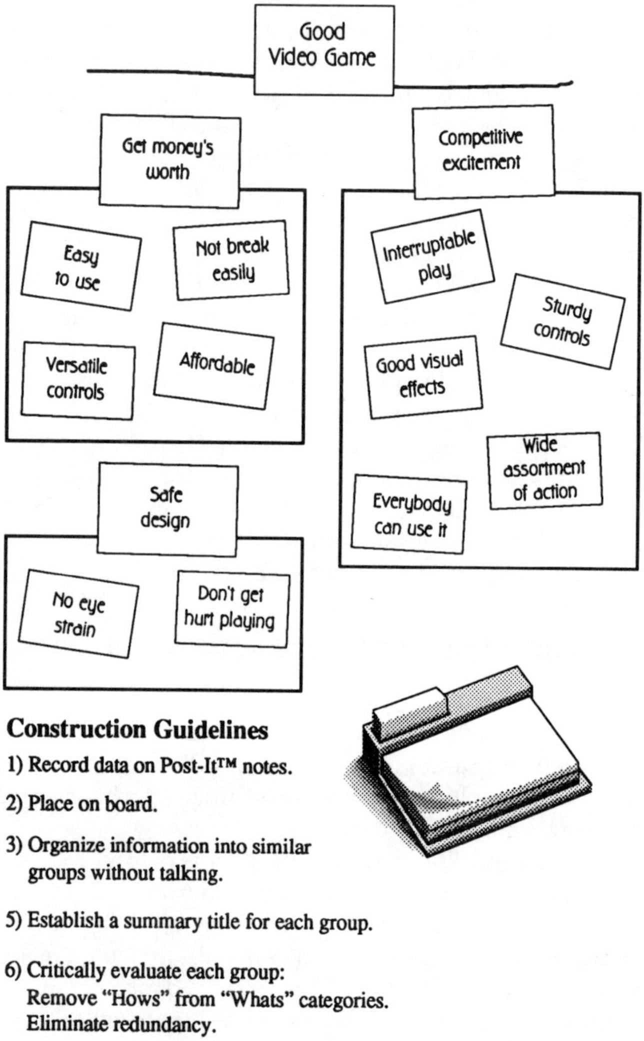

Figure 4-6 Affinity Diagrams using the development of a video game as an example.

diagram. The purpose of the tree diagram is to create a complete list of requirements or objectives but in a more categorical manner than simple affinity diagramming. Essentially the individual elements of a tree diagram can be traced back to a central grouping or concept idea, like the branches of a tree lead back to the trunk. Tree diagramming also allows the visual creation of relationships at various detail levels and can help the group to determine if, in the initial Core QFD Matrix creation, they want to go down to the lowest level of detail, or tackle the matrix at a higher level of grouping. The tree diagram can also reveal any gaps or holes in the sequence or logic in the grouping itself, or in the determination of the customer needs.

Tree Diagrams

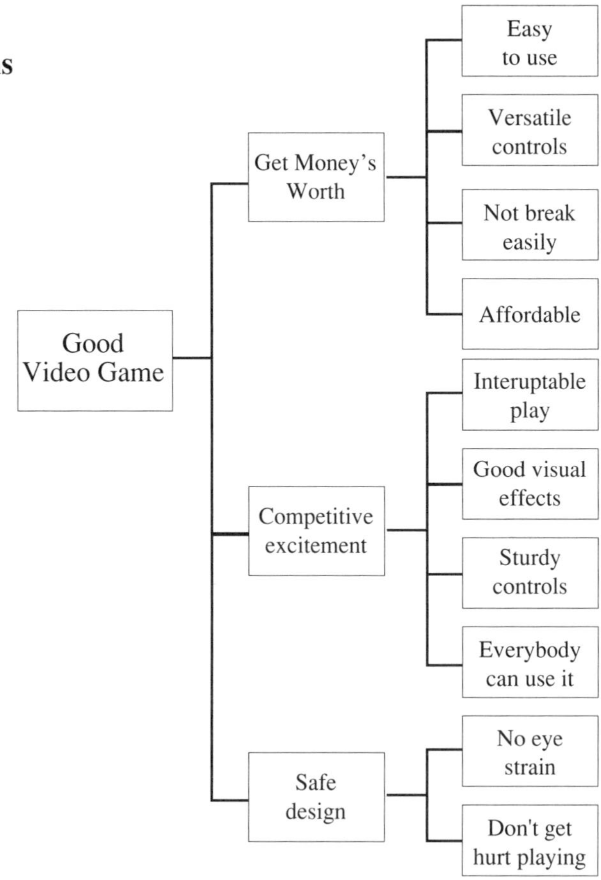

Purpose:

Create complete list of requirements / objectives

Construction Guidelines:

1) Translate the Affinity Diagrams into a hierarchy.

2) Start hierarchy on left side of chart.

3) Place more detailed items in secondary levels or tertiary levels depending on relationship.

4) Establish and record definitions of categories as necessary.

5) Examine tree for gaps and consistency in detail between levels.

Figure 4-7 Tree Diagrams using the development of a video game as an example.

4.3.2 Developing the detailed Core QFD Matrix

The eight elements of the Core QFD Matrix can be further divided into some subelements for purposes of calculating the mathematical relationships between the various customer needs and technical solutions. Figure 4-8 refers to each of the eight sections and subelements using numbers and letters for easy reference in this chapter. By understanding how to create the core QFD Matrix, the reader will have the understanding to develop the four Matrices in the Integrated QFD Approach™ that were introduced in Chapter 3 (Figure 3-8), and are analyzed in detail in Chapter 7.

The subelements of the Core QFD Matrix are not fixed constants impervious to alteration. As the reader becomes more familiar with the structure and logic behind each element of the matrix, the possibilities of adapting some of the subelements and calculations to specific situations should become apparent. The Core QFD Matrix is just that — a core to be added to or modified as required by circumstances.

The following describes how the matrix is completed and from where the information comes. Some facilitation hints are given based on actual situations. However, Chapter 10 discusses in more detail how to facilitate these sections more effectively. The reader should approach the following material with an eye for understanding the mechanics of generating the information for each of the eight sections of the Core QFD Matrix. Appendix A provides a quick reference guide to performing the mechanics of the Core QFD Matrix.

4.3.2.1 Determining the customer needs (Section 1)

The customer needs section contains the information gathered by the team to represent the key elements the customer wants. After brainstorming and using affinity and tree diagrams, the CVM team came up with the customer needs contained in Figure 4-9. The team placed in Section 1 of the matrix the three basic attributes that the video game needed to meet: (1) get money's worth, (2) provide competitive excitement, and (3) safe design. These needs were then broken down into subcategories on the left side of the matrix. Having defined their initial customer needs, the next step for the team was the second element of the Core QFD Matrix: prioritizing the customer needs and doing competitive benchmarking.

4.3.2.2 Prioritizing customer needs and competitive benchmarking (Section 2)

The next task for the project team was to prioritize the importance to the customer of the identified needs, and do competitive benchmarking against two of Compulsive Video Mania's competitors. Section 2 of the QFD matrix is divided into nine subsections, represented in 2A through 2I below and in Figure 4-10.

2A. Degree of Importance — The first step in prioritizing needs is represented by 2A-Figure 4-10. Column 2A is the relative degree of importance of each of the customer needs. The question being answered by the team is: "How important is this need to the customer?" The scale of ranking

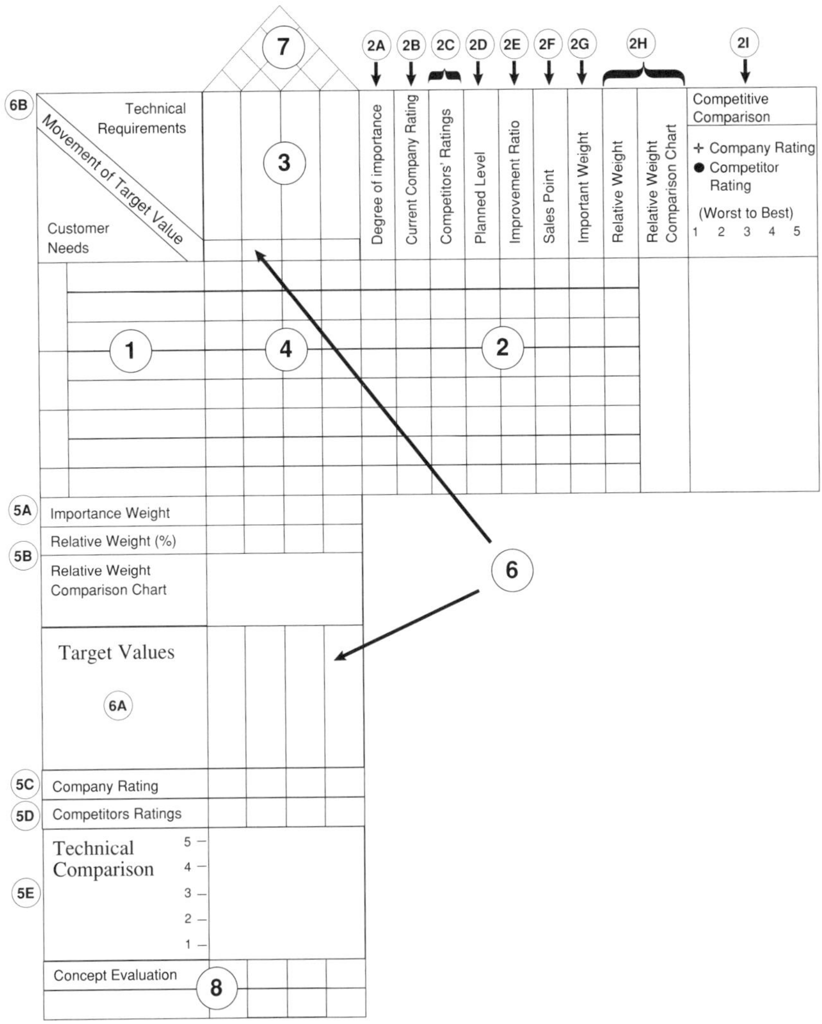

Figure 4-8 Detailed Core QFD Matrix guide.

is 1 to 5, with 5 being the most important. When ranking the customer needs, the team first determined which of the customer needs would be *the* most important. As you can see, the range of ranking went from 2 for "versatile controls" and "interruptable play" to the most important elements (5) that were "wide assortment of action" and "good visual effects."

When ranking the degree of importance, it is not as important to have a range of customer needs ranked 1 to 5, as it is to have directed discussion to truly determine which of the customer needs ranks the highest and which are not so important in the total scheme of customer needs. One way to determine this relative ranking is to ask the team what they would give up among the customer needs if they could maximize another customer need. Another

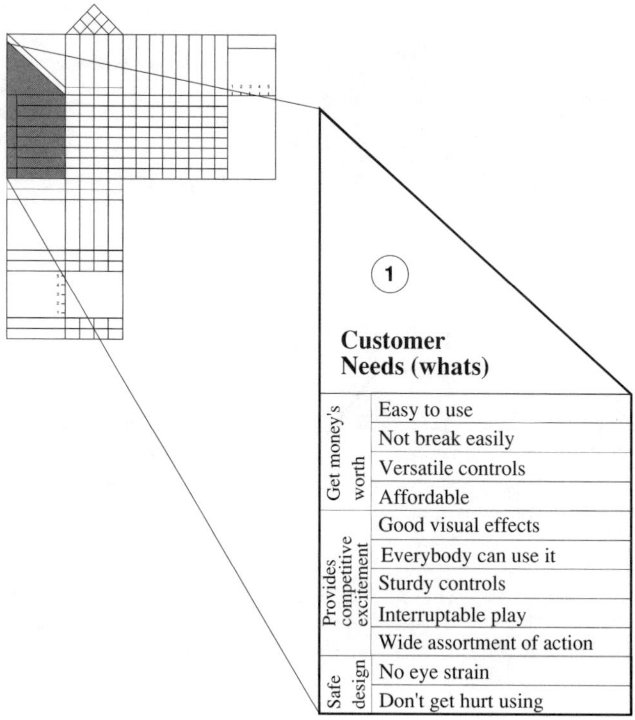

Figure 4-9 Video game Core QFD Matrix example — Section 1: Identify Customer Needs.

technique is to ask the customer, if present, what is most important. Sometimes customers feel overwhelmed and outnumbered by the team members and do not voluntarily discuss their preferences, and failure to bring them into the discussion can lead to catastrophic failure. One of the authors was involved in a QFD with the customer present, where the focus of the QFD was to map into the QFD matrix the specifications originally given by the customer. At the end of the day and after a grueling amount of work, the customer pulled the author aside and said, "You know, I really do not want what's in there (the QFD) — What should I do?" The author offered to intervene and let him talk specifically to the team to describe his idea of the customer needs. After describing eight mission goals to the team, he said that what he wanted was for the team to tell him what kind of performance was required for these goals, and what kind of system concepts would meet those needs within the proposed budget. The next day the group started again, but the QFD facilitator (not the author) ignored the customer's request stating that if it was not in the "specs" it was not a real need. The team finished the original QFD and lost the contract.

2B. Current Company Rating — The next step is to fill in Column 2B which ranks the company's performance or ability to meet this current need. The question being answered here is: "How well do you think your company

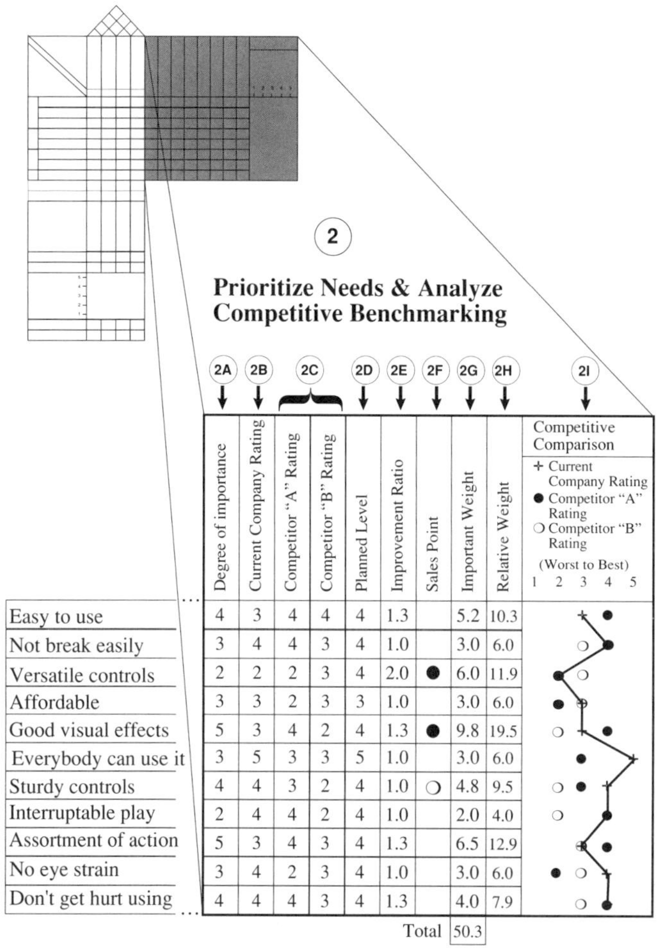

	Degree of importance	Current Company Rating	Competitor "A" Rating	Competitor "B" Rating	Planned Level	Improvement Ratio	Sales Point	Important Weight	Relative Weight
Easy to use	4	3	4	4	4	1.3		5.2	10.3
Not break easily	3	4	4	3	4	1.0		3.0	6.0
Versatile controls	2	2	2	3	4	2.0	●	6.0	11.9
Affordable	3	3	2	3	3	1.0		3.0	6.0
Good visual effects	5	3	4	2	4	1.3	●	9.8	19.5
Everybody can use it	3	5	3	3	5	1.0		3.0	6.0
Sturdy controls	4	4	3	2	4	1.0	❍	4.8	9.5
Interruptable play	2	4	4	2	4	1.0		2.0	4.0
Assortment of action	5	3	4	3	4	1.3		6.5	12.9
No eye strain	3	4	2	3	4	1.0		3.0	6.0
Don't get hurt using	4	4	4	3	4	1.3		4.0	7.9
							Total	50.3	

Figure 4-10 Video game Core QFD Matrix example — Section 2: Prioritize Needs and Analyze Competitive Benchmarking.

meets this need?" Again, a 1 to 5 ranking is utilized, with 5 indicating that the company fully meets that customer need at this time and that there is no room for improvement. The CVM team has ranked their company's ability to meet the identified customer needs. This can be done in a variety of ways, including third-party interviews and marketing analysis. A 5 would indicate that the company is completely meeting a need and improvement would be difficult.

2C. Competitors Rating — The next step, which also deals with benchmarking as does Column 2B, is Column 2C: competitors rating. The question that needs to be answered here is: "How well do you think your best competitors meet this need?" This is often one of the most difficult columns to complete in the total scheme of needs ranking and benchmarking. Often it is difficult to determine, first of all, who is a competitor in a particular field and if that competitor currently performs to a certain level of expertise regarding that particular customer need. This is where research and market data gathering can be the most beneficial. CVM compares favorably in most categories to its competition, Competitor "A" and Competitor "B." Because markets are changing so quickly, it often makes sense to analyze the competition at a point projected in the near future. The question to be answered then becomes: "How well do you think your *best* competitor will meet this need by the time you launch this new product, service, or business operation?"

2D. Planned Level — After completing the ratings for competitors, the next step of competitive assessment and benchmarking is to determine the planned level (Column 2D) for your organization. The question the team must answer is: "By the time you launch this new product, service, or business operation, where would you like your company to be relative to today?" Compulsive Video Mania determined that they wanted to increase their ability to meet their customers' needs by one to two steps for every element identified. This step of the competitive assessment can also be difficult because often, at this stage of the development of the QFD matrix, it is difficult to determine not only where the company may be in terms of benchmarking these elements, but whether or not the company can improve in certain areas. One effective way to look at this measurement is to determine where the company should be in order to *stay* competitive in the market regarding the planned level. Therefore, the second question to be answered by the team is: "Do you have or should you get the resources to make this improvement, taking into account how important this need is to the customer and how well your competitors provide this need?" In other words, Compulsive Video Mania determined that they currently have games that are easy to use and that no competitor outshines them in this area. Therefore they would like to remain at the planned level of 4. As with the other measurements in this section, it is important to look at these numbers as a comparative weighting among all the customer needs. It is important to identify exactly what a "4" means in terms of level. If a company assesses itself as being very good in that particular customer need at this time, or if the company assesses itself as needing improvement in that particular customer need, the definition of what is a "4" is often based upon marketing research and data taken.

2E. Improvement Ratio — The next step in filling out the Core QFD Matrix is to determine the improvement ratio (Column 2E), which is simply taking the planned level for that particular customer need and dividing it by the current company rating (2D divided by 2B). This ratio gives a weighted value for those customer needs that would require the most improvement. For example, Compulsive Video Mania feels that they currently perform at a level of "2" for versatile controls, but they feel that it is necessary for their new product to attain a planned level of "4" in their versatility. This would make the improvement ratio 4 (Column 2D) divided by 2 (Column 2B) which equals 2.

2F. Sales Point — Having determined the improvement ratio, the next step is to determine if any of the customer needs, when met, would give a competitive advantage to Compulsive Video Mania. This column is represented by the sales point (Column 2F). Not all elements are sales points. As mentioned earlier, a sales point is often a feature that the customer may not think is needed, but if it were provided it would create the "wow" factor. In other words if the customer saw this feature, perhaps not even anticipating it, the response would be "wow!" Sales points can be divided into three areas: (1) a big advantage which is worth 1.5 (represented by a filled circle), (2) a slight competitive advantage which is equal to 1.2 (represented by an open circle), or (3) no competitive advantage at all which is equal to 1 (represented by no symbol). As a rule of thumb, with a matrix consisting of approximately ten to fifteen customer needs, there should be no more than one, or possibly two, sales points that give a big advantage, and one or two sales points that provide a slight advantage. These are represented on the QFD matrix as symbols for a quick visual assessment of which elements provide sales points.

2G. Importance Weight (Customer Needs) — The next step is to determine the importance weight of the customer needs. This is represented in Column 2G and is calculated by multiplying the value of Column 2A (Degree of Importance) times Column 2E (Improvement Ratio) times Column 2F (Sales Point) of each customer need. This calculation takes the three most important elements and multiplies them for a final weight.

2H. Relative Weight (Customer Needs) — The next step is to normalize the importance weights in Column 2G as represented in Column 2H, the relative weight. This is done by taking the sum of the total of Column 2G and using it as a denominator for each value in Column 2G and then multiplying each column by 100 to normalize it. For example, in the QFD matrix created by Compulsive Video Mania, the total of Column 2G equals 50.3. The relative weight for the customer need of "easy to use" is then calculated as follows: (Value of 2G divided by a total of Column 2G) times 100, which is numerically represented as (5.2 divided by 50.3) times 100, which equals 10.3. This relative weight is then used in prioritizing technical solutions, as will be illustrated below. Often the relative weight is graphically represented in a bar chart for quick visualization of driving customer needs.

2I. Competitive Comparison Chart — The last task in Section 2, represented by 2I, is the graphical representation of columns 2B and 2C, Current Company rating and Competitors Rating, respectively. This step is not essential in the creation of the matrix, but permits two analyses: (1) The graphical representation gives a quick view of how your company compares to its competitors in meeting the customer needs; and (2) When later taken in comparison with Section 5 (which graphically portrays how you and your competition compare in meeting technical requirements) is a double-check to see if marketing and engineering or the requirements departments are in step in evaluation of technical and customer needs.

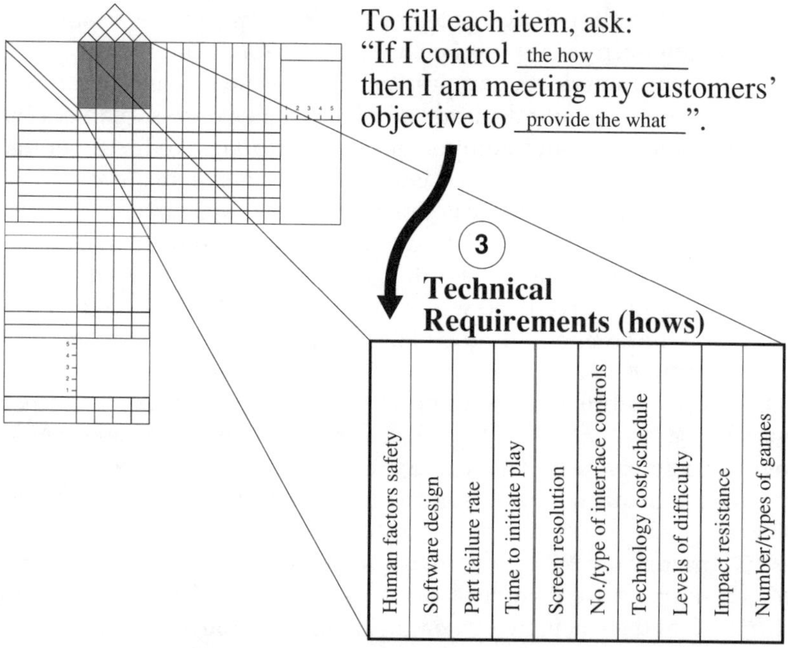

Figure 4-11 Video game Core QFD Matrix example — Section 3: Develop Technical Requirements (products, services, or business performance requirements).

4.3.2.3 Develop technical requirements (Section 3)

Technical requirements for meeting the customer needs can be developed in a manner similar to that used to develop the customer needs themselves (see Figure 4-11). The question the team should be answering is: "What can I control that allows me to meet my customers' objectives?" For example, "If I control the *number and types of games* then I am meeting my customers' need to have a wide assortment of action." Brainstorming and affinity diagramming can be combined with marketing knowledge to determine levels of technical detail. In other words, the following factors should be taken into consideration and used when developing the technical requirements:

- Use the "If I control . . . " question to help determine technical requirements;
- Use existing data and the combined experience from team members;
- Collect all available technical data from sources such as:
  - Statements of work/requests for proposals
  - Related publications
  - Technical interchanges with the customer
- Use brainstorming to identify any additional requirements;
- Use affinity/tree diagrams to organize the information;
- Additional guidelines:

- Address top-level requirements initially (such as systems before subsystems);
- Identify performance parameters;
- More detailed technical solutions are not necessarily better;
- Avoid restating components of particular products, services, or organizations;
- At the initial level try not to include parts or mechanisms, but focus on systems or technical solutions;
- Establish definitions for the technical solutions that are understood by all team members.

For Compulsive Video Mania, the technical requirements are listed across the top of the Core QFD Matrix. These technical requirements focus on product requirements and do not take into consideration the service and business operation elements as a result of the original scoping of the QFD project by the team. In addition to the standard technical requirements, there may also be special requirements such as government requirements, company policies, or other compliance-related requirements. For example, in the environmental industry there are often government agency requirements that must be met when considering design solutions.

4.3.2.4 Correlating technical requirements to customer needs (Section 4)

Section 4 of the initial matrix as represented by Figure 4-12 is the relationship area of the Core QFD Matrix. In this section, technical requirements are correlated to individual customer needs to determine strength of relationship and impact on the need. These relationships are numerically represented in Section 4. In order to determine the strength of a technical requirement to satisfy a customer need, four measures are utilized:

- *Strong relationship,* which has a numerical value of nine and is represented by a filled circle;
- *Moderate relationship,* which has a numerical value of three and is represented by an open circle;
- *Weak relationship,* which has a numerical value of one and is represented by a triangle; and
- *No relationship,* in which case the cell is left blank.

The three symbols representing the various degrees of relationship are allegedly derived from the Japanese horse betting system, with a strong relationship (a filled circle) equating to a "win," a moderate relationship (represented by an open circle) meaning "place," and a weak relationship (the triangle), equalling a "show." This horse racing analogy is also useful when facilitating the creation of a QFD matrix. It is often easier to rank correlations if the team considers whether or not the technical requirement is a "winner" for meeting the particular customer need.

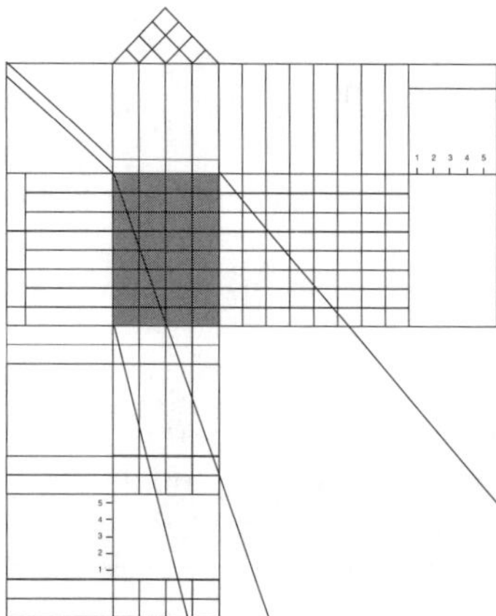

● High Impact
❍ Medium Impact
△ Low Impact

Information is evaluated by asking how each Technical Requirement meets the Customers' Needs.

		Human factors safety	Software design	Part failure rate	Time to initiate play	Screen resolution	No./type of interface controls	Technology cost/schedule	Levels of difficulty	Impact resistance	Number/types of games
Get money's worth	Easy to use		●		●		❍		❍		
	Not break easily	❍		●						●	
	Versatile controls		●	❍			●				
	Affordable	❍	●	●		●	❍	●	❍	△	❍
Provides competitive excitement	Good visual effects	△	❍			●		●			
	Everybody can use it				❍	△	❍		●		
	Sturdy controls			❍						●	
	Interruptable play		●								
	Assortment of action		●		△	❍	❍	●	❍		●
Safe design	No eye strain		●			●	●		●		●
	Don't get hurt using	●		❍		❍				❍	

Figure 4-12 Video game Core QFD Matrix example — Section 4: Correlate Technical Requirements to Customer Needs.

Another important tool in the team's evaluation process for correlating technical requirements and customer needs is to use the following sentence:

> If I control (the technical requirement) it will (significantly/moderately/slightly/or not) impact my customers' needs to (insert specific customer need).

This approach allows the team to determine whether or not the technical requirement correlates to a specific customer need. For example, in Figure 4-12 the technical solution of "software design" (Column 2) when compared to the first customer requirement of "easy to use" would be done as follows:

> If I control **software design** it will *significantly* impact my customers' need to **make it easy to use**.

and conversely,

> If I control the **screen resolution** it will *not* impact my customers' objective to **make it easy to use**.

The corresponding cell has no correlation in it because the technical requirement of screen resolution would not affect the ability to have the game easy to use. However, the technical requirement "time to initiate play" (Column 4 in Figure 4-12) would very definitely have a strong or "winning" effect on instructions being easy to use. Therefore, a filled circle is placed in that particular cell.

The selection of the numerical values 9/3/1/blank is a relative weighting so that strong correlations are valued heavily while correlations that are nonexistent or weak carry little value. Some users of QFD have substituted other values for the 9/3/1/blank system such as 5/3/1/0 and 4/2/1/0 for strong/moderate/weak/no relationship. The Japanese prefer the 4/2/1/0 or the 5/3/1/0 numerical values. The authors believe the 9/3/1/blank system better identifies drivers to the system and causes less arguments over distinctions that may not really matter, as well as providing greater separation between the technical solutions; this highlights the critical issues. Through sensitivity analyses the authors have discovered that the relative prioritization of technical solutions does not change significantly with any of the above numeric schemes. When facilitating a team through a QFD exercise, it is often important to spend a lot of time discussing a 9 vs. a 3 or a 1 relationship, but spend little time discerning between a 3 relationship and a 1 relationship, because of the lesser overall significance.

The symbols are used so that a visual examination of the chart can yield a quick evaluation of important technical requirements and important customer needs. In addition, evaluation of this section of the matrix, if done properly, can signal to the team tall poles or areas of focus, as well as gaps in either technical requirements to meet customer needs, or perceived technical requirements that do not add or correlate to any customer needs. This can be determined as follows: if a row for a customer need is blank or perhaps has only a single correlation among the technical requirements, the team may re-examine this customer need to first determine whether or not it is a real customer need and secondly whether or not there are any technical requirements that would meet that customer need. Similarly, any columns of technical requirements that are blank indicate either a technical requirement that fulfills no customer needs, or, possibly, a perceived need for a technical requirement but the customer need has not been properly identified.

In this way, Section 4 of the Core QFD Matrix operates in two ways: (1) as a mechanism to *correlate* customer needs and technical requirements, and (2) as a mechanism to *reevaluate* or check the results of selecting the requirements and customer needs.

4.3.2.5 Prioritizing technical requirements (Section 5)

Determining the importance weight and the relative weight of the technical requirements is represented in Figure 4-13 by references 5A through 5E of the figure. These values are calculated as follows:

5A. Importance Weight — The importance weight of the technical solutions is calculated as the sum of the value of the relationship (9, 3, or 1) multiplied by the relative weight of the customer need. For example, in our video game analysis the importance weight of the fourth column "time to initiate play" is represented as the sum of the following customer needs times their relative weights (Column 2I).

Easy to use	(9 × 10.3)	=	92.7
Everybody can use it	(3 × 6.0)	=	18
Wide assortment of action	(1 × 12.9)	=	12.9
Sum	124	=	123.6

Each of the columns is calculated in the same way so as to determine the relative weight of each of the technical solutions as compared to the customer needs.

5B. Relative Weight (Technical Requirements) — Once all of the columns have been totaled to determine the importance weights (5A) then the relative weights can be calculated through a normalization as was done in the calculation of the relative weights for the customer needs. In the video game example, the sum of the importance weights equals 2542. In order to determine the relative weight of each of the technical requirements columns, one merely has to divide the importance weight by the total weight, which will yield a percentage. Normalization of that number is then accomplished by multiplying each of the relative weight percentages by 100 as illustrated in Figure 4-13.

Once the relative weights of the technical requirements have been determined, it is important to examine the results to determine the key technical requirements that can help satisfy the customer needs. The relative weights

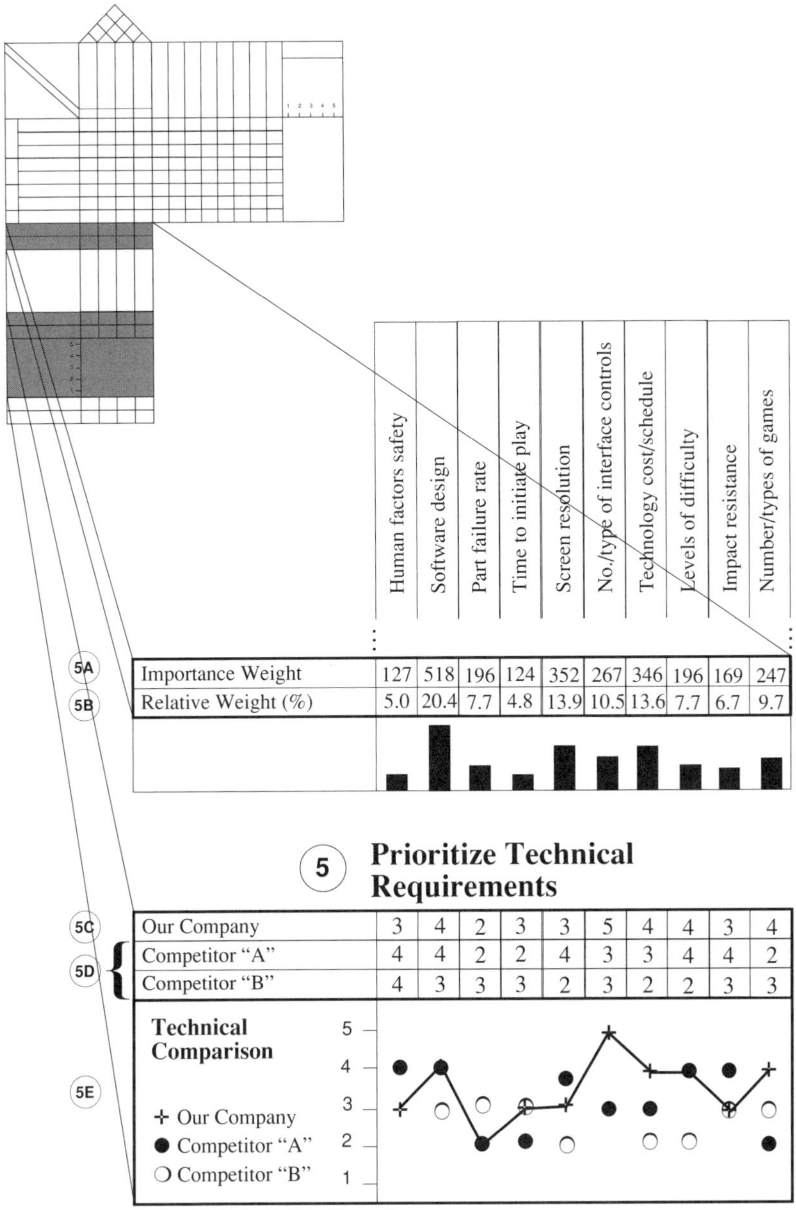

Figure 4-13 Video game Core QFD Matrix example — Section 5: Prioritize Technical Requirements.

are usually presented graphically through bar graphs to provide quick analysis of the key technical requirements. In the video game example there are three technical requirements which rank very high when compared with the customer requirement. Those three technical requirements are (1) software design, (2) screen resolution, and (3) technology cost/schedule. When examining customer needs for the video game, the design team knows that the top 30 to 40% of the technical requirements are going to probably drive the design and ultimate manufacture of the video game in terms of satisfying customer needs. By selecting the highest ranking technical requirements the design team would want to investigate them further and determine how the development of these elements are going to basically drive the design. The design team does not ignore the other elements of the design in favor of these elements, but rather investigates these candidates rigorously and examines the overall data to determine what can be done in these three areas to make sure the customer needs are satisfied. Additionally, if there are certain technical difficulties in any of these three areas, the design team knows that they will have difficulty in satisfying customer needs and that there may be a need for improvement in these areas through investment of either time or money.

5C, 5D, and 5E. Competitive Rating (Technical Performance) — Another element of Section 5 of the matrix is to determine how well both you and your competitors are meeting the technical requirements (based on a scale of 1 to 5, with 5 being the highest). As with the competition comparison represented in 2I of Figures 4-8 and 4-10, the technical comparison in Section 5E of Figure 4-13 will give a rough estimation of whether or not you currently possess some of the technical requirements in-house in order to meet the customer needs defined in the product or service. It is important to benchmark against your competitors in this area to determine if there is a technical gap, or a technical superiority, which again gives the team another element for focus to provide a superior product and meet those customer needs. In Figure 4-13 the numerical values of 1 through 5 have been transferred into symbols for visual comparison of the video game manufacturer and its two closest competitors. Although this is not necessary when taking a look at the initial Core QFD Matrix, teams often translate information into a visual, as well as numerical, representation for quick overall comparison. Also, many people interpret data better if they see it visually as opposed to examining information based solely on numbers.

Technical benchmarking is accomplished much in the same way as benchmarking satisfaction of customer needs. In order to determine how you compare to your competitors in certain technical areas, a company can use techniques such as laboratory testing, bench testing, and prototype simulation, and use both internal and external experts to support any rationale for determining capability. Companies, especially those in software or high-tech fields, often possess a great amount of data on their own technical capabilities, as well as the competitors' or the industry's capabilities. This information is often informally as well as formally exchanged, and also can be determined by comparing how well a company meets standardization criteria (UL, ISO 9000, etc.)

Comparing this technical requirement benchmarking with customer needs benchmarking also acts as a check between marketing and engineering. If the competitive benchmarking of a customer need shows that the competition is better at meeting that need than the company, and the competitive benchmarking of the technical requirements fulfilling that need are ranked higher than the competition, either marketing is not recognizing and selling a strength it has to offer to the customer, or engineering is overrating technical capability.

4.3.2.6 Set specific targets (Section 6)

6A. Target Values — After the technical solutions have been determined, and relative importance and weight have been assigned to them, the next step is to determine target values for those technical solutions. This step is often overlooked when conducting the Core QFD Matrix. Technical solutions, as mentioned above, are those elements that you plan on controlling in order to meet the customer needs. In order to control something you must also be able to measure that element; and if you are going to measure it there should be a target value that you want to achieve. Establishing the target values can focus a design team's efforts, as well as provide additional information about the design parameters.

When a team determines target values, they first should try to assign specific target values to as many of the requirements as possible. In assigning the target values it is important to define specific goals or ranges for designers, engineers, or individuals determining the service parameters. If a team spends sufficient time in setting target values, they can establish targets for trade studies and analysis. They can later use techniques such as Experimental Design to determine the validity of the target value as well as the methodology for reaching that target value. Figure 4-14 illustrates sample target values developed for the video game. Often these target values can be based upon information gathered in the original market surveys, or in the initial benchmarking of the technical requirements against competitors or the industry. For example, the time to initiate video game play is targeted at an average time of 1 min. This would be based upon such elements as the customer's impatience in wanting to start playing a new video game, as well as the current accepted market time frame for initiation of play on video games. Target values can also be based upon industry standards or requirements.

Target values can be a range, as well as a specific technical target that must be hit exactly. When considering the development of target values, the design team should also consider the results of designed experiments that define optimum values, historical or operator data, statistical process control data, and other benchmarking tools. It is important that each target is measurable and that the team can agree on the methodology for that measurement. These targets should be established at levels that are equal to or better than the competition, but should also take into consideration resource allocation and time commitment. There have been cases when a design team has examined target values that they feel are critical for certain technical

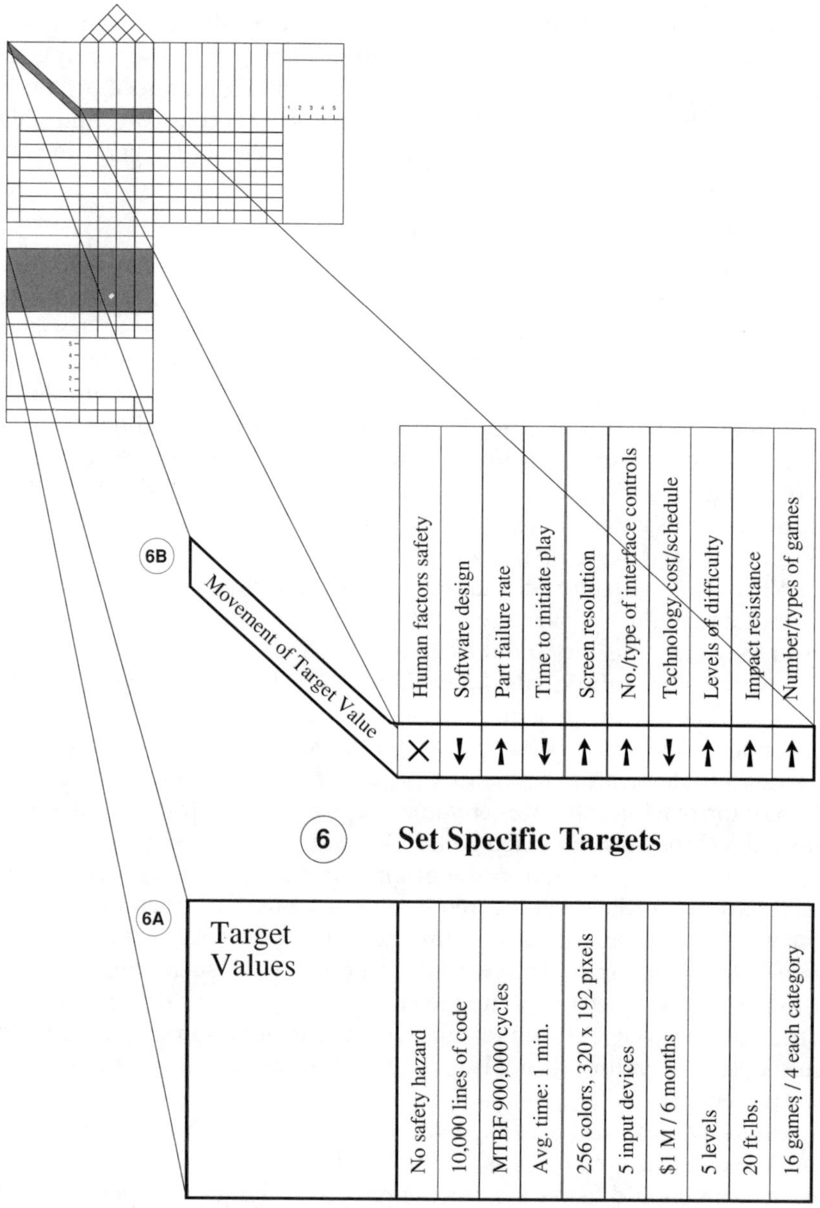

Figure 4-14 Video game Core QFD Matrix example — Section 6: Set Specific Targets.

requirements. However, upon further examination and investigation, the team discovered that these technical target values could not be attained, which ultimately meant abandoning the project. Although this can be discouraging, it also prevents unnecessary and wasteful investment of precious resources on a project that would ultimately end in failure because of an inability to meet a technical requirement that drives a design or service.

Finally, the design, product, or service team should keep in mind that these technical values are often only the *initial* target values that should be reassessed after additional data has been gathered. Target values often can be revised based upon detailed analysis such that they still meet the perceived customer requirement allowing the project to continue.

6B. Movement of the Target Value — Along with determining the actual target values, the team should also consider the movement of those target values for improvement or optimization of the design. This is represented in Figure 4-8 as 6B and is shown for the video game in Figure 4-14, located at the top of the chart directly under each technical solution. There are three entries to be made in this particular column: (1) ↑ indicates an increase in the desired direction of the target value, (2) ↓ indicates a decrease in the desired direction of the target value, and (3) an "X" indicates that a specified nominal value must be attained. In the video game example the team would want to decrease the actual time to initiate play because that would allow the customer quicker access to actually playing the game. Therefore a ↓ is placed under "time to initiate play" (Column 4). It is important to remember that these arrows are based on the target values that have been set in Section 6A. As a second example, the first technical requirement, "human factors safety," would probably best be represented by the meeting of United Laboratories requirements. Therefore, an "X" is placed in Column 1 indicating that certain minimum standards must be met.

4.3.2.7 Identify technical interactions (Section 7)

Figure 4-15 illustrates how to use the "roof" of the Core QFD Matrix to examine the correlations among the technical requirements themselves. The roof of the Core QFD Matrix objectively measures the interaction between the technical requirements and provides early recognition of positively and negatively correlated technical solutions as determined by the design or service team. The roof of the Core QFD Matrix is one of the more advanced tools that can be used to substantiate intuitive suspicions that focus on increasing the performance of a certain technical area can have both positive and negative impacts on other technical requirements. From a mechanical viewpoint, the roof is completed using four symbols represented in Figure 4-15. Those four symbols represent a range of correlation from strongly negative to strongly positive correlations. In the video game example, technical elements such as part-failure rate and impact resistance are correlated using the strongly positive symbol. This means that if you improve the impact resistance of the product, you would most likely improve the ability of that element to resist failure. Likewise, making the video game more failure-resistant may also positively correlate to greater impact resistance. On the other end of the spectrum, trying to increase screen resolution (which would most likely increase cost and possibly impact schedule) has a negative correlation to technology costs and schedule. This means that when a design team is looking at reducing cost or shortening the schedule they have to take into consideration that this may affect their ability to increase screen resolution with new technology developments. Similarly, if the number and type

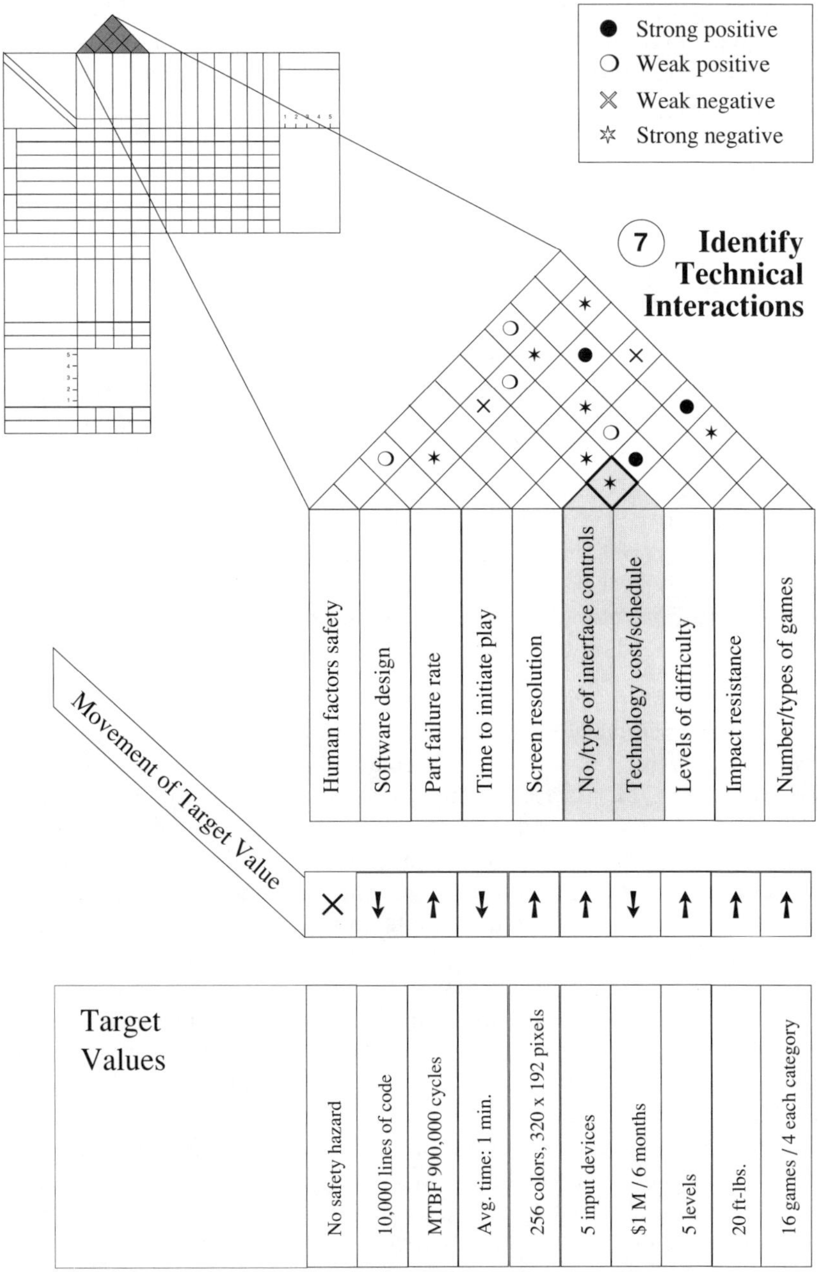

Figure 4-15 Video game Core QFD Matrix example — Section 7: Identify Technical Interactions. This example further illustrates the interaction between "Number of types of interface controls" and Technology costs/schedules.

of interface controls increase, it would have a significant negative impact on the ability to reduce the use of new technology as illustrated in Figure 4-15. By identifying these correlations early in the process, the team can competently discuss tradeoffs and potential "gotchas" in the technical design that may not otherwise be apparent until later in the design process and after the expenditure of potentially large sums of money.

When conducting facilitation to complete the roof of the Core QFD Matrix, it is best to use the most knowledgeable members of the team for the particular technical solutions in order to best identify interrelationships between the technical solutions. This part of the matrix can be one of the most difficult to fill out because it requires the anticipation of problems and potential tradeoffs between departments and product characteristics; therefore, it is often helpful to fill out the entire matrix, with all team members, and then use a subset of the key team members to complete the roof. This gets the team moving in the various areas before having a follow-on session when additional information has been gathered on the actual technical requirements.

4.3.2.8 Concept evaluation (Section 8)

After the first seven sections of the Core QFD Matrix have been fully determined, the next step is to develop various design concepts that take into consideration many of the technical solutions that have been developed in response to the customer needs. This step is for more advanced applications of QFD and is usually conducted by a team that is familiar with the rest of the Core QFD Matrix, and continues with the Integrated QFD Approach™ previously discussed in Chapter 3 (Figure 3-8). In order to complete Section 8 of the Core QFD Matrix the team must develop system alternatives to be evaluated and compared for determination of how well technical requirements are met. Team members may compare their initial best concepts with the requirements to analyze preliminary ideas for major technical gaps and for competitive advantages. However, the authors recommend further development of these concepts by developing functional flow diagrams and brainstorming concept options to meet the technical needs. This will also refine optional system alternatives that can be re-evaluated in the concept evaluation step of the Core QFD Matrix.

A major criticism of other QFD methods has been that these other systems only map what team members already know. The addition of functional analysis and concept brainstorming, further explained in Chapter 7, provides the means to stimulate team creativity in developing products, services, and business practices. These steps should be performed prior to considering Section 8 complete. Utilizing Section 8 of the Core QFD Matrix is a key element in full systems integration of the QFD analysis.

As mentioned earlier, a product or service concept can be evaluated against all other technical requirements as related to the customer needs. When developing a new product or service, there are often multiple ways or concepts that can be developed to meet both the overall customer needs and the technical requirements of the product or service. These concepts are

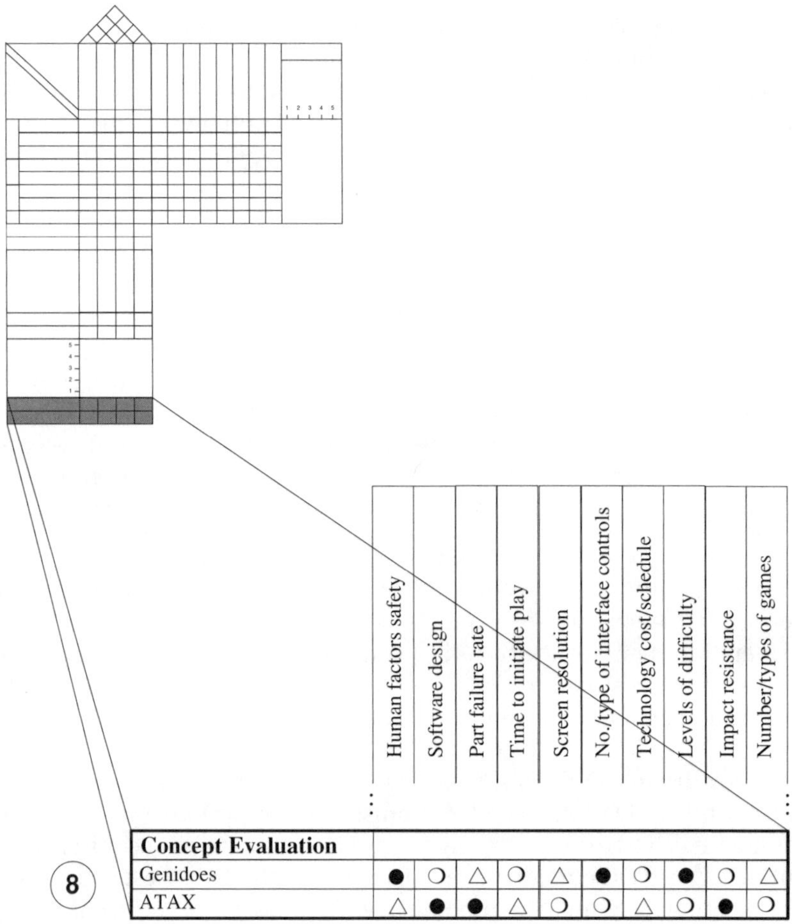

Figure 4-16 Video game Core QFD Matrix example — Section 8: Concept Evaluation.

developed in response to potentially conflicting requirements as identified in Section 7 (the roof) and the technical requirements themselves.

For the purposes of this example, we will assume that CVM has conducted further analysis and developed two system alternatives: ATAX and Genidoes, two potential game systems (see Figure 4-16). The ATAX concept design for the video game maximizes the ability to meet Software Design, Part-Failure Rate (low), and Impact Resistance. However, doing so means that the cost has gone up and the system only ranks a "1" (a triangle) in this category. This may mean that ATAX increases the cost to such an extent that it may not be commercially feasible. On the other hand, the Genidoes concept selection may go for a more middle-of-the-road technical solution when considering Software Design (a "3"), but significantly decreases the cost of the product making it more appealing to a larger audience. Neither system meets the requirement of Video Screen Resolution with a "9," which may mean that neither concept maximizes satisfying one of the key technical requirements. This may be because of the cost factor, but it is a serious consideration and more systems may need to be evaluated in order to determine if there is one that delivers screen resolution without being cost-prohibitive.

Section 8 allows the design or service team to measure each of the various concepts selections, in total, against the customer needs. When doing this the team can determine if a single concept may be high in all areas except one or whether another concept meets more of the customer needs than other selected concepts. Often, there is also the possibility of selecting elements from each of the concepts to obtain the most effective systems design that satisfies the highest-ranked customer needs and leaves no gaps in meeting the technical requirements.

4.3.2.9 The complete Core QFD Matrix

Figure 4-17 assembles all of the information developed for the CVM example in the eight sections of the Core QFD Matrix. This figure demonstrates just how compact the matrix is and how much information it can hold. However, just compiling this information does not mean the team is through evaluating the matrix. The next step is to evaluate the information for completeness and accuracy. This step is discussed in Chapter 5.

4.4 Automating the completion of the Core QFD Matrix

The Core QFD Matrix can often most effectively be completed by using a computerized tool that does all of the calculations given above, and also allows for modification of the various elements of the Core QFD Matrix. This allows teams to tailor the matrix to specific processes, or to use different values or rankings. There are several software programs to address this need, but the authors have found that one program, QFD/Capture (manufactured and sold by International TechneGroup Incorporated), has excellent flexibility and adapts well to the Core QFD Matrix application.

4.5 Chapter summary

1. The Core QFD Matrix is the heart of the Integrated QFD Approach™. It is a product, service, and process planning matrix used to depict customer needs and design, service, and business operations requirements, to target goals, as well as for evaluation of competitive product, service, or business operations.
2. The important point in gathering data for the QFD activity is to understand the needs of the customer fully, and that is often best done by asking the customer directly. This is why more and more companies are bringing the customer into the design and development process — making the customer a part of the QFD activity.

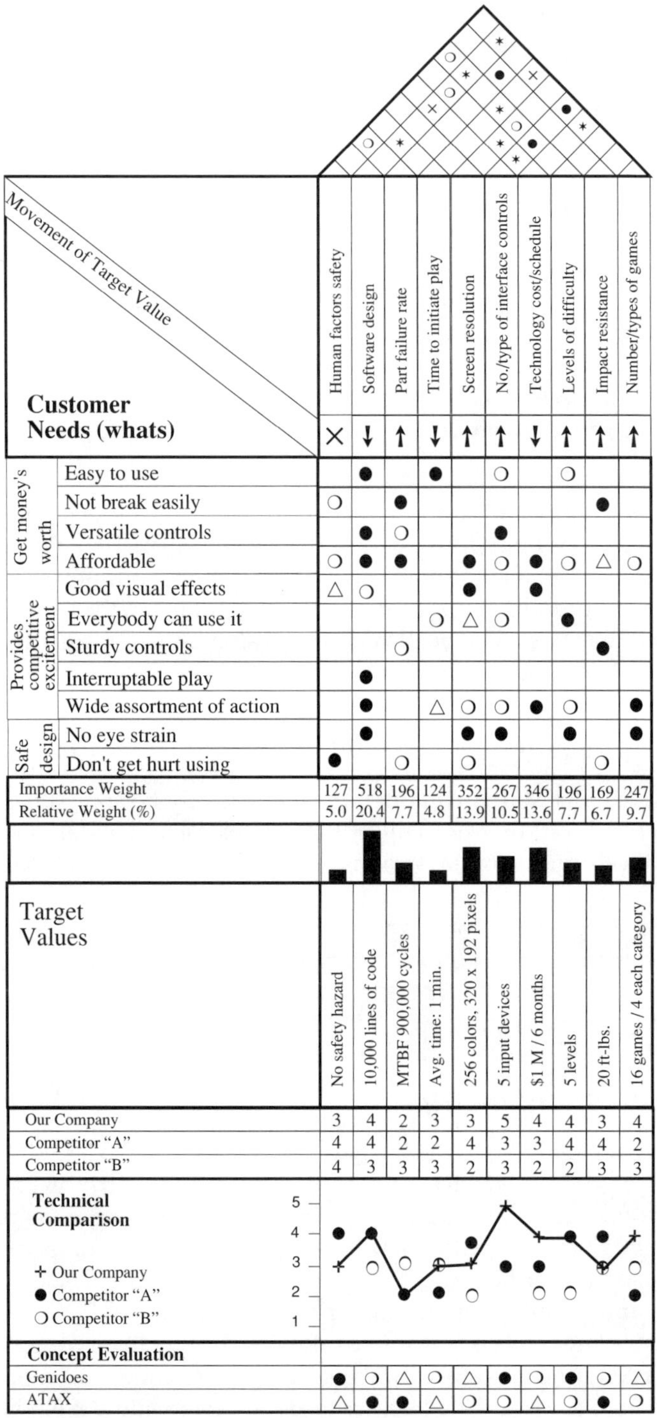

Figure 4-17 Video game Core QFD Matrix example.

Degree of importance	Current Company Rating	Competitor "A" Rating	Competitor "B" Rating	Planned Level	Improvement Ratio	Sales Point	Important Weight	Relative Weight
4	3	4	4	4	1.3		5.2	10.3
3	4	4	3	4	1.0		3.0	6.0
2	2	2	3	4	2.0	●	6.0	11.9
3	3	2	3	3	1.0		3.0	6.0
5	3	4	2	4	1.3	●	9.8	19.5
3	5	3	3	5	1.0		3.0	6.0
4	4	3	2	4	1.0	❍	4.8	9.5
2	4	4	2	4	1.0		2.0	4.0
5	3	4	3	4	1.3		6.5	12.9
3	4	2	3	4	1.0		3.0	6.0
4	4	4	3	4	1.3		4.0	7.9
						Total	50.3	

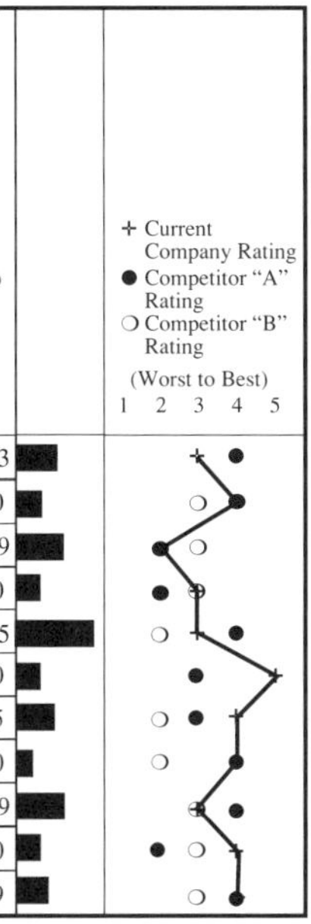

Figure 4-17 (continued)

3. Supporting techniques and tools in developing a Core QFD Matrix include brainstorming, affinity diagraming, and tree diagraming. These methods stimulate the team's creativity, develop and organize information quickly through a structured process, and provide visualization of relationships at various detail levels, which can reveal any gaps or holes in the sequence of information.
4. A detailed reference guide to completing a Core QFD Matrix is provided in Appendix A.
5. The Core QFD Matrix has common elements with all the QFD matrices used in the Integrated QFD Approach™. These steps are highlighted in the following eight steps:

Step 1. Identify Customer Needs — Defines customer needs in a hierarchical fashion.

Step 2. Prioritize Needs and Conduct Competitive Benchmarking — Provides a relative prioritization of customers' needs based on the importance to the customer, competitive assessments, in-house improvements required to meet the customer needs, and the unique attributes of any specific need that could help capture the market.

Step 3. Develop Technical Requirements — Defines the product, service, and business operations in a hierarchical fashion that can be controlled to satisfy the customers' needs.

Step 4. Correlate Technical Requirements to Customer Needs — Provides an indicator of how each technical requirement impacts the ability to meet the customers' needs.

Step 5. Prioritize Technical Requirements — Provides a numerical and graphical representation of prioritized technical requirements to be used as a basis for developing an integrated plan of action to improve products, services, and business operations.

Step 6. Set Specific Targets — Defines measures for a successful product, service, or business operation that can be integrated into product specifications, service, and business procedure guides.

Step 7. Identify Technical Interactions — Defines conflicting and redundant performance system goals. These interactions are further analyzed in system trade-off studies and process improvement analyses.

Step 8. Perform Concept Evaluation — Products, services, and business operation concepts can be evaluated against technical requirements and related to customer needs. Each independent element of a particular concept can be evaluated for strengths and weaknesses as compared to the performance targets set during the QFD analysis.

6. Once the Core QFD Matrix is complete, it should be analyzed for completeness, adequacy and consistency, which is discussed in Chapter 6.

chapter five

Analyzing the Core QFD Matrix

The difficulty is not in catching the Tiger;
It is what to do with it when you've got it by the tail!

5.0 Ensuring an accurate matrix

After a team has developed the Core QFD Matrix, the work is hardly over. The next step in using the matrix as shown in Figure 4-8 is to evaluate the information on which the team based the prioritized technical requirements and customer needs, and create an integrated plan — right?

Wrong.

If you have carefully followed the steps to construct the matrix, and assembled all of the information to be considered, the next step is to conduct a cross-check of the matrix itself to see if the information you have put into it and the priorities it yielded make sense. The matrix should not be used as a substitute for critical thinking, but as a lens to focus a team. Judgment should never be surrendered for the sake of a method or tool. If precautions are not taken to ensure that biases or lack of knowledge (both technical and about the customer) have not corrupted the analysis, the results will be limited. It is axiomatic to say that no approach is 100% foolproof, but no method of analysis is useful unless fundamental checks can be employed to determine whether or not the majority of the analysis can be relied upon for accuracy. The QFD matrix must also be capable of analysis for completeness, bias, and adequacy. Therefore, it is necessary to take a look at the format of the matrix before interpreting the data to create an integrated plan (the interpretation of the matrix and how it fits into an integrated plan is discussed in Chapter 6).

5.1 Evaluating the Matrix for completeness

The reader has seen how the Core QFD Matrix is an extremely effective tool for prioritization of customer needs, prioritization of technical requirements, and identifying the various relationships between the two. It is also an extremely effective tool for organizing a team, developing a strategy for a new product or service, or re-evaluating processes for process improvement. But how do you know if the matrix is complete? Does it include all of the necessary customer needs? Has the team considered all of the key technical requirements?

The first level of analysis to determine whether or not a developed QFD matrix is complete involves analyzing the matrix for any gaps in either customer requirements or technical requirements. The need for properly identified customer needs has been touched on in Chapter 2, but the importance of understanding and analyzing the Core QFD Matrix cannot be overlooked. Chapter 2 discussed some of the ways to identify customer needs, but the Matrix itself can give information on whether or not those customer needs have been met. After the completion of the first five sections of the Core QFD Matrix (up to determining the relative importance of technical requirements) the product design or service team should analyze the matrix for any "gaps" in customer needs (whats) or technical requirements (hows).

In order to determine whether or not there are any customer requirements that have not been addressed by technical requirements, a team should analyze each "customer need" row, and be suspicious of any customer needs which are not correlated to any of the technical requirements so far identified. A blank row should immediately signal the team that they have identified a customer need that has not been addressed by any technical requirements or solutions. Figure 5-1 illustrates a blank row in a Core QFD Matrix. In conducting this analysis it becomes very apparent the reason for utilizing symbols for correlation of the strength between the technical requirements and the customer needs as opposed to merely filling in numbers in Section 4 of the matrix. At a quick glance, the team sees an empty row, or a row with only one or two triangles or circles in it (meaning that the technical requirements meeting the customer need are weak, at best). A blank row means that:

- Further discussion and analysis should be conducted to determine whether or not the customer need is properly identified, or whether this is a customer need at all;
- A technical requirement has been identified as a customer need making development of further technical solutions difficult; or
- No technical requirements have been identified to support the customer need (in most cases this is the problem).

This kind of gap in the development of a QFD matrix is easy to spot, but sometimes requires rethinking the customer needs and technical requirements in order to find a remedy.

A similar gap can often be identified when analyzing the technical requirement columns. If a column is blank, or has only a few low-level correlations to customer needs, the likelihood is great that the technical requirement is a red herring. The matrix is useful in detecting superfluous requirements that are often thought to be essential to the customer. If the team has carefully conducted the needed research to develop accurate customer needs, then the failure of a technical requirement to meet any of those

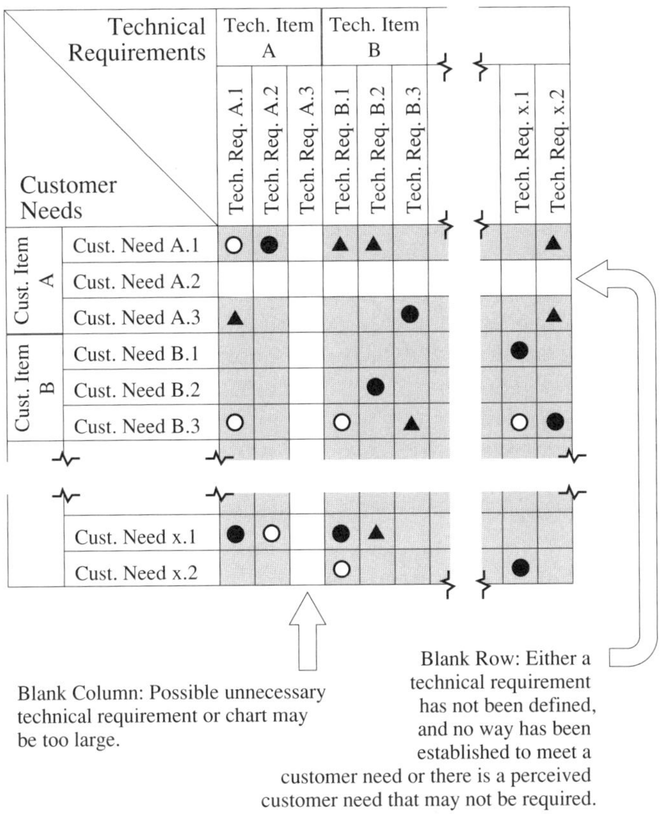

Figure 5-1 Analyzing the Core QFD Matrix for completeness (part 1).

needs usually means the technical requirement should be either defined in greater detail or dropped. However, care should be taken to also consider the following points:

- There is supporting documentation for all the customer needs;
- The matrix may be too large, including too many levels of technical requirements for the level of detail of customer needs, making it difficult to establish significant correlations;
- The team may not have a common understanding of the definition of the technical requirement, making correlation to the customer need difficult; and
- The perceived technical requirement is actually a "what" (customer need), and therefore independent from the other customer needs.

Figure 5-1 illustrates what a matrix may look like when a technical requirement is unnecessary. It is possible that the matrix may be too large and include too many technical requirements for the current level of analysis when using the Integrated QFD Approach™ on more complex product developments. Engineers and technically oriented individuals have a tendency to go right to the details of a design to solve a problem. The team should be sure to analyze a technical requirement in the framework of the tree diagram created for all technical requirements in order to determine if it belongs at a lower level of detail.

When two or more customer needs are correlated exactly to the same technical requirements, the possibility exists that the customer needs have been broken down to a level that is not useful in the initial matrix development. Figure 5-2 illustrates a matrix exhibiting this situation. It has been the experience of the authors that two customer needs rarely exhibit direct correlation to the same technical requirements. The customer needs should first be examined to determine if the hierarchy is sufficiently defined. For example, customer needs such as user-friendliness and operational flexibility may be perceived as identical at this level of detail. As a result, the two needs may be equally evaluated to technical requirements. A lower level of detail may mean that the two needs are really a lot more disparate than could be determined without the subcategories.

Such similarity of technical correlation to customer needs may also indicate that insufficient analysis of the customer needs exists. This scenario often occurs when a team does not want to "sacrifice" perceived customer needs that are actually very similar in the functions they perform. It may also indicate that the technical solutions that have been chosen are too generic to provide meaningful distinction.

Figure 5-2 also illustrates how a matrix may appear when a customer need is met only minimally by technical requirements. As mentioned above, the result is the failure to meet a customer need adequately, and the technical requirements should be re-evaluated.

5.2 *Evaluation for adequacy and bias*

After determining whether or not there are any gaps in the technical requirements satisfying the customer needs, the next step is to take a look at whether the initial matrix is adequate in evaluating the task selected by the team. Sometimes a matrix "looks" sufficient on first examination: there are one or two technical requirements to meet each customer need; the team feels that appropriate research was conducted in order to define customer needs; and the team arrived at the Core Matrix construction through Section 4 without a lot of disagreement or bloodshed. If this is the scenario, than the Core Matrix may still be inadequate! QFD is a dynamic process. Arrival at a matrix that has only one or two technical requirements for each customer need (no matter how strong the correlation) is indicative of a problem, such as defining technical requirements as customer needs.

Figure 5-3 illustrates a scenario that occurs when technical requirements are categorized as customer needs. The matrix indicates that there is only one or two technical requirements for each customer need. This immediately

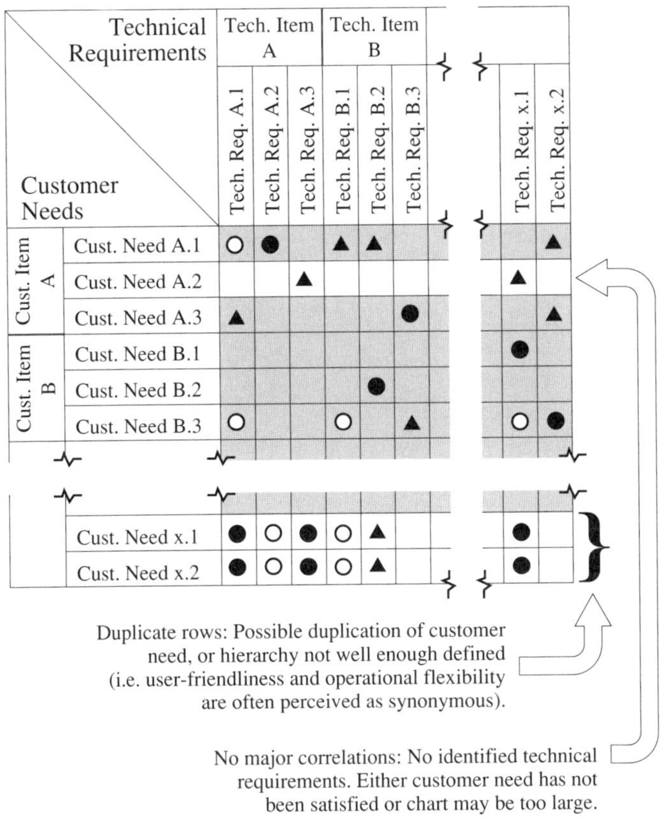

Figure 5-2 Analyzing the Core QFD Matrix for completeness (part 2).

becomes suspect because the essential nature of QFD allows the construction of a matrix to show how customer needs can be satisfied by several technical requirements, and technical requirements can satisfy several customer needs. A seemingly unique correlation of customer need to technical requirement should alert the team that there has not been sufficient analysis to define the actual customer needs, and the team should go back and decide whether or not the customer needs are truly "whats" as opposed to technical design or service solutions (hows). This situation occurs most often when the team is working exclusively from a "requirements" document provided by the customer or developed in-house. It also occurs when teams are changing or working from "old" or existing designs.

A perfect example of this involved two teams organized by a manufacturer to redesign an existing product to meet new customer requirements. One team had been working for 6 months designing a new system, and one had just been assembled to re-examine the problem. Both teams were to complete a QFD analysis, led by one of the authors. The team that had been together for 6 months dashed through the QFD mapping creating a matrix

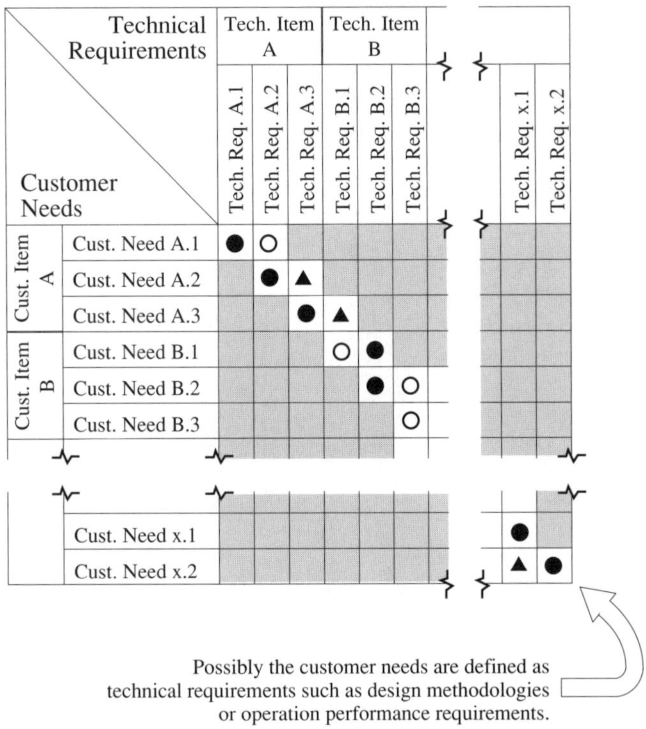

Figure 5-3 Analyzing the Core QFD Matrix for adequacy (part 1).

similar in appearance to Figure 5-3. They used preconceived notions of customer "specifications" and concluded that the design they were working on already was the best. The second team that had been together for only 2 weeks did a step-by-step analysis that included redefining the customer needs into what they believed to be product characteristics, but not necessarily design "specs." After 3 days the second group came up with *three* new product concepts and proceeded to begin patent procedures on one that was only one tenth the cost of the original product.

Why did this happen? The group that had been together for 6 months insisted that a technical requirement in the specification was a customer need, and mapped it to a "requirement" which limited options for further consideration. They accepted a single solution and "reversed" development of the customer needs. The "new" group approached the problem from a different angle, asking *why* the customer wanted the element in the specification, and how the total system was to be used. The technical requirements then became the means to achieve the end operational needs. The groups then approached the customer with the new concepts asking for a redefinition of the specifications, and both the customer and the manufacturer found a solution that was more affordable and easier to install and maintain. This is an example of bias toward a design or approach overshadowing the first team's ability to see the problem from a fresh viewpoint.

The error in using technical requirements as customer needs can be illustrated through an example using snacks. If a group of people want snacks, characteristics should be brainstormed as opposed to asking for specific snacks. Sometimes the group needs to be prompted into thinking differently about requirements and needs. If a group member asks for a chocolate bar, that's exactly what they get. If a member asks for a *candy* bar, they may get a variety of candy bar snacks that could include a chocolate bar. If the group asks for *something sweet* a multitude of possibilities arises including cookies, chocolate bars, pastries, ice cream, etc. The *real* customer need can then be defined as something sweet because it has fewer technical limitations and opens the door for many options. Customer need identification is best conducted when the choices for a solution are not limited or chosen too soon.

Another scenario that should create suspicion by the team of their analytical methods and identification of customer needs and technical requirements is represented in Figure 5-4. Visual examination of the represented QFD analysis shows a "clumping" of technical requirements to customer needs in the upper left corner of the correlation matrix. Such a clumping usually indicates that the requirements (both technical and customer) are not properly understood and that the hierarchy of each requirement has not been properly identified. The team should re-analyze the affinity and tree diagrams supporting the development of the customer and technical requirements.

Figure 5-4 also reflects a scenario when customer needs are taken to too great a detail. This focus results in technical requirements satisfying only one customer need, creating a bias or overemphasis on that need. The bottom half of Figure 5-4 demonstrates that a problem exists when one customer need (Customer Need x.1) is a subrequirement of a tree diagram, but heavily correlated to all technical requirements, whereas a similar requirement (Customer Need x.2) is not. At this point, the customer needs should be re-evaluated to determine if that hierarchical category has been taken down to a level of greater detail than the rest of the customer needs identified. It is important to identify whether or not all customer needs are at relatively the same level of detail. If a single customer need has been broken down into, for example, ten subelements that are at a greater level of detail than the rest of the customer needs, emphasis placed on a certain detail level will be in improper proportion to the other customer needs. This causes the QFD analysis to be skewed heavily in favor of that particular element, when removing that detail would reveal an element without as great an emphasis in comparison to the other customer needs. The same logic also applies if a single column (technical requirement) is overly emphasized in comparison to all of the customer needs, as illustrated in Figure 5-5 (Technical Requirement A.3).

Figure 5-6 demonstrates a scenario that at first looks like a "good" Core QFD analysis has been performed. A good distribution of correlations exists between the technical and customer need; but closer examination shows that there are far too many weak relationships and this would be extremely

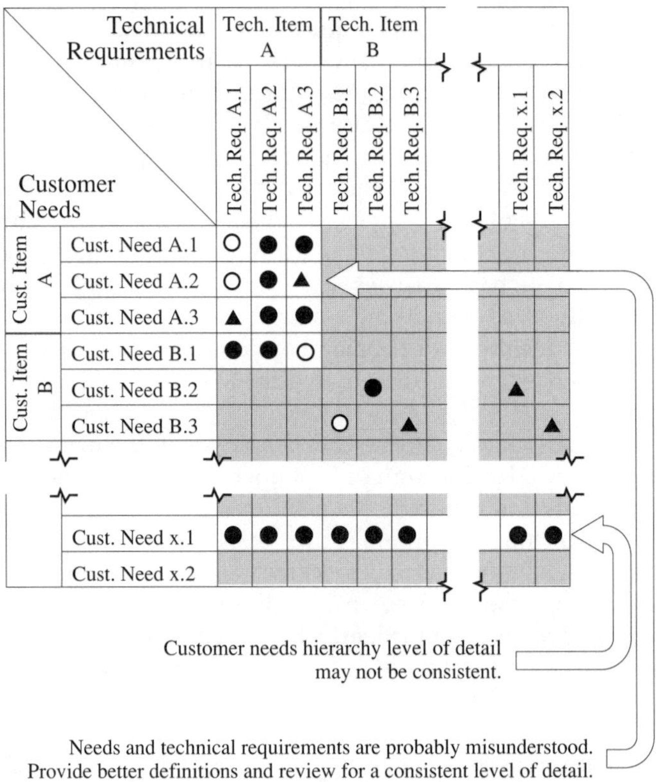

Figure 5-4 Analyzing the Core QFD Matrix for adequacy (part 2).

unusual when conducting a QFD analysis. Typically, the development of a QFD matrix errs on the side of too many high correlations between many of the elements as opposed to many weak correlations. If the team determines that there are far too many weak correlations, then two possibilities exist: (1) that no clear technical requirements have been determined for the customer needs, and (2) the technical requirements have not been clearly understood. A matrix such as this would dilute the results even if taken to the next level of detail. If the team is struggling with identifying technical requirements that relate strongly to the customer needs then the team should focus on each customer need and suggest technical requirements that they feel would strongly solve that customer need. After doing this for each customer need, the team should then re-evaluate if any of these technical requirements are redundant, or if there is another level of detail that can be addressed.

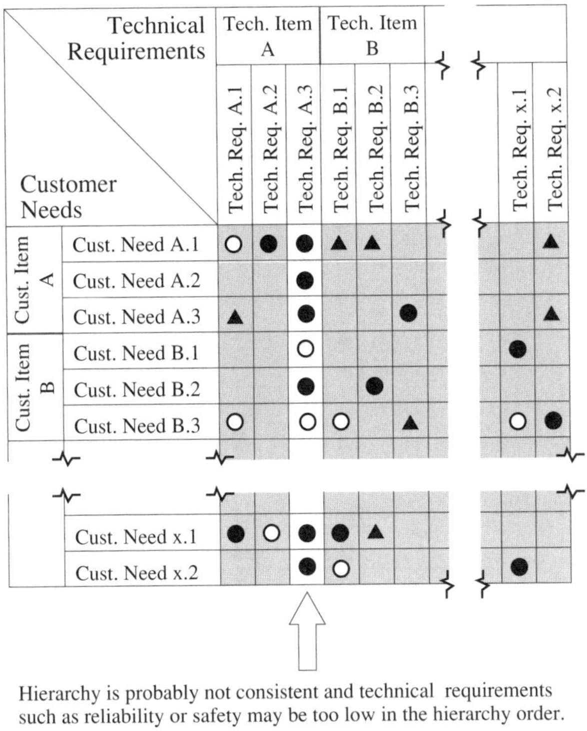

Figure 5-5 Analyzing the Core QFD Matrix for adequacy (part 3).

5.3 *Using competitive benchmarking to check marketing and technical perceptions*

As mentioned in Chapter 4, comparing the graphical representation of the Customer-Needs Competitive Comparison (Section 4.3.2.1 referring to 2I of Figure 4-10) with the Technical Competitive Comparison (Section 5E of Figure 4-13) not only offers benchmarking, but also a check to see whether marketing and the technical disciplines are on the same page. This is not an "apples-to-oranges" comparison. The technical elements do not map one-to-one with customer needs, but the mapping indicates a general "trend" that can give a basic reading of perceptions. If the comparison of your company's technical and customer benchmarking against the competition is relatively the same (Figure 5-7) then marketing and the technical departments are operating on roughly the same perceptions.

If, however, the evaluation of how well your company meets the customer needs is generally higher than how well you meet technical requirements compared to other companies, then marketing may be overselling the

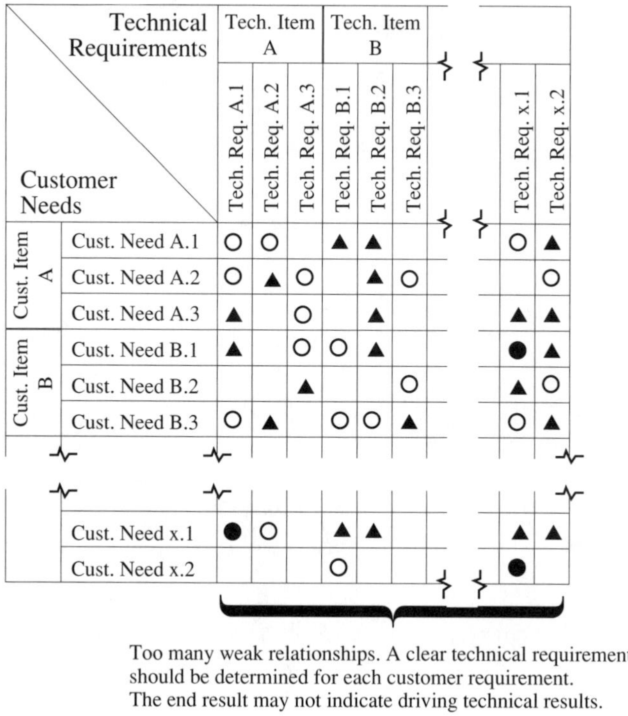

Figure 5-6 Analyzing the Core QFD Matrix for adequacy (part 4).

company's technical capabilities. This is important to know because if marketing oversells the technical capabilities of the company, customers may be dissatisfied with products or services that do not meet the level requested, or the company may not be able to deliver on customer needs. But this is not the only answer for a discrepancy in perception of competitive benchmarking. If marketing feels that the company's technical abilities are meeting customer needs, the technical evaluation may be out of line with reality, which could lead to emphasis on resources to improve a technical capability that may not need improving.

Conversely, if marketing perceives a lower capability to meet customer needs than competitors, and the technical evaluation shows a generally higher capability than competitors, one of two things may be true: (1) marketing may be underselling the strengths of the company, or (2) the technical departments may not have a clear picture of the company's technical competitive position.

The bottom line is that any discrepancy in these two elements — customer-needs competitive benchmarking and technical benchmarking — needs further analysis and input to resolve.

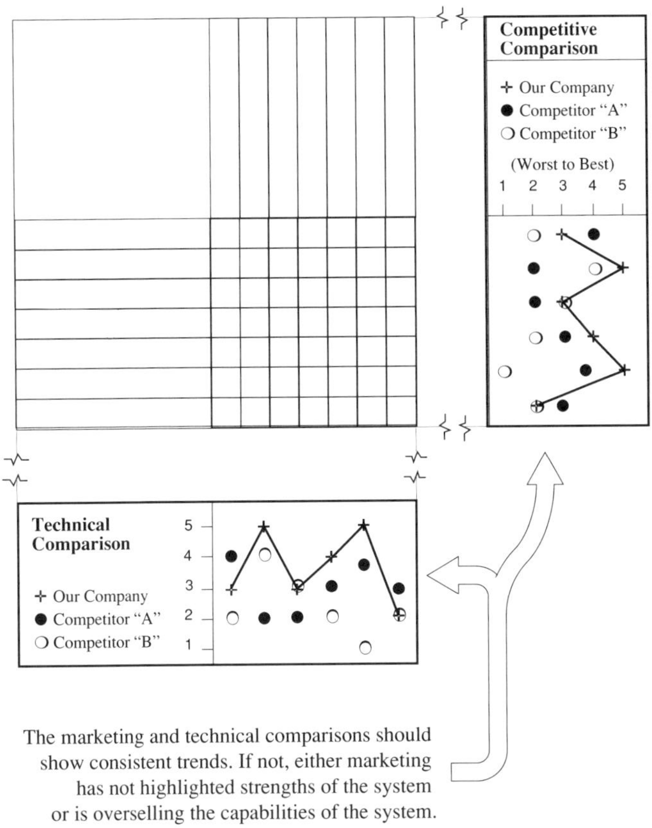

Figure 5-7 Analyzing the Core QFD Matrix for consistency.

5.4 Checklist of relationships

The authors hesitate somewhat to give a checklist of what to look for in evaluating the adequacy of a QFD matrix because checklists are often used as substitutes for clear analysis and reasoned thought. However, checklists can be useful in getting the team to conduct its first analysis and to suggest potential problems that can be found in the QFD matrix. With this caveat, the authors would recommend that the checklist represented in Table 5-1, be utilized as a starting point in analyzing the QFD matrix. No checklist can substitute for experience or familiarity with the subject matter upon which the QFD analysis is performed. But the checklist does give some basic guidance when analyzing the QFD matrix. The previous examples of potential problems identifying an incomplete or biased QFD matrix are not meant to be an exhaustive list, but merely illustrative of thought processes used to check the efficacy of the QFD analysis.

Table 5-1 Parameter Checklist for Analyzing the Core QFD Matrix

❑	Blank Rows or rows with weak relationships.
❑	Blank columns or columns with weak relationships.
❑	Are customer requirements consistent with technical requirements in competition assessment?
❑	Are there possible technical parameters that might serve as discriminators with competition?
❑	Do conclusions seem realistic based on comments from members outside team?
❑	Do the marketing and technical comparisons indicate similar trends?

As a facilitation suggestion, the authors have found that one of the most effective methods of checking the completeness and adequacy of a QFD analysis — especially on large projects — is to develop the initial product or service matrix using two teams with similar skills, but not allowing the teams to communicate during the construction of the initial matrices. This way, not only is there an element of competition between the two teams, but there is also the ability to cross-check analysis and utilization of data in developing the initial matrices. The authors would recommend this approach only if there is sufficient time and resources to conduct the parallel efforts, and the team leaders feel it would be justified to obtain a parallel view of the product or service development process. At a minimum the team should share their results with experts in the field to validate the outcome.

5.5 Chapter summary

1. The QFD Matrix should be analyzed for completeness, bias, and adequacy.
2. Completeness — key indicators that the matrix is not complete include:

   - Blank rows or columns; and
   - Rows and columns with only weak correlations.

   Potential reasons/solutions:

   - No technical requirements have been identified to meet the customers needs;
   - A technical requirement has been identified that is not necessary;
   - A perceived customer need has been identified that is really not a customer need;
   - Matrix may be too large; and

- Terminology of customer need or technical requirement not well understood by team members.

3. Adequacy and Bias — key indicators that the matrix is biased or inadequate:

   - Limited technical solutions (one or two for each customer need) in a diagonal pattern;
   - A densely populated area of high correlations or "clumping";
   - A row completely filled or nearly filled with strong relationships;
   - A column completely filled or nearly filled with strong relationships; and
   - A matrix filled with weak relationships and very few strong relationships.

   Potential reasons/solutions:

   - Customers' needs defined as technical requirements;
   - Level of detail or hierarchy in the technical or customers' requirements inappropriate;
   - Matrix too large and too detailed; and
   - Misunderstanding of terms.

4. The team should use experts in the field to validate the results.

chapter six

Developing an Integrated Plan from the Core QFD Matrix

Even if you are on the right track, you'll get run over if you just sit there.
Will Rogers

6.0 *The next step*

The Core QFD Matrix is not an end in itself. It is part of an integrated system for creating repeatable success. Once a Core QFD Matrix has been completed for the top level evaluation of a product, service, or process improvement, it should not be discarded; and team members should not be left alone with their assigned tasks to attempt implementation of the priorities and insights developed during the QFD process, or without the benefit of a plan of action. Once the Core QFD Matrix has been completed and checked for consistency, bias, and accuracy (Chapter 5), the results of the matrix should be carefully analyzed to determine how to proceed with an integrated plan — which is also the next action in the Integrated QFD Approach™.

But before looking at the results of the matrix and how to work them into an integrated plan, we should consider a parable about planning.

6.1 *The dam too far*

Back in ancient times (before cellular phones) there was a farming community in a land that was rich and fertile. The community was called Bizillion and was in the valley of the great river Commercia. This river provided abundant water for the community and over the millennia had deposited the rich, loamy soil that made the valley a very productive crop-growing region. The only drawback to living in the valley was that every 20 years the river flooded, destroying the village, flooding the cropland, and basically making life miserable for one or two seasons.

Now the villagers were an industrious and creative lot. Shortly after one of these flooding events, they held a town meeting and decided that something needed to be done. Well, the village elders assembled a multidisciplinary team from the populace consisting of farmers, merchants, road builders, and laborers. This team was given the task of developing a technical solution for saving the village and preventing future floods. The team members worked furiously for 2 months brainstorming solutions to this problem. After much effort, a design was selected for damming the river upstream where the valley narrowed. The village elders carefully scrutinized the design and gave the go-ahead for the project. The team was asked to accomplish the dam before the next flood, 17 years away.

The team was very motivated at first. They had selected the most important technical requirements for building the dam (using an ancient form of QFD) and knew they had the best design to meet the requirements set out by the elders. Unfortunately, the team did not get underway immediately. Each team member had commitments to its group. The farmers on the team had to begin planting for the next crop; the merchants were busy trading to replace some of the lost profits from the last flood; the road builders were reconstructing the washed-out roads; and the laborers were busy rebuilding structures. These were all very important activities, and the team members felt obligated to their individual groups to take care of these commitments before working on the dam. After all, they still had quite some time before the next flood came.

A year came and went. The team met several times during the year to revisit the design and reaffirm their commitment as a team. The team was optimistic they could build the dam in time, but were unsure of the exact steps to accomplish this task. In general the farmers needed to gather some straw and prepare mud for the dam; the merchants needed to round up some wagons and horse teams to haul supplies; the road builders needed to map out the construction of the dam and how to handle the flow of the river during the construction phase; and the laborers would be needed to move stones, dig earth, and generally labor. These steps were part of the technical requirements for the dam, but there was no real task definitions or schedules, no estimate of materials, and no evaluation of human labor needs or schedule.

Another year passed.

The team met a few more times, but the urgency of the dam always took a back seat to more pressing needs: the village was growing and more food had to be grown, homes had to be built, and roads were under construction.

Several more years passed.

The team continued to meet, but morale was slipping. Everyone felt they were attending meetings, but no progress was made. Some of the team members became disgruntled and stopped coming to the meetings entirely. A little work was accomplished (a nearby quarry was located for the rock), but all-in-all, not a lot of progress was made.

Several more years passed.

Some of the team members had died by this time, and the others had gotten older and were less excited about the work involved in the dam construction. Then the elders called on the team leader to check on the status of the work. When the team leader spoke to the elders about the lack of progress, the elders panicked. The next flood was only 2 years away! The team had to meet immediately and start work on the dam.

The team leader quickly assembled what was left of the team. Several team members had died or moved to other villages and the team leader had to recruit additional team members. These team members did not have the benefit of knowing how the design was developed or what had transpired before. No one anticipated having to work on the dam immediately and there was much grumbling among the various groups who had to send additional workers to help with the dam.

The team did a quick analysis of what was necessary to build the dam according to the old design and realized such an undertaking would require 5 years to complete. They only had 2 years. So the team modified the design and accepted a less massive structure that could be completed in the 2-year period.

Work began immediately, but problems soon hindered progress. Manloading was difficult because of the other needs of the village. The dam site was several miles away from the village and attacks by wolves were not anticipated, so additional men had to stand guard over the workers. It was difficult to find enough straw because of the short time frame and some of the livestock had to be sacrificed, creating a shortage of milk for the children. Things were in terrible disarray, but the tenacious workers managed to complete the revised dam prior to the spring flood. (Fortunately there was no environmental impact statement required or they delay would have prevented completion!)

The villagers were tired and worn from the ordeal, but a celebration was planned to honor the deed. The day of the celebration saw the spring monsoonal rains come down in torrents, testing the dam's strength. Unfortunately for the villagers, who were gathered in the central meeting hall, the new design did not hold and the backed-up waters of the river cascaded through the valley at a greater volume than ever before, carrying the village (villagers and all) away.

Although the flood was certainly a disaster, enough of the villagers had managed to scramble out of the flood to come back to the valley, determined to rebuild. The first thing the new governing elders did was to make a proclamation that the dam team would not use the same method to construct the new dam. After all, the method was faulty because it did not result in a dam that held back the waters of the river. The members of the old team who survived the flood were immediately hanged and the village returned to life as before.

Unfortunately, this parable has been all too commonly played out in companies over the last few years. QFD is tried and then abandoned and the system is blamed. One of the causes has been the failure of the team to take an important step after creating the matrix — creating an integrated plan to take action. The remainder of this chapter discusses how to use the information in the Core QFD Matrix to develop an integrated plan.

6.2 *Analyzing the results of the Core QFD Matrix*

The information contained in the Core QFD Matrix sets the stage for the remainder of the work to be done on the product or service. Chapter 4 took the reader on a step-by-step tour through the Core QFD Matrix and the reasoning behind each of the eight sections of the matrix. After completion of the matrix, the following information is now known:

1. A focused goal for the product, service, or process improvement;
2. A concise list of customer needs based on research;
3. A concise list of technical requirements to meet the customer needs;
4. A prioritization of the relative importance of the customer needs;
5. A competitive benchmarking of how well your company compares to its competition in meeting customer needs and providing technical requirements;
6. A prioritized relative ranking of importance of the technical requirements;
7. Target values for measurement for each of the technical requirements;
8. Advance knowledge of potential conflicts between technical requirements; and
9. A method to compare technical adequacy of system concepts when they are developed later.

This information comprises the matrix and documents the team's activities and thought processes in creating a product or service concept. The team now knows which technical requirements are the most critical in meeting customer needs. To better view the relative importance of the customer needs (whats) and the technical requirements (hows) a bar chart should be constructed showing the whats and hows ranked by importance weight. Figure 6-1 illustrates the results of the Core QFD Matrix analysis by CVM, using the information obtained in Column 2H of Figure 4-10 and Row 5B of Figure 4-13. The bar charts in Figure 6-1 illustrate that of all the customer needs, a "wide assortment of action" is the most important, and "software design" is the most important technical requirement, with "screen resolution" a very close second. But what does this mean? Does this mean that CVM should work on software design to the exclusion of the other technical requirements? Absolutely not!

One of the misperceptions about QFD is that it creates a priority system for technical requirements that can exclude certain low-ranking technical requirements — even if they are critical to the customer. QFD analysis does not prioritize customer needs and technical solutions to allow exclusion of the lowest-ranked elements of each. Instead, QFD analysis allows a methodical ranking of these elements so that resources can be allocated intelligently. No one factor is to be ignored. The relative rankings tell the team that in order to meet the customer needs, certain technical elements must be met. In

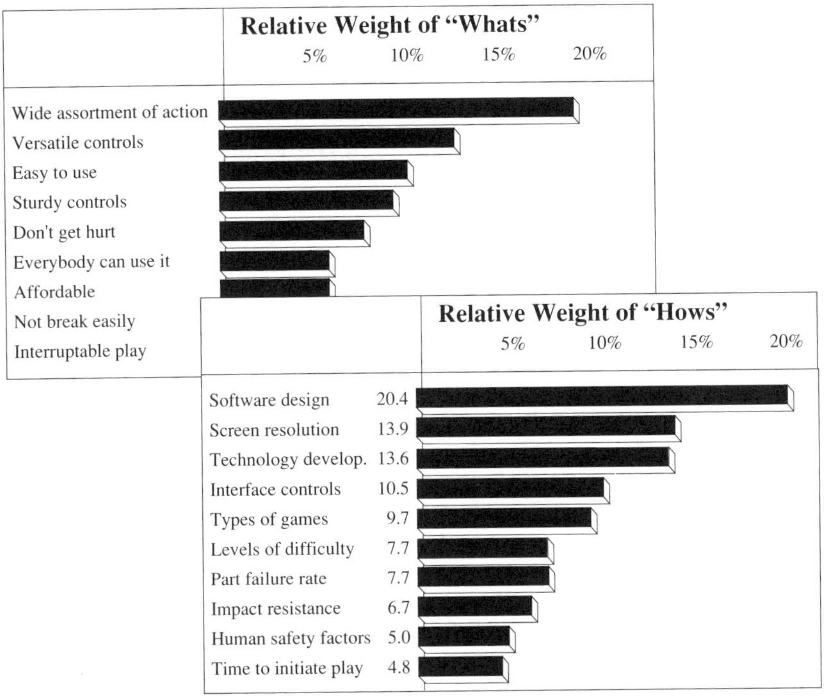

Figure 6-1 The results of the QFD matrix provide a prioritization of technical requirements that can be translated into a plan of action.

addition, these elements have targets that need to be achieved. By prioritizing the technical requirements, the team now knows where to start further analysis and what requirements are critical, thereby needing to be addressed first.

In the case of CVM, failure to address the need for good screen resolution will mean that their product will suffer a significant decrease in meeting the customer needs in several important categories. Likewise, technology development is another key technical requirement for meeting customer needs and should be focused on. The ability to prioritize these elements communicates to the team how to spend resources, allot time, and focus energy.

Sometimes, the completion of the Core QFD Matrix results in a "no go" or abandonment of the project because it highlights areas of technical ability beyond the company's capabilities. This happened in one case involving a government contractor and a defense contract. After completing the QFD analysis, the customer and the contractor realized that not only was there no real market for the product, but the cost of completing the project far outweighed any gain or benefit. The project was stopped and the money was reinvested in a different project more aligned to meet the long-term goals of the customer. This is another benefit of QFD — prevention of capital investment into projects with too high a risk or little chance of success.

6.3 Translating prioritized technical requirements into an integrated plan

Even if a team has developed a sensational Core QFD Matrix showing customer needs, technical requirements, and interrelationships, it serves no purpose if the matrix cannot be translated into action steps. In talking to individuals from companies who have done QFD analyses but experienced some sort of failure or setback, the most common ailment seemed to be a lack of ability to *use* the information developed in the QFD analysis. The authors are only too familiar with these stories told to them by employees and management of small and large corporations. They describe how a team was pulled together and a "vision" created for a project or new product or service, often using some form of a QFD matrix; but the project either failed or did not accomplish the goals originally set forth by the team, resulting in disillusion with the process. In almost all of these cases, one of the key elements that prevented success of the team was a failure to develop an integrated plan based on the QFD or planning tool that gave specific tasks not only to departments, but also to team members. The team can prevent this. Immediately after completion of a valid Core QFD Matrix, the team can develop an integrated plan to set specific tasks and assignments based on the prioritized rankings in the matrix. However, how does the team translate the matrix information into tasks?

6.3.1 What is an integrated plan?

There is a large amount of literature available about developing planning skills, developing work breakdown structures (WBS), and conducting project management. It is not the intent of this book to reinvent planning and project tools that many companies are already successfully using. Rather, the key to an effective plan for project management is knowing what task to take on first and how to get to the desired outcome. For the purposes of this book, an integrated plan is a plan that is tied directly to the technical requirements that are developed out of a QFD matrix. If the QFD matrix and the Repeatable Success Method™ are part of the road map for product and company success, an integrated plan really represents the vehicle that travels that road.

An integrated plan can be something as simple as a checklist of tasks that come out of the initial QFD analysis, or a full-scale project management plan complete with WBS, manpower hours, and resources. The focus of developing an *integrated* plan is to create tasks and task ownership by using the matrix to establish relationships and provide guidance. This is important even if the QFD identifies gaps in customer knowledge or technical ability. In facilitating QFD analysis, the authors have found that often an integrated plan is not developed until after a detailed QFD matrix is completed because the team members feel that there is either missing information or insufficient detail to develop a plan. Unfortunately, when this happens the failure to set forth specific tasks for team members to accomplish (such as gathering additional information) leads to an ultimate failure of implementing the priorities and elements identified in the QFD matrix. The authors would recommend that in facilitation, even if the results of the initial QFD analysis show that there is insufficient data to go forward with the design or new service, a plan should be developed to remedy the lack of information or knowledge, and prepare for the next implementation step after assembling the additional information.

6.3.2 Translating the Core QFD Matrix into specific activities

Those who are familiar with developing any kind of plan for a project know the difficulty in first defining work priorities and responsibilities for accomplishing specific tasks. Using QFD to first prioritize the technical requirements jump starts the process of developing an integrated plan.

After the development of the initial matrix, the team possesses a prioritized list of the major technical elements that need to be explored. Sometimes these technical elements represent technical subsystems, or — in the case of a service project — they can represent a combination of products, core processes, and management or support processes. In either case, the team possesses enough information to set forth tasks associated with each of the prioritized technical requirements and develop action items for further work.

Development of an integrated plan can be done in many ways. The important thing is to take a methodical approach to identify the task, identify the resources to be applied to the task, and define inputs and outputs in order to complete the task. The following represents one approach for developing an integrated plan from the QFD matrix.

6.3.2.1 Identifying the tasks

Figure 6-1 ranked the technical requirements from highest to lowest priority. As a general rule, the authors have found that this list can be divided into thirds for the additional analysis that is part of the integrated plan:

- **Top Third** — These requirements should receive the most analysis, such as modeling, experimental design analysis, bench testing, and other more detailed methods of examination.
- **Middle Third** — These requirements, although important, are not the project drivers found in the top third. The level of analysis and tasks undertaken should reflect this position, and existing information, trade studies, and other less intensive and costly methods of analysis are recommended.
- **Bottom Third** — These requirements are not to be ignored, but should be looked at more for potential problems. A less intensive assessment should be used on these requirements because they have the least effect on meeting the customer needs.

The above criteria are guidelines only, and should be treated as such. QFD analysis is not an excuse for abandoning good engineering or technical judgment (or common sense, for that matter!) to assess available resources.

Having prioritized the technical solutions, the team should then take each technical requirement and assign it to a person on the team whose skills most closely supports the task subject matter. That person then becomes the team coordinator and interface for that particular technical element and task. Sometimes, technical solutions can cover several areas of expertise or knowledge, and it is important to assign several individuals to that particular technical solution based upon experience and knowledge. However, there still should be a single coordinator of all the knowledge. Having assigned the technical solutions, the next step is to define the activity that must be undertaken to support each particular technical solution. This activity then becomes the process or task that must be undertaken. To do this a "Task Sheet" is used to identify the following task elements:

- **Task Title** — Usually the technical requirement title or an element of it.
- **Responsibility** — The person or persons who "own" the task and are responsible for its completion and interfacing with other team members and functional areas.
- **Description** — The task description should be in sufficient detail so that all team members know what the task is and what it does.
- **Inputs** — What information, tests, reports, or interface is needed to accomplish the task. This will often help determine the order for task execution in creating the total plan. Essentially, what is needed before this task can be accomplished?
- **Outputs** — What are the expected outputs of the task and what other area will receive the output.
- **Required Resources** — This area is composed of Resources (physical and staff), Skills (of individuals to complete task), Total hours; Man-hour Loading, Duration (time); Earliest Start Date, and Latest Start Date.
- **Training** — Are there any training requirements for the task to be completed, or will there be any on-the-job training? What is the experience needed?
- **Performance Measure** — The performance measure is the task measure or analysis that will be achieved at task completion. The information provides guidance to answer the question: "How do we know when this job is complete?" This might be the performance target for the technical requirement, or the ultimate outcome of the task.

These elements of a Task Sheet are shown in Figure 6-2. Using CVM as an example, the Task Sheet is completed for the Task Title "Technology Development." Technology development was the third highest technical requirement identified by the CVM QFD team, and, therefore, it was identified

Task Title: Technology Development

Responsibility: Nick B.

Description:

Identify new technology developments.
Assess cost, risk and schedule.
Consider new graphics capabilities, S/W development, the ability to create new games & control systems.

Inputs:	**Outputs:**
Customer requirements Previous warranty claims Customer complaints	Concept & where possible prototype for customer evaluation

Required Resources:	**Skills:**
• 2 486/33 MHz computers with design / cost & risk S/W • Prototype lab & materials • 4 people with indicated skills	• Research • Design • Prototype craftsmanship • S/W code development

Total hours: 1,200 hrs.

Man-hour level: 5

Duration: 1/4/94 to 6/14/94

Earliest Start Date: 12/1/93

Latest Start Date: 2/3/94

Training:

Possibly S/W case tools training

Performance Measure: • Customer feedback
• Enjoyment or action games • Time to initiate play
• Ergonomic factors for controls.

Figure 6-2 Illustration of a task form sheet used to support the development of an integrated plan for the video game.

as a requirement that was one of the "drivers" to meeting the customer needs. Consequently, it receives the appropriate importance in the integrated plan.

6.3.2.2 *Define the task*

The first step in translating requirements into activities is to define the task, or tasks, associated with the requirement. In this example, the CVM team has defined the task as identifying new technology developments and assessing those developments for cost risk and schedule. The team will also consider new graphic capabilities that are consistent with the customer needs identified earlier.

The defined tasks can be as complicated as necessary to accomplish the goal of meeting the technical requirement. Sometimes they are only studies to assess further analysis, or they can be design and bench testing, for example. The task description should validate target parameters and the sensitivity of performance of those targets. The description must also describe interfaces with other systems and subsystems, and any analysis to be performed. Some of these interactions can be identified by using the "roof" of the Core QFD Matrix.

6.3.2.3 *Inputs and Outputs*

A task that fits into an integrated plan cannot exist in a vacuum. Information or other inputs are required in order for the task to be accomplished. Likewise, once the task is accomplished there is an outcome or output that often supports another task. The importance of developing the integrated plan is to identify what information (inputs) is needed to support that task and what is the outcome of the task. Having identified inputs and outputs, two things can happen: (1) the relationship between the various tasks that are attributed to the technical requirements can then be identified, and (2) the order for completion can be developed. Task plans can then be scheduled according to the overall time frame of the product development or project, and man loading and resource assignment can also be developed. Figure 6-2 shows the inputs and outputs for the CVM task "Technology Development." At this level, the team needs inputs from customer needs and previous warranty claims. The output will be a concept or prototype where possible.

6.3.2.4 *Required resources*

The team must know the schedule, man-loading, and skill needs in order to function effectively. This section of the Task Sheet should reflect a best estimate of these elements based upon the information available, along with the initial schedule and cost estimates. By indicating earliest and latest start dates the team can also better schedule all of the tasks in the plan.

6.3.2.5 *Training*

This section identifies any training needs that may impact the ability to meet task completion. In today's business environment, training should always be taken into consideration on any new product or service. Knowledge and skills are constantly changing and this is an element that should not be ignored.

6.3.2.6 *Performance measure*

The performance measure is a critical piece to the Task Sheet. It details how the team will measure completion of the task, and how the task ties back into the customer needs and target goals identified as part of the matrix. A team should always be measuring progress and outcome, because if these cannot be measured the team has no ability to determine if the technical requirement is being met.

6.3.3 Linking the task to the requirement and the requirements to each other

As shown above, the tasks for the plan are taken directly from the prioritized technical requirements. Figure 6-3 reinforces that each technical requirement must be evaluated for additional analysis, but in accordance with the importance to the project as defined by the QFD analysis. No task should exist without being linked to some requirement.

The next step is to link all of the requirements into a plan of action with a defined system start and stop. Figure 6-4 shows how the interactions between tasks can be represented using the Task Sheets. The inputs and outputs can be used to define interactions between tasks, but the team should also use the "roof" (Section 7 of Figure 4-15) of the Core QFD Matrix to define positive and negative interactions and establish communication lines between activities. The benefit of using this method to establish links is prevention of the "gotchas" in developing a product or service because of conflicting requirements. The team can therefore stay in communication while completing the tasks outlined as a result of prioritizing the technical elements.

Completion of the integrated plan using the Task Sheets results in the creation of a schedule for completion, represented in Figure 6-4 by the Gantt chart at the bottom. The creation of a schedule is almost as important as the plan itself. The team members need to put the elements of the integrated plan into the time frame of the project completion in order to know if the tasks they have scoped are capable of being completed in the time allotted.

Sometimes the creation of the integrated plan leads to pleasant surprises when considering schedules. A team at a clutch company that performed a QFD analysis on its internal bid and proposal system was convinced that it would take 6 months to conduct a complete implementation of the technical requirements needed to revise the process. After developing the integrated plan and laying out the specific tasks associated with the plan, the team concluded that the analysis needed to take the next step in process improvement would take only 3 months — and that was a schedule including some margin for error!

6.4 The integrated plan and the project resources

The integrated plan (IP) developed at this stage of the Integrated QFD Approach™ is meant to be a working document. It should be easily changed as the team begins analysis and investigates the parameters of the tasks defined in the plan. The IP can become the interfacing document for the

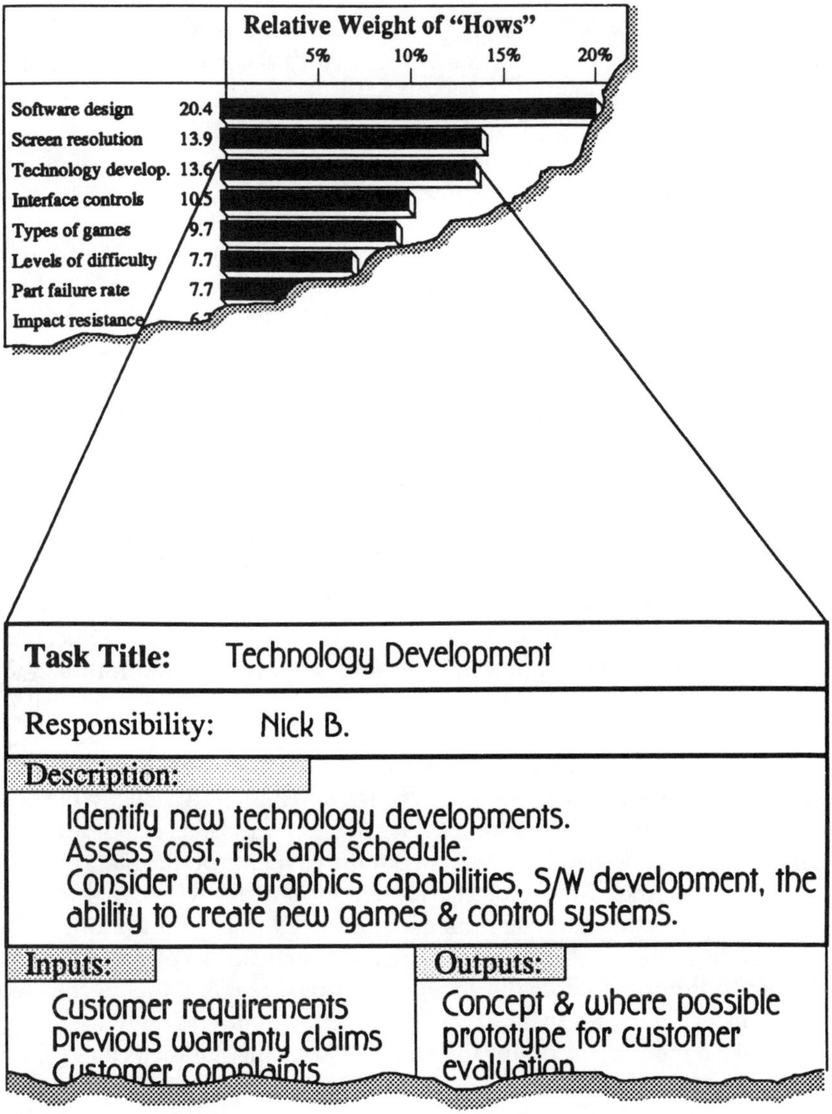

Figure 6-3 Each video game technical requirement is evaluated for the need and extent to perform additional analysis or trade-off studies to further define the system concept as a means to meet the technical requirement targets or further refine the system sensitivity.

project team because it defines the interfaces necessary to carry out the tasks. Regular meetings should be held to keep the team updated on the progress being made toward the final goal.

One of the most important benefits of the IP, if it is maintained through the project (as indicated in Figure 3-8 of Chapter 3), is its ability to become

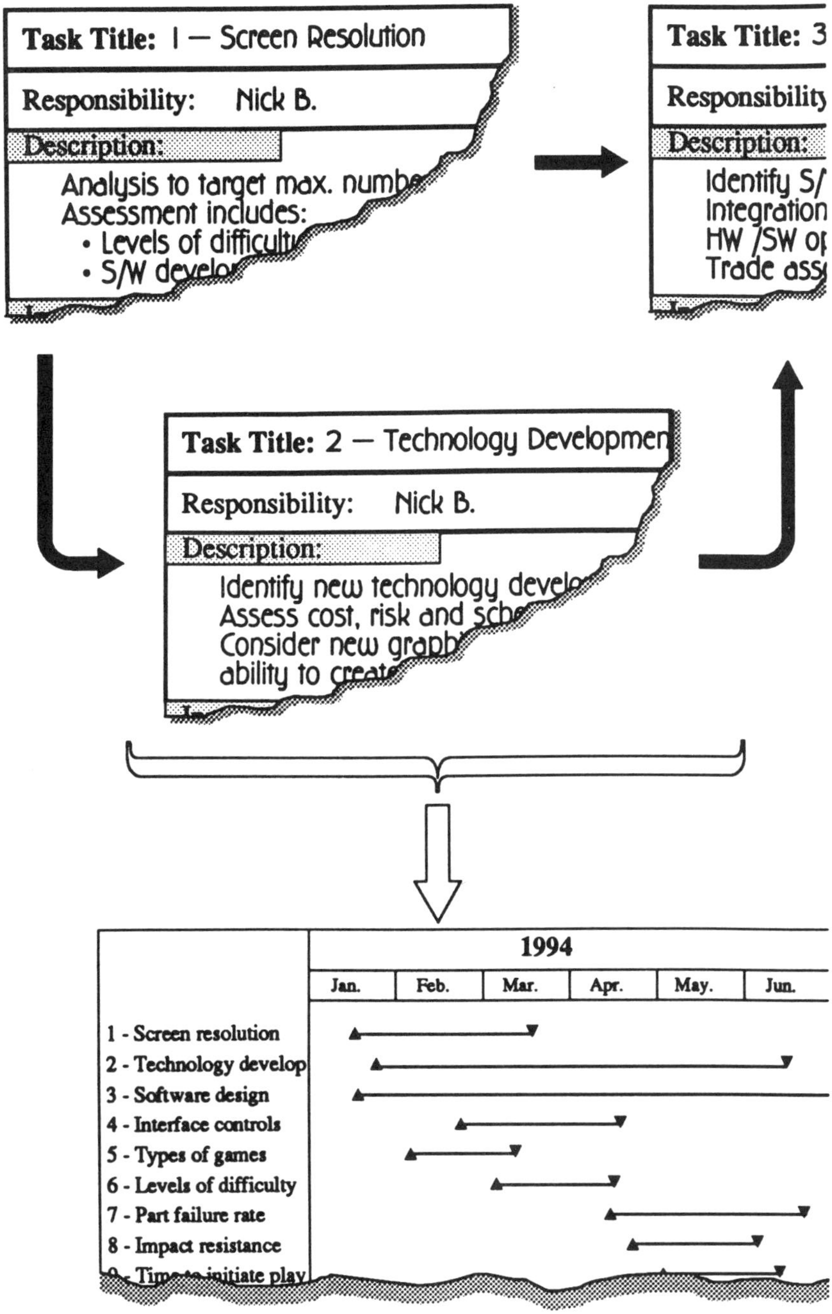

Figure 6-4 An illustration of how the coordination of information flow and interrelationship between the translated QFD tasks is translated into an integrated plan and schedule.

a resource-scoping document. The IP removes a lot of the guesswork in defining the amount of work to be done on the project and the resources (both human and physical) necessary to carry the tasks to completion. The IP allows the team and management to determine what can and cannot be accomplished within the time, cost, and physical resource parameters.

The creation of the IP will prevent teams from experiencing disasters similar to that experienced by the poor villagers of Bizillion. But why worry, that's only a fable — *right?*

6.5 Chapter summary

1. In many cases teams are pulled together and a "vision" is created for a project or new product or service, often using some form of a QFD matrix, but the project either failed or did not accomplish the goals originally set forth by the team, resulting in disillusion with the process. In almost all of these cases, one of the key elements that prevented success of the team was a failure to develop an integrated plan. Success comes from developing an integrated plan based on the QFD or planning tool that gives specific tasks not only to departments, but also to team members.
2. Immediately after completion of a valid Core QFD Matrix the team can develop an integrated plan to set specific tasks and assignments based on the prioritized rankings in the matrix.
3. The first step in developing an integrated plan is to identify the tasks based on the prioritized QFD lists.
4. The second step is to define the tasks using task sheets which include the following information:
    - Task Title
    - Required Resources
    - Description
    - Training
    - Inputs/Outputs
    - Performance Measures
5. The last step links the tasks to each other and creates an integrated project schedule.

chapter seven

The Integrated QFD Approach™ and Robust Designs

The whole is greater than the sum of its parts;
But which parts do you use?

7.0 Wholes, holes, and parts

Chapter 3 introduced the concept of the Integrated QFD Approach™ and how QFD is part of a larger analytical approach called RSM™. The key concept of QFD from the authors' viewpoint is to view it as a part of a "whole" process that translates customer needs to actionable plans. The actual construction of the QFD Core Matrix is only one step in the total application of an integrated approach to problem solving. Too often the authors have seen QFD viewed as an end in itself, with the resulting first matrix neatly filed away after completion, often never to see the light of day again. Sometimes this is the result of an unclear picture of what to *do* with the matrix once it has been established. How do you take a technical requirement such as "software design" or "high video screen resolution," for example, and translate it to lower levels of analysis for parts, labor, and manufacturing controls? How do you design a "robust" product? What is a "robust" product?

Chapters 1 to 3 explained the need for QFD and a larger methodology for implementing and managing change, and defining requirements. Chapters 4 to 6 detailed the fundamentals of the Core QFD Matrix, its interpretation and the basics of an integrated plan. The first part of this chapter uses the CVM example from Chapter 4 to illustrate three elements in detail: how various "parts" of the Integrated QFD Approach™ are developed; how they relate to each other; and the overall purpose of translating customer needs into requirements. The proper use of all of these parts will yield a product that is much more useful and effective than any individual element, and will also prevent the failure to fill any "holes" in the business, product, or process areas.

The second part of this chapter addresses the use of statistical analysis tools, including Statistical Process Control and Robust Design, to address the issue of variability within the requirements identified by the QFD analysis. Robust Designs essentially are designs that can tolerate variance in operating conditions without compromising the needed requirements. This is another example of how QFD uses other methods or applications to prevent "holes" in customer needs or technical requirements.

7.1 The nine steps of the Integrated QFD Approach™

Most QFD methodology incorporates a concept of breaking the customer needs down to individual parts and processes. The Four-Phased approach described in Chapter 3 used four matrices to translate the needs down to process control planning (see Figure 3-6). The Matrix of Matrices approach uses a series of submatrices geared toward certain elements of manufacturing, test, design, and processes. However, it relies on the team to make the decisions as to which matrices are to be chosen, and presents a highly complicated approach not usually attempted at the level of detail necessary to trace requirements in a comprehensive, controllable fashion (see Figure 3-7). The Integrated QFD Approach™ breaks the customer needs into four distinct matrices: System Matrix, Concept Matrix, Process Matrix, and Control Matrix. The Integrated QFD Approach™ not only shows what matrices should be developed for a product, but what are the service and business operations elements that also need to be developed in order to support the product element. In this way the entire organization can be focused on the project, and a true "systems" approach yields an integrated result.

Figure 7-1 shows the functional flow of the nine steps of the Integrated QFD Approach™ (Figure 7-1 is the same as Figure 3-8). The nine steps are as follows:

1. Develop System Matrix
2. Develop Initial Integrated Plan
3. Develop Functional Flow
4. Develop System Alternatives
5. Evaluate System Alternatives
6. Develop Concept Matrix
7. Develop Process Matrix
8. Develop Control Matrix
9. Refine Integrated Plan

These steps were briefly described in Chapter 3, and the Customer Needs Matrix (Step 1) was illustrated by example in Chapter 4 using the customer needs for the video game being developed by our imaginary high-tech company, Compulsive Video Mania. That example focused on the product area of the video game; however, there are three types of requirements that can be integrated into a System Matrix: Product, Service, and Business Operations. Step 2 — Developing the Initial Integrated Plan — was examined in Chapter 6 in some detail to illustrate how a preliminary integrated plan could be developed from the initial customer needs matrix. These first two steps of the Integrated QFD Approach™ are essential in developing the insight into design, processes, and products; however, they are only the beginning to a

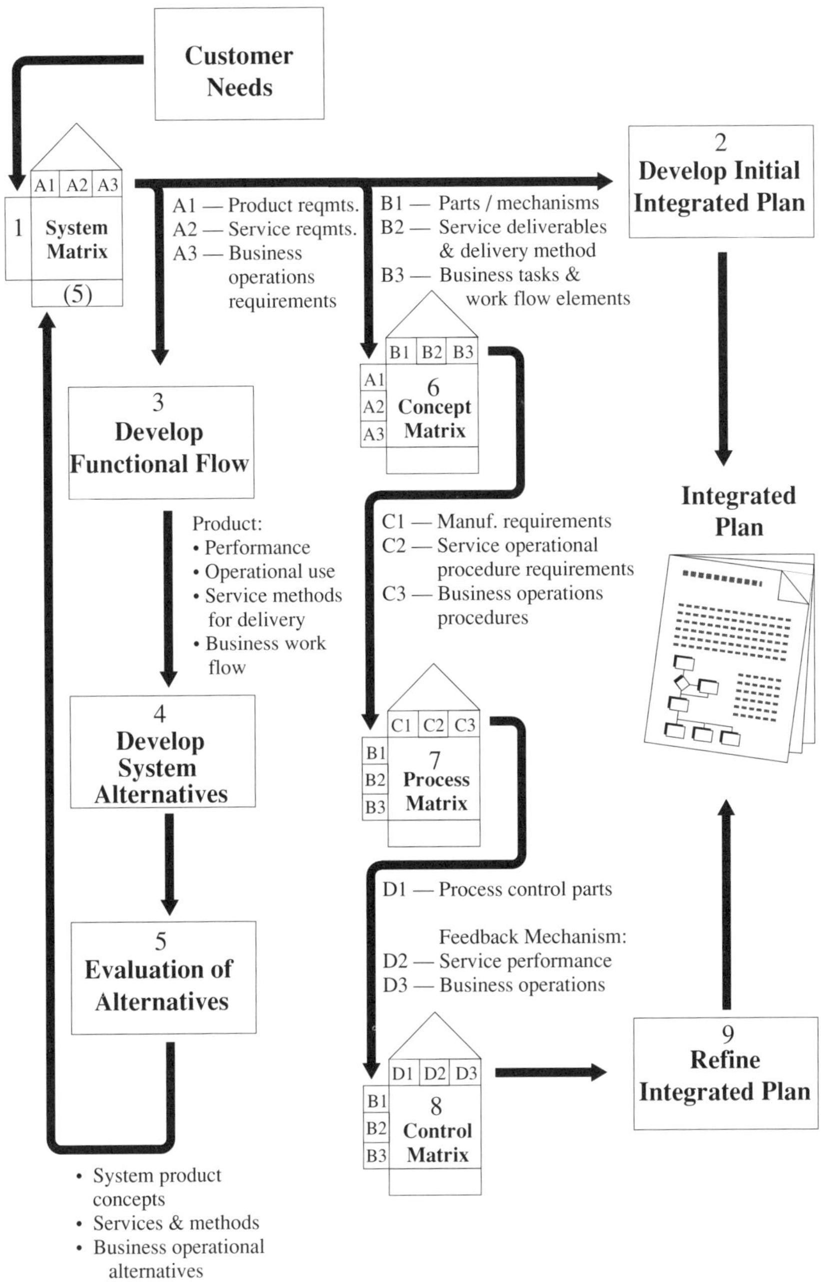

Figure 7-1 The nine steps of the Integrated QFD Approach™.

successful implementation of QFD and integrated planning, coupled with other tools, to create traceability of customer needs down to parts, processes, and operation details. This is where the power of QFD can be used to achieve the dramatic results in improved quality, productivity, and management.

This next section takes the reader through the nine steps of the Integrated QFD Approach™ illustrating the source of the information for each step, and how the results of each step provide the input for the next application. In this way, the reader can understand the logic and flow of the approach. For the purposes of this illustration, elements of Product Requirements, Service Requirements, and Business Operations Requirements have been selected to carry through the nine steps of the Integrated QFD Approach™ based upon some of the information already developed on the video game example.

7.1.1 Developing the System Matrix

Chapter 4 rigorously defined the steps to complete a Core QFD Matrix for customer needs focusing on the product aspects of the video game example. The Core QFD Matrix represented elements common to all QFD matrices in the Integrated QFD Approach™. As the Core QFD Matrix is integrated into the Approach, the features of the System Matrix will seem familiar.

The product requirements selected to meet the customer needs are only part of the equation when considering a total QFD implementation. When analyzing the customer needs for a product, for example, the service requirements and the business operations requirements should also be taken into consideration (this is also an important step when conducting "business re-engineering," and is more fully discussed in Chapter 9). Customer needs are not limited only to the product itself, nor does the development of the product happen without interaction with service functions (such as technical support) and impact on the infrastructure of the business operations (such as the product development process). Therefore, a thorough QFD should encompass service requirements and business operations. Likewise, if the purpose of the System Matrix is to analyze the creation or improvement of a service, the analysis should consider the impact on any "product" delivery and business operations. Each of these three areas — Product, Service, and Business Operations requirements — should be included in the initial System Matrix developed in Step 1 for a full QFD analysis.

Recognition of the three requirements elements in Step 1 can avoid confusion in how to meet and prioritize customer needs later. At the 1993 TQM Conference in Colorado Springs, Colorado, the authors discussed the concern of one high-tech company about the difficulty of implementing QFD because of what appeared to be validation problems. The individual charged with facilitating the QFD said that when he did the initial QFD with upper-level management, the focus was on the "business operations" and the ability of the company to meet customer needs. When attempting to further implement the initial priorities with the technical teams, there was a distinct dichotomy between "product" and "service" priorities depending on the technical bent of the team. The facilitator felt that QFD could not be validated because the priorities varied depending on the group conducting the initial analysis.

This is not an uncommon complaint heard by the authors about QFD. However, it is an unfair one. Ask an engineer what is most important and he or she will tell you it is design; a company executive may tell you it is new product development or marketing; and the marketing representative may say it is service. They all may be right, from their own perspective. This is why it is important to bring these viewpoints together to analyze the total scope of a QFD analysis. Dividing the customer needs into three categories assists in the prioritization of the elements.

This concept is illustrated in Figure 7-2, which takes two of the product requirements of the video game example from Chapter 4 and adds two service and two business operations requirements. The matrix represented in Figure 7-2 is a partial one that will be used for demonstrating the following application of the Integrated QFD Approach™ and does not reflect an entire System Matrix. The goal is to take the System Matrix example through the remaining eight steps using the elements of each requirement group. The matrix in Figure 7-2 shows that the service and business operation requirements can be mapped against the customer needs on the left side of the matrix for a complete prioritization involving all of the elements that will impact the customer needs at this top level. In Figure 7-2 the three types of requirements — product, service, and business operations — are not only compared to the customer needs that have been prioritized for importance weight and given realizable target values, but they also can be correlated in the "roof" of the matrix to understand interactions. For example, the business operations requirement of the Product Development Process will have a strongly positive correlation on the product requirement of Software Design. This may be because in order to shorten the design cycle time to 6 months (a target goal for the Product Development Process), the number of lines of code to be written may have to be reduced. Reduction in the number of lines of code is a target value for the Software Design requirement.

Grouping the requirements into product, service, and business operations categories requires cross-functional analysis beyond the narrow scope of the product and takes into consideration full integration of support functions. Once the initial System Matrix has been completed in this fashion the QFD team has determined the following:

- The interaction of customer needs with product, service, and business operation requirements;
- The prioritization of the customer needs;
- The relative importance of the technical requirements;
- The target values for the three groups of technical requirements;
- Competitive benchmarking of the customer needs;
- Competitive benchmarking of the technical requirements (not shown in Figure 7-2); and
- Correlation between all technical requirements.

Having accomplished this important initial analysis, the next step is to develop the initial integrated plan.

7.1.2 *Developing the initial integrated plan (Step 2)*

Chapter 6 explained the importance of having an integrated plan to implement and track the progress of the project after initial determination of the System Matrix and top-level technical requirements. An integrated plan is the method by which the QFD team can assign and take responsibility for

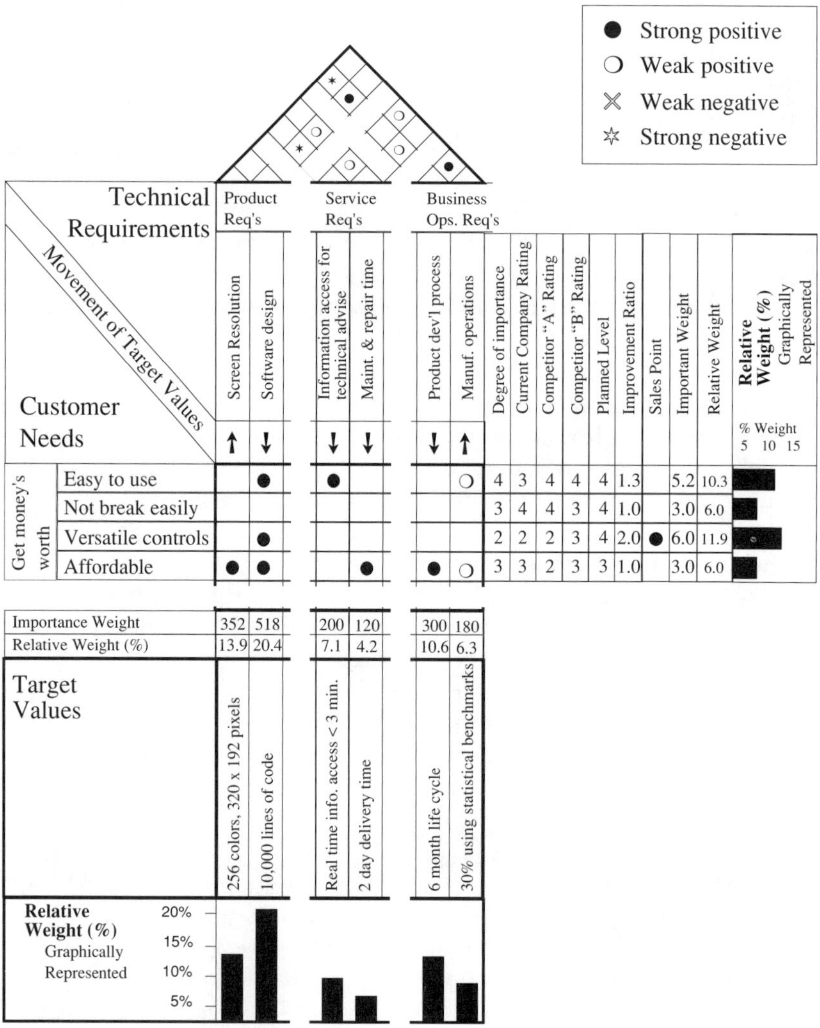

Figure 7-2 Step 1. An example of an expanded system matrix to include Product, Service, and Business Operations requirements using the video game example.

accomplishing the project tasks in a controlled manner. But the initial integrated plan developed after Step 1 in the Integrated QFD Approach™ is a key component for completing the remaining matrices to be developed and in connecting the functional elements of the project. This section will create an example of the integrated plan using the information developed in Figure 7-2, but will take it a step beyond Chapter 6 by integrating the requirements into product development procedures and company procedures.

The initial integrated plan begins with charting the prioritized System Matrix. Figure 7-3 shows how translating the customer needs of the video game example into a graphical representation by priority allows an understanding of the key elements of the project from a customer satisfaction viewpoint. The next step is to translate the three categories of technical requirements into a graphical representation of priorities. Figure 7-3 illustrates that the order of importance for the video game example, using the six sample requirements, is (1) Software Design; (2) Screen Resolution; (3) Product Development Process; (4) Information Access; (5) Manufacturing Operations; and (6) Maintenance and Repair. These results will give the team its first level of priority in order to allocate resources to deliver the best product (and service) to the end user. The translation of the information from the QFD System Matrix to graphical representations of priority is not necessary, but the authors have found that by taking this step the team, and those supporting the team, will more quickly grasp the level of importance of the technical requirements. Some people are more visual learners than others, and graphing the priorities is an excellent method of quickly communicating this information that can become lost in the complexity of the QFD matrix.

The next element of the integrated plan (initially illustrated in Chapter 6) is to brainstorm and analyze the individual technical elements in order to create task sheets for inclusion in an overall schedule and plan. The task sheet is critical to defining the task, describing the activities involved, assigning responsibility for task completion, defining inputs, and defining outputs (this was covered in detail in Chapter 6). Figure 7-4 shows sample task sheets for two of the requirements for the video game being developed by Compulsive Video Mania: Software Development and Product Development Process. In developing the task sheet for Software Development, responsibility for task completion is given to "Brent C." The task is then described in detail such as "Identify operational steps for using and maintaining the video game," etc. The task description should contain all elements and action steps for completing the task and be developed by the team members possessing expertise in the particular task. Tasks may also be broken into some subtasks depending on the size of the project.

The next key step is to determine the inputs and the outputs. Chapter 6 described how important it was to link the tasks, and that in order to determine the sequence of tasks and interdependencies, the inputs and outputs needed to be well-thought out. For Software Development, the QFD team described some of the inputs as Customer Requirements and Software specifications from previous products, building upon the information available

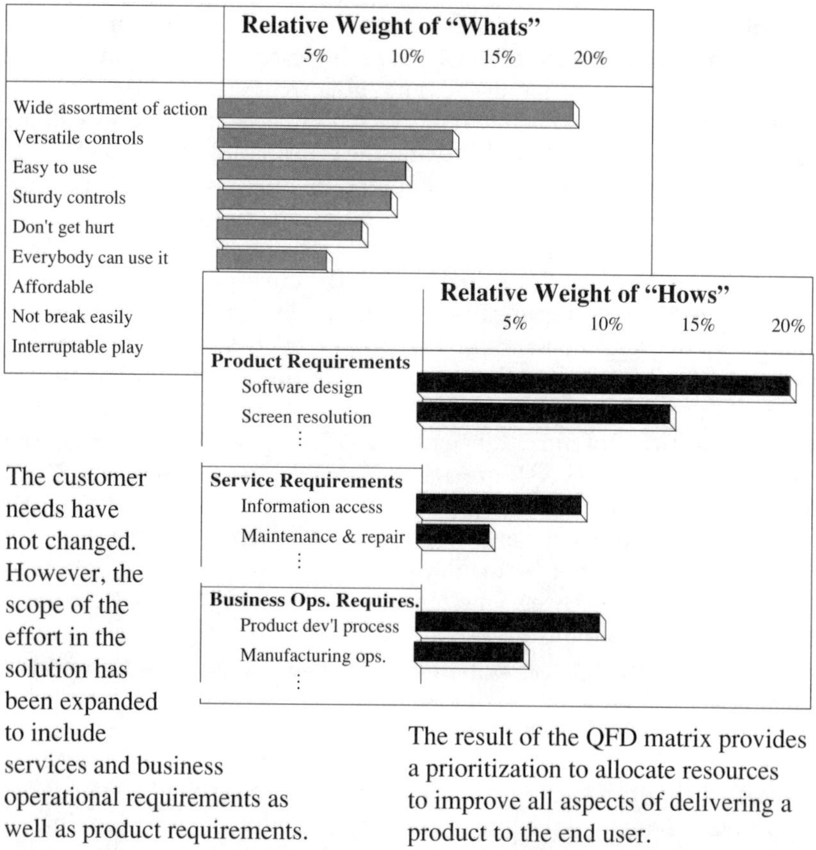

Figure 7-3 Step 2. Translating the System Matrix into an Integrated Plan. Begin by understanding the relative importance of resulting product, service, and business operations requirements.

from the previous efforts. Outputs were items such as Software Development Plan, Code Development, and Pilot Demonstration. These are not all of the possible inputs and outputs but indicate how the task sheet is used to begin mapping the process of developing the entire product and related service and business operations. Figure 7-4 gives an example of the task sheet for the Product Development Process and some of the details that would be associated with this kind of business operation in support of the video game development.

It probably would come as no surprise that Code Development would be an output for software development; however, the purpose of developing the integrated plan from the identified technical requirements is to ensure the follow-through on the tasks associated with the technical requirements. The integrated plan is a systems tool allowing insight into the entire development process, and, therefore, identification of specific efforts between functional areas must be coordinated. At this high level of task identification it is not yet

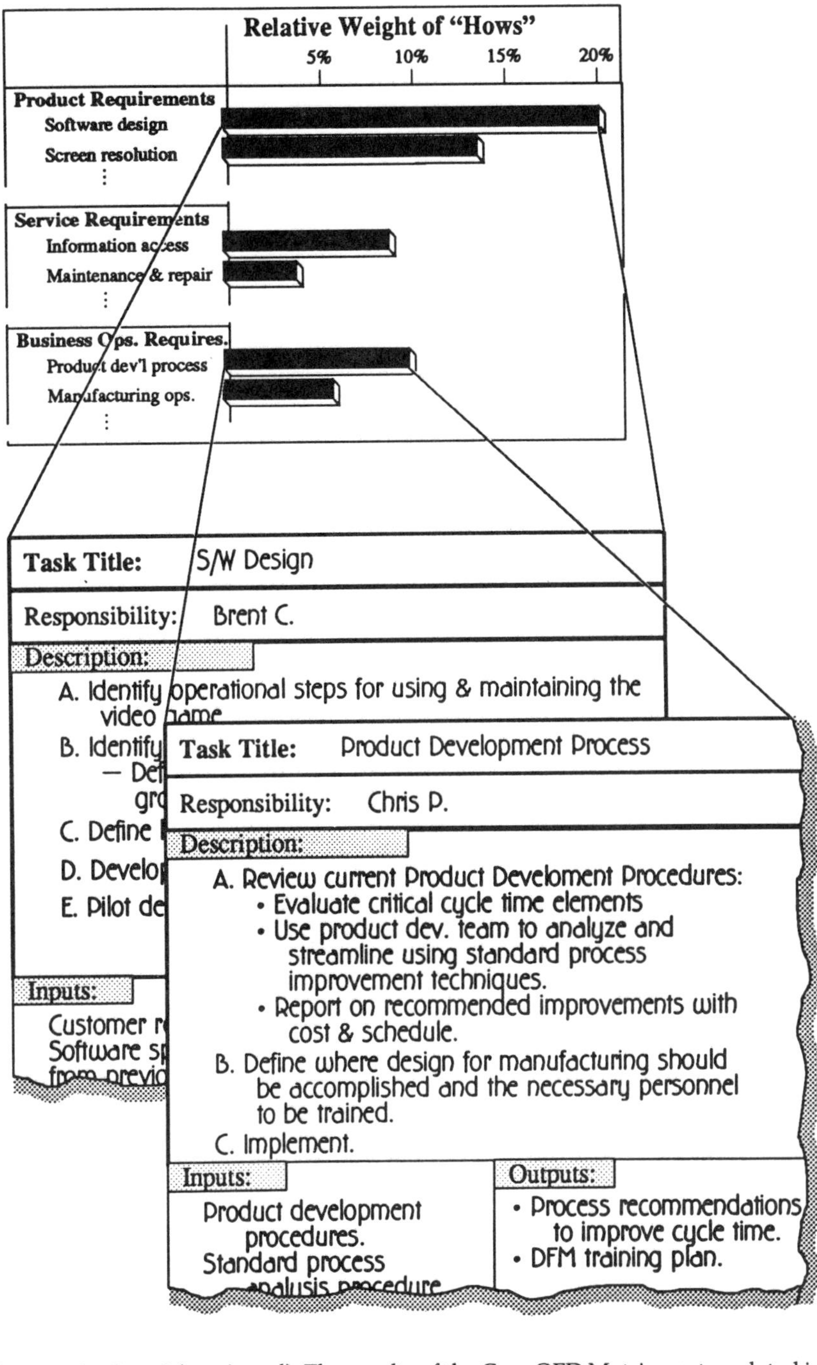

Figure 7-4 Step 2 (continued). The results of the Core QFD Matrix are translated into task plans of action for all aspects of delivering a product to the end user.

possible to assign time and total resources to the tasks, but by the end of the Integrated QFD Approach™ these details can be placed into the plan.

Another benefit of developing the integrated plan is to see how the elements fit into the company or project product development cycle where the primary focus of the QFD effort is product development. Most companies have a process for developing a product even if the process is not set down in writing or in a formal policy document. The process usually involves getting from the market analysis to the sale of a developed product, but what is sometimes lacking is the ability to trace the actual development activities to the specific areas of the development process. Development of an integrated plan based upon the priorities and requirements developed from the System Matrix results in alignment of activities within the existing product development environment for a company or project. An example of how the initial integrated plan can relate to the product development process is given in Figure 7-5 using a model for product development and total design developed by Stuart Pugh, an engineering design consultant and professor at the University of Strathclyde (U.K.).[26] The authors have chosen the design method of Stuart Pugh as a representative example of how an integrated plan can be mapped to any process or product development, recognizing that the design and product development process varies widely not only between industries but from company to company within industries as well. Pugh's method is indicative of the European design process and is being used by increasing numbers of companies in the U.S. for product design development.

Pugh's total design strategy is based on the recognition of what he calls a central core of activities that includes the functions to get a product from the initial identification of user need to sale. Pugh states:

> Total design is the systematic activity necessary, from the identification of the market/user need, to the selling of the successful product to satisfy that need — an activity that encompasses product, process, people and organization.[27]

He concentrates on a major element of this core — the design core — which includes market (user need), product design specification, conceptual design, detail design, manufacture, and sales. These concepts are represented in Figure 7-5. By way of illustration, three of the task sheets for the video game example are included in Figure 7-5 and the various task descriptions for each task are lettered and tied to a specific activity within the product development procedures of the Pugh design core. QFD, through requirements identification and prioritization, connects the activities of the integrated plan to the existing activities contained in the product development procedures of the organization. Responsibility for ownership of the tasks described in the task plans can be delegated according to the existing product development process. In the video game example, Part D in the Description under Software Development is "Developing Lines of Code." This task would belong under "Conceptual Design" of the Pugh design core as shown in Figure 7-5.

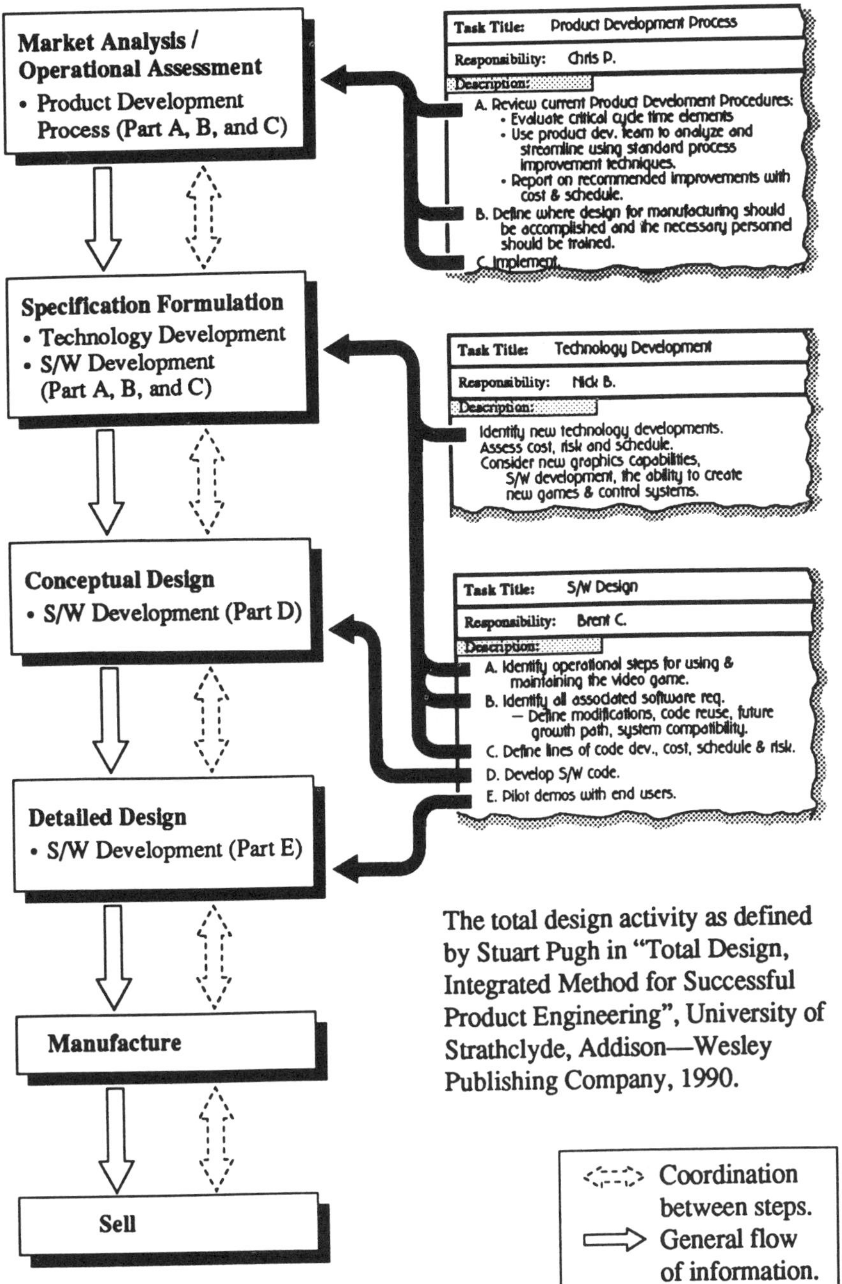

Figure 7-5 Step 2 (continued). The Product, Service, and Business Operational Requirements and Support Task Analysis should be integrated into current company or project product development procedures.

The results of the System Matrix and the associated requirements can also be integrated into current design activity to create a schedule of sequenced events to arrive at the top-level planning for the product design. Using the engineering design process defined by Benjamin S. Blanchard,[28] Figure 7-6 illustrates how the sequence of design activity encompasses the tasks defined by the integrated plan. As the tasks are attached to each sequence of design activities, a schedule is developed to achieve the results. There is a tremendous advantage to developing the task sheets and then integrating the tasks into the existing design process. Not only does the team know the "tall poles" and priorities of the product and design process, but traceability now exists from the individual tasks back to the importance of the customer need. Interfaces between departments and functions are identified, and specific roles and responsibilities result from the connection to existing design processes. The integrated plan can contain a multitude of tasks and activities, but at this level of detail care should be taken not to attempt the actual *design* of the product through the plan. The plan should be at a fairly high level of detail (e.g., a systems level), but can also contain various tasks that are critical to the success of the program and reflect the priorities obtained in the customer needs QFD matrix. Some of the types of information the plan could contain are:

- Product investigation and analysis
- Process analysis
- Design process
- Manufacturing scheduling
- Evaluation of cycle time and streamlining
- Conceptual design
- Prototyping activities
- Scheduling for product completion
- Technology development investigation

The plan should contain all of the types of activities that will be necessary to complete the project and will be the blueprint for the program development. Many of the tasks in the plan at this early stage are at a top-level definition (such as "Define lines of code, cost, schedule, and risk" under Software Development) because the details cannot yet be known. However, the point of the plan's development at this stage is to begin to piece together a flow for completion of the program. The plan will be modified as additional information from the Integrated QFD Approach™ steps is added.

Once the initial integrated plan has been constructed in the manner described in Chapter 6, care should be taken to streamline the plan. Identification of the tasks associated with the requirements identified in the System Matrix will almost always result in some duplication of activities, overlap of responsibilities, identification of automation needs, and questions about execution of the tasks. These and additional considerations when streamlining the plan are contained in Table 7-1. Merely creating the plan does not mean

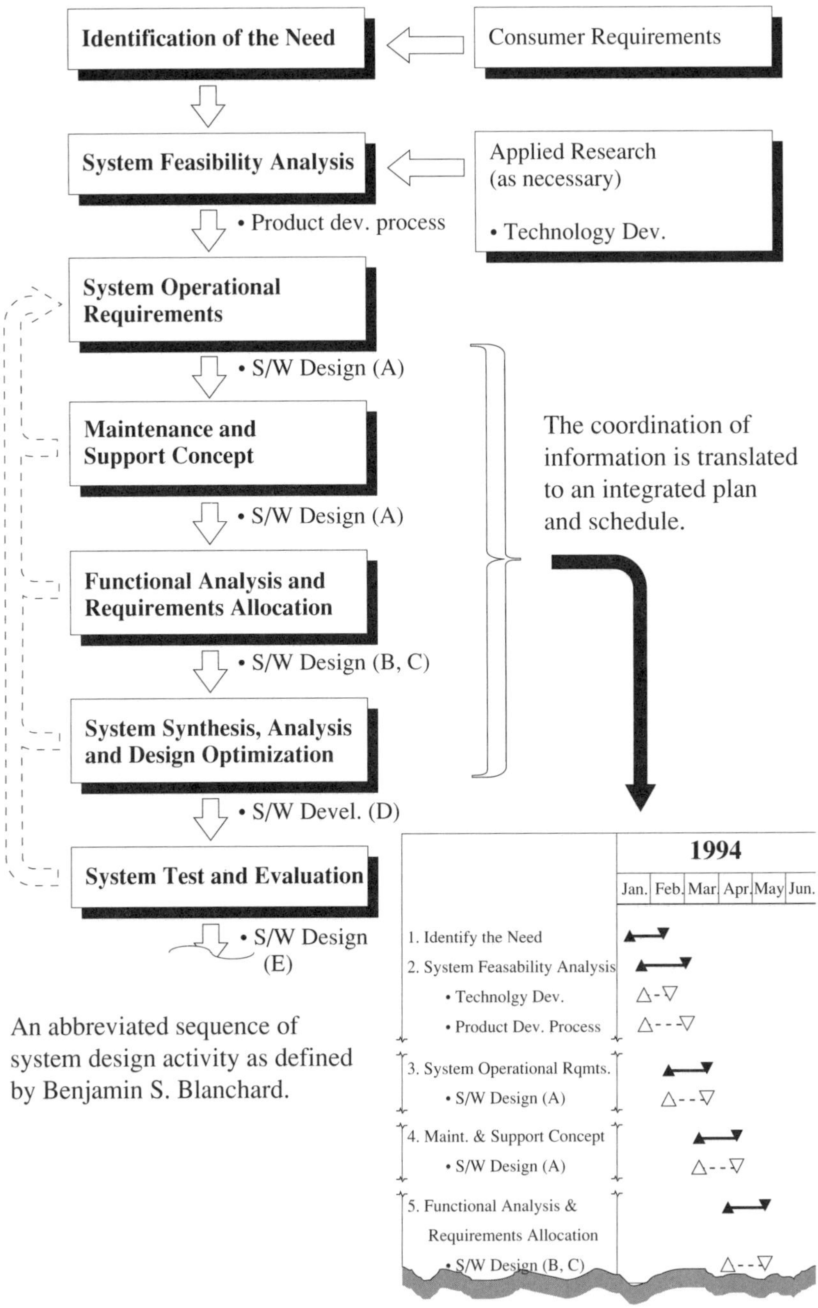

Figure 7-6 Step 2 (continued). An example of the unique QFD solutions developed through the QFD matrix integrated into the Systems Engineering Design Process defined by Benjamin S. Blanchard.

Table 7-1 Considerations when Streamlining the Integrated Plan

• Eliminate non-added value activities or duplication.
• Simplify work flow.
• Make sure roles and responsibilities are understood.
• Identify independent tasks as potential concurrent activities to reduce cycle time of plan.
• Examine how standardizing activities could reduce time and cost i.e. standard formats for documentation, standard equipment, S/W, etc.
• Integrate compliance requirements early in planning cycle.
• Identify potential areas where errors may occur and define strategies to mitigate risks.
• Upgrade personnel and equipment capabilitites to enhance productivity.
• Examine who will use the resulting integrated plan information and tailor the language and type of data to meet their needs.
• Examine possible partnerships between departments or other suppliers to either reduce cycle time or invested costs.
• Look for additional automated, computerized or mechanized methods to enhance team productivity.

that the plan has been cast in stone and all activities have been properly identified and scoped. The initial integrated plan should be a living document and be revised when necessary to accommodate changes in scope, personnel, and task descriptions. It should also allow departments to construct interfaces for communication progress and formats for standardized documentation. The plan is a document for interfacing, and should never be looked at as merely "the schedule."

Once the initial integrated plan has been put together and a schedule has been created, what is the next step? How do you translate the activities in the plan into more detailed functions? Do you now start the design process?

The next step is to start the design process, but not in a conventional manner of putting drafting pencil to vellum or using computer-aided design (CAD) techniques. The design process really starts with understanding the function of the product. Likewise, for service and business operations the next step is to understand the functions of the organization in supporting the product design and delivery.

In the video game example, the QFD team assembled a well-defined platform from which to launch their design effort. From the initial concept of an interactive action game that puts the players in the action, they used a market analysis to develop the System Matrix, which yielded the initial product, service, and business operations requirements. Using the requirements an integrated plan was constructed which will guide the team through product development and manufacturing. The question to be answered next is: What will our product do? Often this is where product development starts, without the benefit of QFD or an integrated plan. Although this is an important element of design, the authors will show why functional analysis is really the third step in the process, and how it plays back into the QFD analysis.

7.1.3 Developing the functional flow (Step 3)

Before illustrating a functional flow for the video game example, it is important to understand what a functional flow is and how functional analysis provides the information to create functional flows.

Arthur E. Mudge, in his book *Value Engineering: A Systematic Approach, Part I,*[29] states that a function is "that which makes a product work or sell." Functional analysis is another tool that has existed for some time and has been used by the aerospace industry extensively. But functional analysis really gains its power when coupled with prioritization and brainstorming associated with QFD. Functional analysis is really part of the tailored process flow using QFD so that alternative methods to satisfy technical requirements — and, therefore, the customer needs — of a product can be methodically developed, analyzed for best concepts, tested for target accuracy, and then put under statistical process control.

It is not the purpose of this book to go into great detail on functional analysis. The purpose of this section is to show the basics of functional analysis in relationship to the design process and QFD.

In general, functional analysis takes a process and divides it into its functional characteristics of how that process is performed. Often, flow charts or block diagrams are utilized in defining function and processes. One of the best ways to view a function is illustrated in Figure 7-7. A function can be described as an activity that is expressed by a verb and a noun. That activity, as shown in the figure, acts upon inputs (described as materials, energy, and signals) and yields an output (shown as the altered materials, energy, and signals).

To understand the application of functional analysis it is best to use examples of functional analysis in simple systems. Functional analysis will come into play often in the development of a product, and its first application here is at a systems level to understand what a product will do. When defining systems functions, some guidelines should be kept in mind to facilitate the analysis. Table 7-2 shows several general guidelines for defining system functions. The first is to express the function using a verb and a noun. This allows the system definition to be more representative of the activities associated with the product. It also forces the analyst to examine those activities rather than just name elements of the product.

The second rule is to break down the expression of work and sell functions using different categories of verbs and nouns. A work function is action and measurable. A sell function is passive and nonmeasurable. For example,

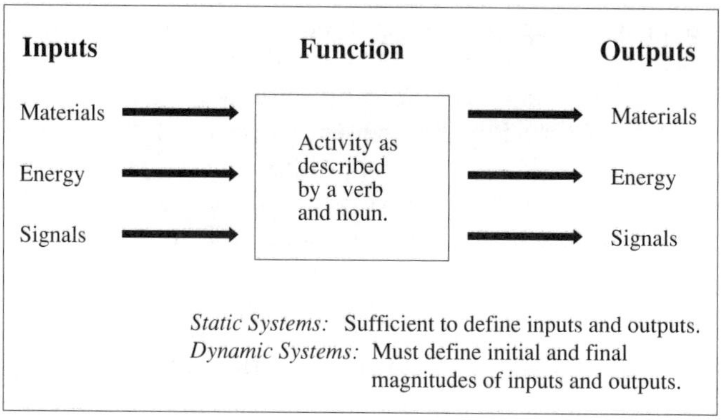

Figure 7-7 The term "function" can be related to an input/output relationship for a system as a means of describing the system task performance.

work functions should be expressed as "collect liquid" or "conduct heat." "Sell" functions would be expressed by using passive verbs and nonmeasurable nouns: "Improve appearance" or "increase features."

Finally, functions should be divided into primary and secondary functions. A primary function is the top-level function or the purpose of the activity. Primary functions would be "store liquid" or "harvest potatoes." Secondary functions support the primary function or purpose but cannot accomplish the primary purpose alone. For "store liquid" secondary functions would be "measure liquid" and "indicate quantity," and for "harvest potatoes" secondary functions would be "lift potatoes," "separate potatoes," and "collect potatoes."

A functional flow is a series of interrelated functions usually expressed as a flow diagram. Using the "harvest potatoes" example, Figure 7-8 illustrates how a primary function can be decomposed into secondary functions, all of which contribute to the task of "harvest potatoes." In the example, the input to the function is potatoes and ground, and the outputs are "good" potatoes, "bad" potatoes, debris, dirt, and rocks. Between these there are five secondary functions: Lift Potato/Dirt Mixture, Sift Mixture, Separate Mixture, Sort Potatoes, and Collect Potatoes. The functional flow diagram in Figure 7-8 allows examination of the system-level, potato-harvesting process for the functions that comprise the top-level function.

Now let's return to the video game example. The Compulsive Video Mania QFD team knows that a customer need is to "Get Money's Worth," and that a subcategory to that is "Versatile Controls" (Figure 7-2). The team also knows that Software Design would be a technical requirement, as would Number/Type of Interface Controls. Software Design and Number/Type of Interface Controls would have Task Sheets associated with them and the integrated plan would reflect activities to be undertaken for completion. Selecting "Number/Type of Interface Controls" as the concept for functional analysis, a flow chart of what the controls must do can be constructed. By understanding what the controls for the video game must do, the requirements

Table 7-2 Considerations when Defining System Functions

System Function Definition Guidelines

- Express functions as a verb and a noun.
- Express work functions with action verbs and measurable nouns, i.e. "collect liquid" or "conduct heat"
- Express sell functions with passive verbs and non-measurable nouns, such as "improve appearance" and "increase features"
- Divide functions into primary and secondary levels. A secondary level is a function that does not accomplish the primary purpose, but does support it. For example:

The primary function is:	***Secondary functions include:***
"store liquid"	"measure liquid" "indicate quantity"
"harvest potatoes"	"lift potatoes" "separate potatoes" "collect potatoes"

for the number and types of controls for the user interface will be understood.

The mechanics of developing functional flows through the use of functional analysis is sometimes compared with other "black arts" such as software development. Intuition, experience and creativity are usually the requirements, and it is difficult to pin down those who do functional analysis to specific approaches. The authors believe that the same team that develops the QFD matrix should also be primary in developing the initial functional flow for the product, service, and operational areas. Company experts in the area of functional analysis (usually these people are found in the systems engineering area) should also be used. The purpose for doing functional analysis at this stage is to develop the initial concept analysis for the product. The functional analysis should be at a "systems" level and should not be taken to greater detail. As will be demonstrated in Step 5, the basis for conducting functional analysis now is to develop system concepts for continued analysis and development in the Integrated QFD Approach™.

Figure 7-9 is a functional flow diagram for the interface controls for the video game being developed by Compulsive Video Mania. The functional flow diagram for the interface controls was the result of a brainstorming session by the QFD team members looking at the function of the controls in terms of the action or activity to be captured by the game. At this point, the team is not actually considering *how* to make the controls, or *what* are the parts or components of the controls. The team is concerned with what the

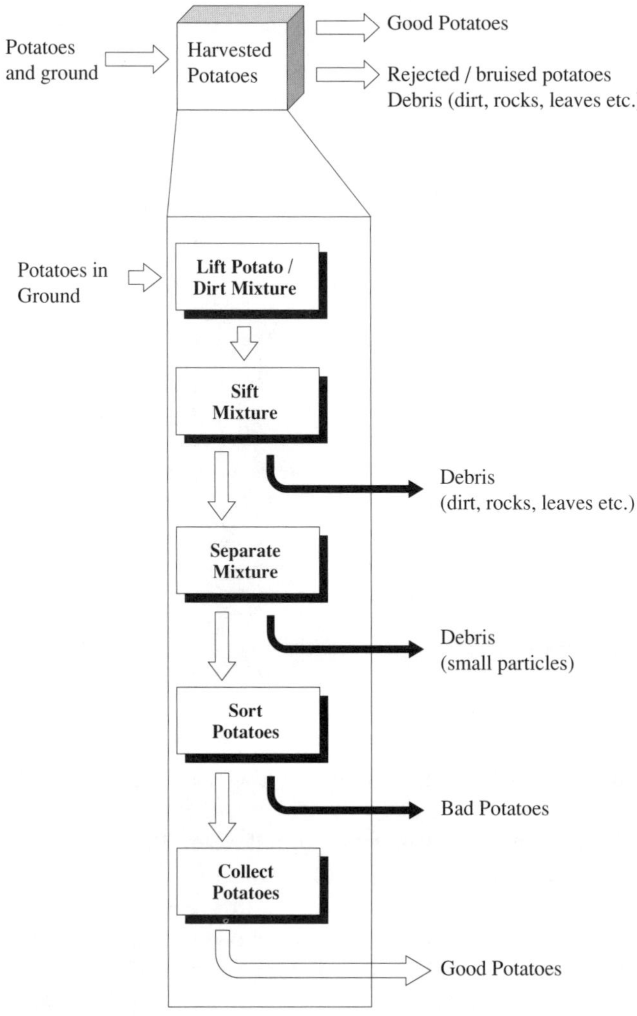

Figure 7-8 An example of a functional flow diagram.

controls do and the function they perform. The top-level functional analysis could easily be: Start game...Play Game...End game. The functional flow in Figure 7-9 takes the concept of how the controls will be used to create a flow of the activities represented by subfunctions, and the type of input (in this case a signal) and output for the functions. The player will select the type of game, start the game, and then select certain features for action. Those features (represented in items 4 to 11 in Figure 7-9) will be activated by the interface controls. At this point no determination has been made as to *how* the controls will accomplish these features. By not defining the types of controls, the team has left open options for technically meeting the functional attributes, and have not locked the design into a specific technical solution.

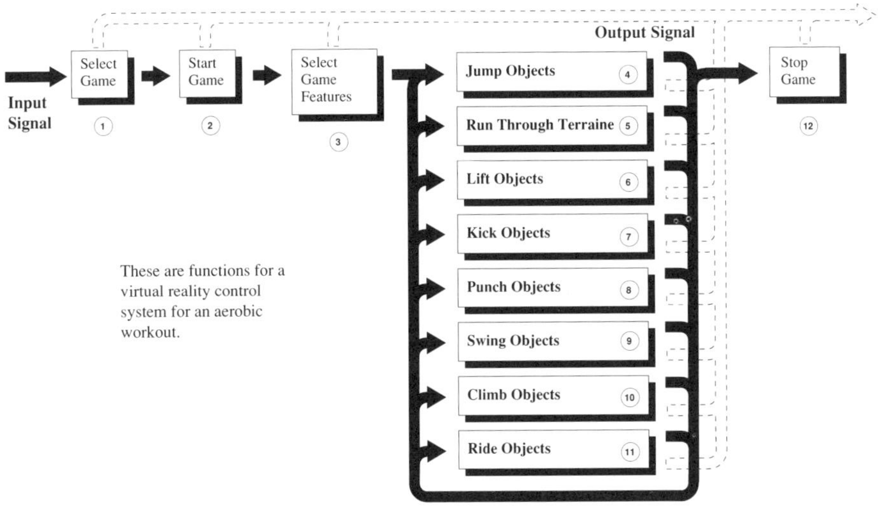

Figure 7-9 Example of a Functional Flow Diagram for the system controls of the video game.

This is just one example of the functional analysis to create functional flow diagrams used to understand what the video game would do, and what the performance characteristics would be for the game. In addition, if the team was also conducting a QFD analysis on the Service and Business Operations Requirements, functional flows would be constructed for top-level systems in these areas as well. The result of conducting functional analysis at the systems level would be a complete picture of what the product, service, and support would look like for the video game concept. The next step is to develop system concept alternatives for each function.

7.1.4 *Developing system alternatives (Step 4)*

Application of functional analysis to system-level designs helps to understand the purpose or attribute of a product, but does not describe how to create the attribute. For example, in Figure 7-9 Function 4 of the control mechanisms requires "Jump Objects." Jump Objects is a function but *how* to go about achieving that goal has not yet been described. This is why the next step in the Integrated QFD Approach™ uses brainstorming to develop system alternatives to meeting the performance goal of jumping objects. Engineers and designers are used to developing methods to meet the performance goals of a product, and this step of exploring alternatives is often used when looking at alternative solutions. However, the organized concept development of using functional analysis first to understand what a product should do and then using brainstorming to understand methods of accomplishing that goal prevents settling on a single design concept that may not be the best solution, or may not work at all.

Function 4.0	Function 5.0	Function 6.0	Function 7.0	Function 8.0	Function 9.0	Function 10.0	Function 11.0
Jump objects	**Run through terrain**	**Lift objects**	**Kick objects**	**Punch objects**	**Swing objects**	**Climb objects**	**Ride objects**
Pad with height detection system	Tread mill	Pulley system	Boots	Gloves	Baton with various weights	Escalator system	Stationary bike
Single mechanized rubber element moves according to height	Circular path	Free weights	Shoes	Single feature attachment	Tennis racket with attachable "heads"	Ladder with handles	Attach own bike to syste
Multiple predetermine size & height of objects	Single path power pad	Nautilus type	Electrical cloth overlay			Step blocks	
Mounted detection height system attached to rail	Tread mill / escalator mix						
Detection system attached to boots	Electrical cloth overlay on path						

Figure 7-10 Concept alternatives to support a video game workout system.

Figure 7-10 shows how concept brainstorming can work to develop several options for meeting the functionality of the video game control interfaces. Figure 7-10 takes Functions 4 through 11 as examples and maps alternative concepts for meeting the functions. Function 5.0, "Run through terrain," lists five alternative methods of meeting this performance goal: (1) a treadmill, (2) a circular path, (3) a single path with a power pad, (4) treadmill/escalator mix, and (5) an electrical cloth overlay with no path. These are five methods of meeting the functionality of running through terrain; and although cost, feasibility, and constraints may eliminate one or more of the methods, the importance in mapping alternatives through brainstorming is the creation of many solutions to a single problem. Alternative method brainstorming should be viewed by the team as a delimiting process and not a restricting process. Free exchange and discussion of the concepts should be encouraged but no concept should be excluded because of a concern it may not work or is technically difficult to achieve. This step in the process has yet to reach the level of actual design solidification. Concepts are just that: concepts to be explored. The benefit of creating several solutions to each function will become more apparent in Step 5 when the concepts are evaluated against the technical requirements of the QFD matrix.

Figure 7-10 lists alternatives for only a small section of the total product concept and does not take into consideration the Service and Business Operations Functions at this point. In doing a full functional analysis and alternative concept analysis each phase of the product, service and business operations elements should be explored fully using the QFD team as a coordinating panel, if not the actual evaluating entity.

Once the functions have been brainstormed and the team feels a sufficient number of concepts have been developed to satisfy the functionality of the product, the concepts would be combined into "systems" concepts. Creativity and experience can combine at this point to link together various concepts that will make an entire product idea to be examined further. The team should construct several concepts for further exploration and analysis. Figure 7-11 pictorially shows how two alternative systems developed by the CVM QFD team will function. System 1 is based on stationary application. The video game would be at a single location and the elements chosen would conform to the requirement for a stationary location. The team selected a sensor pad for jumping (Function 4.0), cloth overlay on running shoes for sensing running (Function 5.0), etc. System 2 is a "portable system" that can be used outside on a running path incorporating features such as free weights, step blocks, and a holographic visual overlay or heads-up display.

The designers at CVM have come up with some high-tech concepts for their interactive video game! Figure 7-12 tracks each element of the two systems against the functional requirements. Each system concept contains one of the functional solutions developed during the concept brainstorming. However, the reader should not assume that only one or two systems could be developed by using this process. Often the number of systems capable of satisfying the functional elements can be huge when considering the combinations. Because of this possibility, the team should employ as much experience in comparing concepts as possible to narrow down the potentially viable combinations.

Do companies use this methodology for brainstorming concepts? Absolutely. Many firms are familiar with the concept of using system engineers and designers to develop concepts to design solutions. Often engineers are confronted with a multitude of possibilities to solve a problem, such as binding two parts together: should they use adhesive, welding, or fasteners? The answer may depend upon the other components in the system as well as cost and environmental conditions. This level of analysis would be at the part or process level, and for the purposes of the concept selection, such detail is not yet desirable or possible.

Functional analysis and development of alternative concepts are also used to develop system concepts for the service and business operations areas. The functions involved in servicing the product would be analyzed and decomposed in the same method as the product analysis, taking into consideration how the customer would be supported by service and the interaction with the product areas. Business operations would include accounting, purchasing, product development, and other support functions for the video game development.

The actual development of various concepts can be a difficult and somewhat time-consuming method. To assist in the brainstorming of concepts, "Concept Stimulators" or directed questioning of parameters and uses can aid the team in opening up creative horizons. Figure 7-13 lists some concept stimulators and how they might raise additional concepts about different

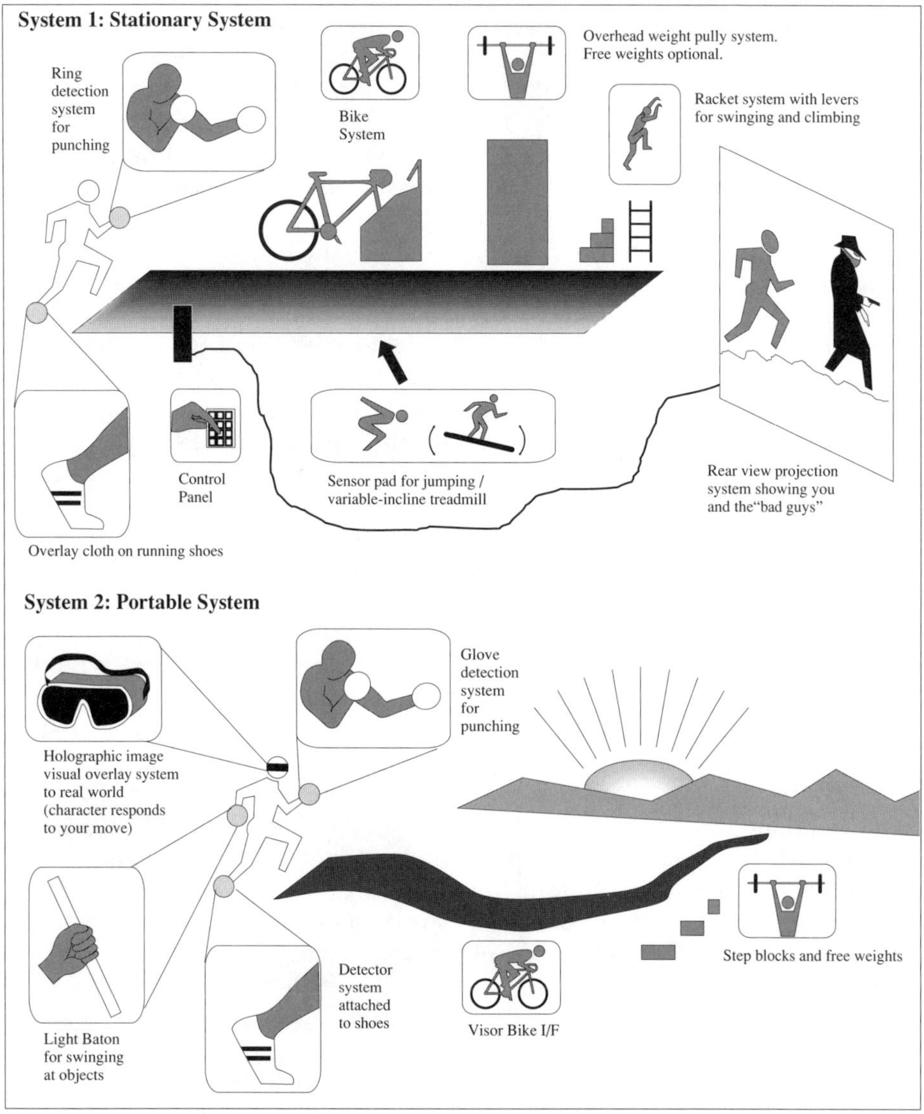

Figure 7-11 Alternative concepts for a video game aerobic workout system.

aspects of the video game example. The point of the concept stimulators is to encourage the QFD team to look at the concepts in a slightly different application or use to increase the potential for new ideas. Most of the concepts in Figure 7-13 would apply to detail or part applications, since this is usually the area of detail where the team can bog down in technology application.

By analyzing all of the functions of the product and its support mechanisms the QFD team would be able to construct various systems to best meet the overall function of the product and satisfy performance measures. The

	Function 4.0	Function 5.0	Function 6.0	Function 7.0	Function 8.0	Function 9.0	Function 10.0	Function 11.0
System 1: Stationary System	Jump objects Mounted detection system	Run through terrain Escalator/tread mill with variable terrain	Lift objects Overhead pulley system	Kick objects Cloth overlay on running shoes	Punch objects Finger holder	Swing objects Tennis racket with attachable heads	Climb objects Escalator/tread mill	Ride objects Attach own bike to system
System 2: Portable System	Detection system attached to shoes	Electrical cloth overlay	Free weights at end of route	Electrical cloth overlay	Lt. cloth gloves with electrical impulse	Light baton with signal impulses Belt with holster when not in use	Step blocks at end of route	Attach bicycle to system resistance S/W I/F

Figure 7-12 Step 4. Develop system alternatives from concept.

Additional Concept Stimulators:	**Concept Changes:**
Can dimensions be change? • Thinner? • Larger?	• System 1 Screen: Thinner material, larger diameter with higher resolution; white reflectant
Can the quality be changed?	• Vary baton weights through attachments
Can the order be changed?	• S/W program — designed for either hight intensity aerobics followed by circuit weight training or vice versa, independent
Can the elements be changed?	• Lengthen / shorten activities real-time through control panel
Can there be a change in cause and effect?	• If program too easy, can be self modulating to work participant harder
Can there be a change in character?	• Remove hills from program if desired
Can the form be changed?	• Head set for System 2: Firmer & padded, with no rubbing & material designed to absorb sweat • Running path for System 1: Material that minimizes foot & leg damage
Can the motion be changed?	• Slower / faster dependent on input control / real time
Can the state or condition be changed?	• All elements evaluated for lighter materials while maintaining strength
Can the use be adopted to a new market?	• Nursing home / medical out-patient therapy • Childrens activity centers (use monsters)

Figure 7-13 Step 4 (continued). Refine system concepts through Concept Stimulators.

development of several system configurations can then be evaluated to determine which is the best. How do you determine the "best" system or combination of systems? — By using the criteria already developed in the System Matrix containing the customer needs.

7.1.5 Evaluating system alternatives (Step 5)

Figure 7-1 shows how functional analysis and development of system alternatives follow the completion of the first system-level matrix. Step 5 in the Integrated QFD Approach™ evaluates the system alternatives for meeting the technical requirements, which, in turn, correlates directly back to the customer needs. Chapter 4 briefly discussed concept evaluation in Figure 4-1 and Section 4.3.2.8, covering the eighth part of the Core QFD Matrix comparing two different design concepts to the technical needs. Steps 1 through 4 of the Integrated QFD Approach™ show where the concepts originate.

Having developed two competing concepts, the next step is to evaluate them against the technical requirements of the matrix developed in Step 1. Figure 7-14 evaluates the two concepts in terms of meeting the six sample technical requirements of a complete matrix. The three symbols used in concept evaluation are (1) a filled-in circle for indicating that the technical requirement has been exceeded; (2) an open circle for indicating that the technical requirement has been met; and (3) a triangle for indicating that the technical requirement has not been met or that it cannot be met with current technical expertise. By using symbols the team, and management, can quickly see graphically how the two systems compare. Figure 7-14 shows that System 1 (the Stationary Game) meets or exceeds all of the technical requirements whereas System 2 (the Portable Game) only meets one Service Requirement — Information Access for Technical Advice — and fails to meet the other five requirements that are shown on the abbreviated QFD System Matrix. The comparison between the two systems is accomplished by comparing the system concept, developed by the functional analysis and brainstorming, to the target value for each of the technical requirements. The QFD team is essentially determining if the target value is being met (or is capable of being met) by the system concept. If the target value is capable of being met, then the technical requirement is being satisfied; and if the technical requirement is satisfied, then there is an excellent chance of meeting the customer needs and creating a successful product.

Figure 7-15 shows how the system concept is compared with the technical requirement in order to determine if the system exceeds, meets, or does not meet the technical requirement. The matrix in Figure 7-15 places the technical requirement on the right and lists the two concepts at the top. As the QFD team evaluated each technical requirement against the concept, they looked for the ability of the concept to meet or exceed the target value assigned to the requirement. For example, for Screen Resolution System 1 (Stationary) could use an active matrix projection system that exceeds the technical target value of 256 colors at 320 by 192 pixels. However, System 2 (Portable) would use a heads-up display with holographic images. This technology would create "haziness" in the picture projection and new technology would need to be developed or the target value could not be met. For the product requirement of Software Development of less than 10,000 lines of code, System 1 is estimated to require only 9500 lines of code, whereas System 2 (because of the new technology for translating signals) would require a totally new software of approximately 200,000 lines of code.

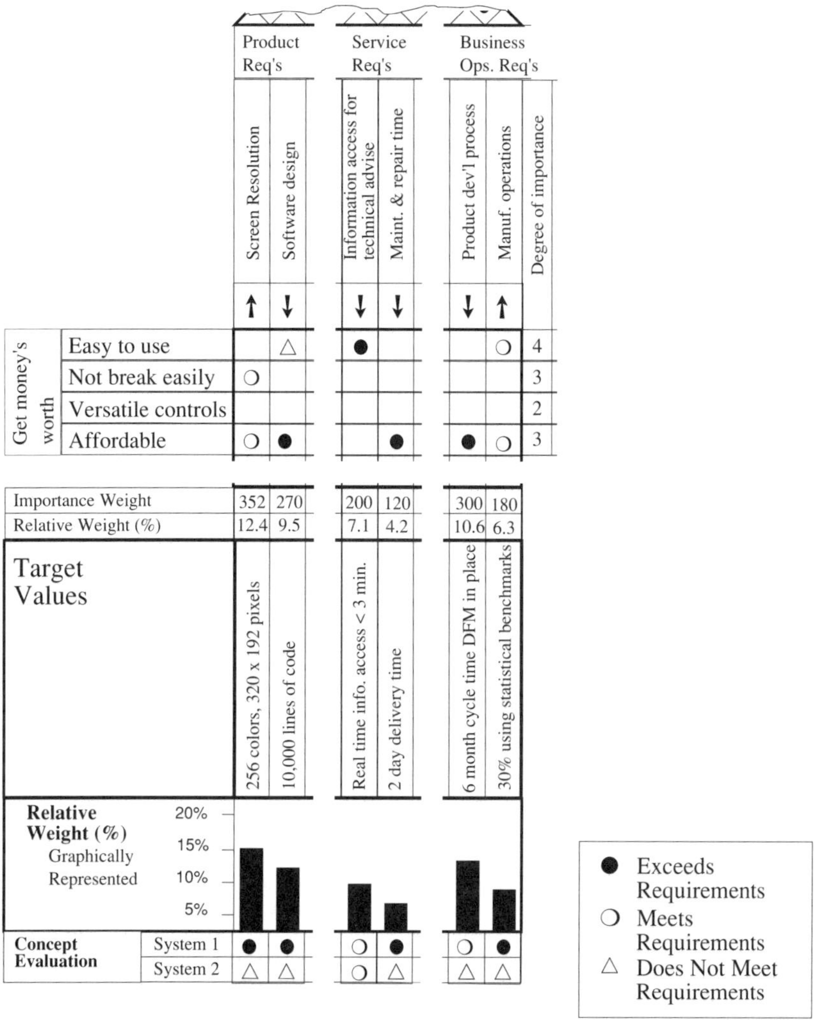

Each concept can be evaluated for its ability to meet the technical needs of the customer as related to products, services, and impact to the company operations.

Figure 7-14 Step 5. Continued evaluation of alternatives.

The information used to evaluate the two systems comes from concept analysis and functional analysis, information gathered as a result of the initial integrated plan (such as research into code requirements after concept identification, etc.) and expertise of the individual QFD team members or information gathered from other resources within the organization. Based on this analysis the conclusion reached by the CVM QFD team is that although

		System 1 Stationary	System 2 Portable
Product Requirements	Screen Resolution	● New screens with active matrix projection system exceed 256 colors with 320 x 192 pixels.	△ Holographic images are still hazy and difficult to see. - New technology required to improve visual images.
	S/W Development (10,000 lines of code)	● Estimate 9,000 lines of code to refine earlier program.	△ Total new code development 200,000 lines of code to translate signal processing system.
Service Requirements	Information Access for Technical Information	❍ Use current satellite communications system.	❍ Use current satellite communications system.
	Maintenance & Repair Time	● Current service capable on 80% of products to be delivered in 1 day, 20% in 2 days.	△ Delivery of head gear has 5 week delay. Inventory control needs improvement.
Service Requirements	Product Development Process 6 Months	❍ With design for manufacturability in place & new CAD systems running this product could be developed in 6 months.	△ • Technology Dev. & Analysis would indicate a 2 year product dev'l process. • Alliances with other companies could minimize this significantly.
	Manufacturing Operations 30% Using Statistical Process Control	● Nearly 60% of all parts to be delivered will be under SPC manuf. ops.	△ New manufacturing operations need to be developed. Only 10% of manuf. ops. anticipated to be under SPC within 6 months.

Figure 7-15 Analyzing the concepts evaluation correlations: comparison of requirements for each system.

System 2 (Portable) had great versatile controls by virtue of the functional analysis, it was not capable of being developed in a 6-month period, would require costly and risky new technology, and would most likely be more expensive than System 1. System 1 meets or exceeds all of the technical requirements, can be developed within a 6-month design-to-production window, and exceeds the critically important Screen Resolution Target Value (see Figure 7-16). Therefore, based on the concept evaluation, the CVM QFD team selected System 1 as the concept design for advancing the new interactive video game.

The advantage of using the QFD prior to doing concept evaluation and functional analysis is now apparent. QFD allows the team to evaluate system designs against the customer requirements and target goals and gives the team yet another indicator to assure on-track development of the product and ultimate customer satisfaction.

Concept evaluation against the technical requirements in this manner may also reveal important considerations in a single system. Some proponents of concept evaluation advocate using a ranking system to compare each technical requirement to the concept much like the 1–3-9 ranking of the technical requirements. Each concept system is then weighed as a whole against other concept systems and the one with the highest ranking is chosen as the concept to take forward into design and manufacturing. The authors advocate against this type of all-or-none comparison. By using the three symbols for comparison, the team can see if a single requirement (which may be a key one as determined by the System Matrix) is not met by a concept system despite its meeting or exceeding all other requirements. For example, if System 1 did not meet the Screen Resolution requirement (which ranked the highest in importance weight of all requirements in Figure 7-14) then even if all other requirements were met the team would know that System 1 could fail as a product because it did not meet customer needs of good screen resolution, affordability, and ease on the eyes. If this had been the case, then the QFD team could first try and use part of another system concept that did meet the screen resolution criteria, or revisit the functional analysis and concept development steps to select alternative methods for meeting the requirement but keep the other parts of the system concept that meet or exceed the other technical requirements.

Concept evaluation allows the identification of potential holes in a concept by tracing the concept to the customer needs through the QFD matrix. Fruitless design investigation can then be avoided, time is conserved, and costs are saved. The cycle of QFD matrix-to-concept evaluation (Steps 1 to 5) should not be viewed as a one-time exercise but as an iterative loop to refine the best concept to meet customer needs. Only after the QFD team is satisfied that a workable and promising concept has been developed should the team move on to the next step in the Integrated QFD Approach™ — The Concept Matrix.

Although System 2 has great versatile controls and a wide assortment of action, the technology required to satisfy customers' needs is not sufficiently developed to be cost effective within a 6 month period. In addition the complexity and advancements of the system requires additional operational streamlining, particularly in the manufacturing of new materials and integration of the highly sensative equipment.

Alternatively System 1 can meet all the necessary requirements to launch the product within a 6 month period time frame. In many cases System 1 exceeds the requirements such as high screen resolution. This would provide great visual effects and could be clearly demonstrated as the feature that sets it apart from the competition, i.e. the technical solution puts the sales point in action.

Figure 7-16 Step 5. Conclusions reached from concept evaluation analysis for the video game systems.

7.1.6 *The Concept Matrix: Translating Requirements Into Product, Service, and Business Task Requirements (Step 6)*

Once the system concept has been chosen the next step is to develop the second or Concept Matrix of the Integrated QFD Approach™. The Concept Matrix maps the technical requirements against the next level of detail in the product development: Parts/Mechanisms, Service Deliverables and Methods, and Business Task Elements. The Concept Matrix begins to develop the design, service, and business operations "concepts" that further define the manner in which the requirements targets will be met. Many QFD analyses proceed directly from the customer benefits matrix to the parts/mechanisms (concept) matrix without conducting functional or concept analysis. Without the additional definition of the product and its use developed by using the previous functional and concept analysis, the team can quickly become bogged down in doing some of that analysis as part of the Concept Matrix. As a result, the second matrix has been known to become large and unwieldy and progress may become questionable. The authors have been told of many QFD analyses that never progress to the second matrix because of the perception of difficulty, time consumption, and conflict that inevitably arise when focus has been diluted. The Integrated QFD Matrix™ allows a focused progression to the Concept Matrix using functional analysis and concept evaluation, thereby eliminating much of the redundant or fruitless evaluation of design analysis of elements at the parts/mechanism level.

The Concept Matrix is divided into ten elements and is developed much in the same way as the initial System Matrix of Step 1. Figure 7-17 is an example of how the second matrix for the CVM product might look, focusing on the development of the concepts contained in System 1 above. Notice that the service and business operations requirements are mapped against service methods and business task elements. This enables the team to not only track the development of the product but also track the development of the service and business support functions and their continued interaction with the product (Figure 7-17 is also an abbreviated matrix, like Figure 7-2, showing only two elements for each of the product, service, and business groups). Although the Concept Matrix appears to be much the same in form as the System Matrix, there are enough differences to warrant a discussion of each of the ten elements of the matrix shown in Figure 7-17. Those ten elements are given below.

1. **Technical Requirements** — The technical requirements for the concept matrix come from those identified in the System Matrix in Step 1 and are placed on the right side of the Concept Matrix (see Figure 4-8, Item 3). However, not all of the technical requirements developed in Step 1 should be used in the Concept Matrix. The authors have found through both experience and analysis that only the highest priority technical requirements (usually about 40% of the requirements) need to be taken to the next level of QFD analysis. These requirements are selected based upon the values obtained in the System Matrix as the Relative Weight of the technical requirements (see Figure 4-8, Item 5B). Because the very nature of QFD is to prioritize and find the system drivers, decomposing all requirements to the next level of QFD analysis is neither necessary nor productive. Most QFD advocates, no matter which method of QFD itself is used, emphasize the utility of keeping the effort focused and avoiding looking at all requirements in subsequent QFD analyses. The authors advocate only using the key drivers for the design for the following reasons:

- The highest priority technical requirements are those that will have the largest impact on the customer needs and drive the ultimate design of the project;
- Focusing on the top technical criteria does save time and resources, which are elements that are usually a significant concern of any project;
- The iterative nature of the Integrated QFD Approach™ allows the QFD team and project personnel to revisit other technical concerns if the design changes; and
- Focusing on the key requirements allows the team to move more quickly through the entire design process and discover potential "show stoppers" early in the areas of highest correlation to meeting customer needs.

Although the authors do recommend that only the highest priority requirements be selected for the concept matrix, QFD is a flexible analysis tool and nothing should preclude the QFD team from selecting all of the requirements for further analysis if the program sees all

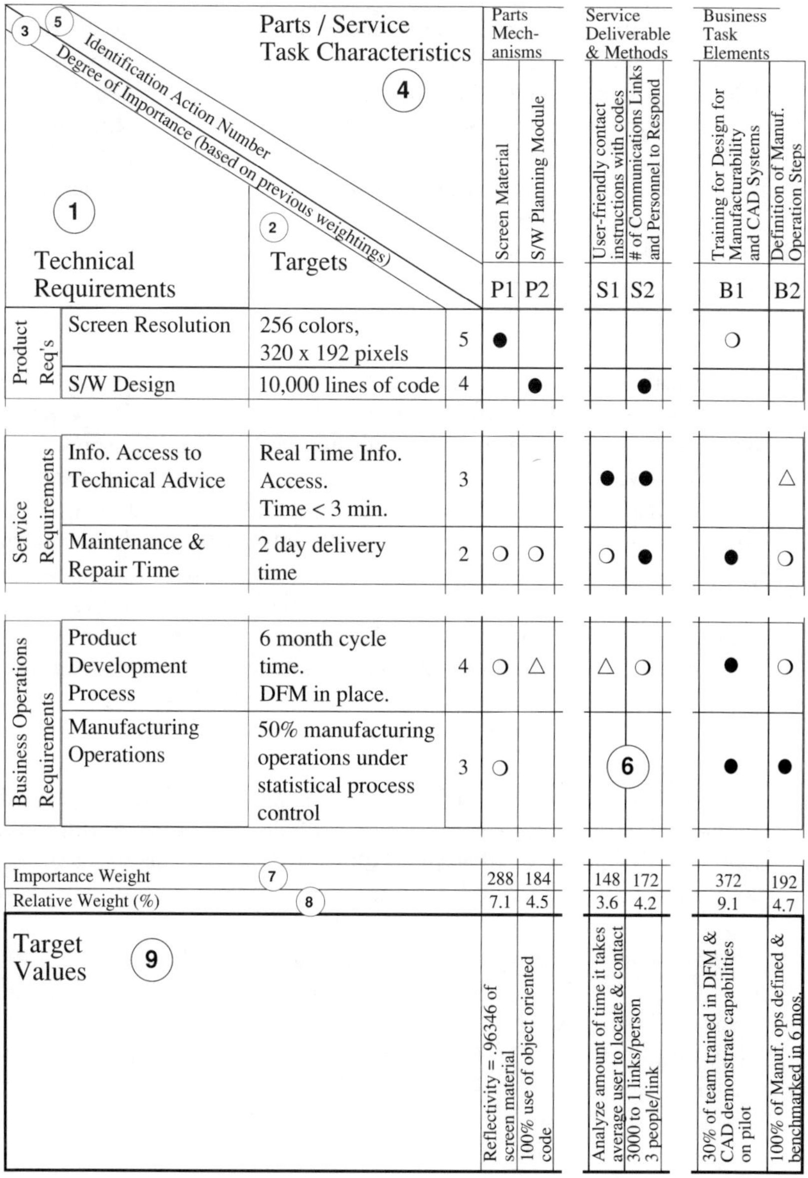

Figure 7-17 The Concept Matrix: translating the technical requirements into product/service and task requirements matrix number 2.

requirements as critical to the success of the design and there are substantial considerations for each requirement such as new technology, close tolerances, or a new untried process. Figure 7-17 assumes that the six requirements in Figure 7-2 are part of the top 40% highest priority requirements, and maps the parts, service, and business task characteristics against them.

2. **Target Values** — The Targets Values for the technical requirements should be included in the Concept Matrix. Requirement Target Values provide a measure that the parts/service/business task characteristics need to support. At the Concept Matrix level of detail, the QFD team must keep in mind that they are seeking to satisfy ever-increasing levels of detail. Placing the Target Values on the Concept Matrix provides this reminder and keeps the team focused on technical requirement definition.
3. **Degree of Importance of Technical Requirement** — The degree of importance of the technical requirement comes from the relative weight of the technical requirement as calculated in the System Matrix (see Figure 4-8, Item 5B). Although the technical requirement relative importance weight is expressed as a normalized number (based on a percentage between 0 and 100), the team should use this weight of the requirement in order to convert the importance to a 1 through 5 value. For example, in Figure 7-2 the requirement of "Screen Resolution" received the highest Relative Weight of 12.4. When used in the concept matrix, "Screen Resolution" should be ranked as a 5 to reflect this importance. Requirements that are close in value, such as "Software Design" (9.5) and "Product Development Process" (10.6) should be ranked the same. In this instance each merits a 4 because they are of next highest ranking to "Screen Resolution." QFD team members should resist the urge to completely re-rank the technical requirements with a voting process for two reasons: (1) The initial QFD matrix has accomplished the ranking based upon detailed information and debate, and (2) Team members may have a tendency to "second guess" their previous decision-making analysis using intuitive analysis rather than empirical reasoning.
4. **Parts/Service/Business Task Characteristics** — The parts/service/business task characteristics are developed using the same tools as those used to develop the original technical requirements. The QFD team should use brainstorming, affinity diagrams, and tree diagrams to develop the part/service/business task characteristics that support the technical requirements. In developing the original technical requirements (Section 4.3.2.3) the QFD team relied on existing technical data, technical interchanges, and brainstorming. The QFD team uses these techniques in developing the characteristics to support the technical requirements. Figure 7-17 shows six characteristics, two for each area of focus, that support the technical requirements. Attention to detail is more important at this level than at the requirement level, and the team should be allowed to focus more directly on detailed solutions to meeting the target values for requirements.
5. **Identification Action Number** — The Identification Action Number is a new addition to the QFD matrix and is used to track each detailed characteristic in an integrated plan. Because the resulting prioritization of the Concept Matrix should also be included in the integrated plan, identification of each characteristic with a task number will create a

more traceable path from the matrix to the integrated plan and facilitate coordination.

6. **Correlation Matrix** — The correlation matrix is developed using the same symbols and values always used in correlating two elements in a QFD matrix: (1) a **strong relationship** is a filled circle and has a numerical value of 9; (2) a **moderate relationship** is represented by an open circle and has a numerical value of 3; (3) a **weak relationship** is represented by a triangle and has a numerical value of 1; and (4) **no relationship** is represented by a blank cell. The relationship between a part, service, or business task characteristic and a technical requirement should be analyzed using a modification of the "If I control . . ." sentence of Section 4.3.2.4:

   > If I control (the part/service/business task characteristic), then I am meeting my technical requirement to provide (what).

   Using this sentence allows the QFD team to determine whether or not the characteristic correlates to a specific requirement, and the extent of that correlation. This is where having the requirement target values on the QFD matrix helps to determine the nature and strength of the relationship. For example, in Figure 7-17 the characteristic of "Screen Material" has a strong correlation to the requirement of "Screen Resolution" because material will have a direct effect on the ability to meet the target values for screen resolution. Using the sentence would yield a statement similar to the following:

   > If I control the Screen Material, then I am meeting my technical requirements to provide Screen Resolution (of 256 colors and 320 by 192 pixels).

   By looking at the parts/service/business task characteristics from a viewpoint of how such characteristics would meet the requirements as defined by the target values enables a clearer understanding of the correlation. Breaking the characteristics into part, service, and business elements will allow the team to more easily define the tasks associated with the elements and their correlation with the requirements. The three distinctions also provide a clearer means for assigning task ownership and integrating them into the functional areas of responsibility for a company. The interrelationship between the three areas of product, service, and business operations is more easily seen and analyzed. For example, under "Business Task Elements" in Figure 7-17 is the subcharacteristic of "Training in Design-for-Manufacturability and CAD Systems." This business task characteristic is not only strongly correlated to the "Business Operation Requirements," but has a strong relationship to the "Service Requirement" of "Maintenance and Repair Time" and a moderate relationship to the "Screen

Resolution." By analyzing these relationships in the concept matrix, the CVM team, and the project management, know that training in design-for-manufacturing and CAD systems will help to support one of the key product requirements and affect maintenance and repair time. This will help to establish communication channels between these tasks and help define tasks and schedules as part of the integrated plan.

7. **Calculation of the Importance Weight** — Calculation of the importance weight is accomplished by multiplying the Degree of Importance (Item 3) for each requirement by the numerical correlation value of the characteristic, and then adding the products for the entire characteristic column (see Table 7-3, Item 7).
8. **Relative Weight of the Characteristics** — The Relative Weight of the Characteristics is calculated by normalizing the Importance Weights (Item 8 in Table 7-3) in the same manner as for the calculation for the Core QFD Matrix in Chapter 4. The Relative Weight can then be used for comparing the importance of the individual parts/service/business task characteristics, and can also be converted into a graphical representation (bar chart).
9. **Target Values for the Characteristics** — Target Values for the parts/service/business task characteristics should be developed using established values or tolerances based on specifications, test capabilities, benchmarked data, or historical information. Time and effort should be taken to properly characterize the targets, and the targets should be capable of being measured. Target Values that cannot be measured prevent the ability to measure compliance. If compliance cannot be measured then there is no control over the element and it will remain unknown if the other factors relying on that measure will be satisfied. Ultimately, because all measures will lead back to customer needs, it will be unknown if the product and service is fulfilling those needs. Target Values should also be quantified wherever possible, especially in the product area. Turning to Figure 7-17, the Target Value for the "Screen Material" is "Reflectivity = 1." The Target Value could have been simply "Bright Screen Material," but by giving a reflectivity value — based upon research, benchmarking, or other data — the team has a specific performance value to achieve in determining the screen material to be used. The business elements should also have measurable target values such as "100% Manufacturing Operations Defined and Benchmarked in 6 Months," which is the Target Value for "Definition of Manufacturing Operations Steps and Benchmark Performance" (see Figure 7-17).
10. **Design Sketches, Flow Charts, and Other Visual Material** — The Concept Matrix contains a place for inclusion of design sketches, flow charts, and other visual or graphical representations that can assist the team and any project members in understanding concepts for parts or mechanisms, service concepts, or operational flows. This section is not mandatory but allows additional areas for interface and understanding between team members and across functional departments and areas.

Table 7-3 The Concept Matrix: Product/Service and Task Characteristics

Inputs:

- Highest priority technical requirements from the System QFD Matrix (approx. top 40%).
- Target values from System QFD Matrix.
- Importance weights from System QFD Matrix.

Desired Result:

- Identification of critical parts and mechanisms, deliverable service items and methods and critical business operation tasks.
- Selection of new or best subsystem concepts.

1. Key Product/Service and Business Operations Requirements
   - Transfer from the Core QFD Matrix (items 3 and 5b of Fig. 4-8).
   - The critical items are usually within the top 40%.
2. Target Values
   - Transfer from the Core QFD Matix (item 6a).
3. Degree of Importance
   - Assign 1 to 5 based on the ratings established from the Core QFD Matix (1 being low and 5 being high). (Relates to item 5b).
4. Part & Mechanisms/Service Deliverables & Methods and Business Task Elements Requirements
   - Use affinity and tree diagrams to brainstorm.
   - Use the following phrase: If I control __"how"__, Then I am meeting my customers needs to provide __"what"__.
   - Allocate requirements to either Parts and Mechanisms, Service Deliverables and Methods, or Business Task Elements.
5. Identification Number
   - Provide an identification number to each part, service or task characteristic for plan moderating and coordination.

Table 7-3 summarizes the ten elements of the Concept Matrix. There is one additional analysis that can be done that is not shown in either Figure 7-17 or Table 7-3: Correlation of the Parts/Service/Business Tasks Characteristics. This analysis would be accomplished by conducting the "roof" analysis that was done in the Core QFD Approach (Section 4.3.2.7 of Chapter 4).

Table 7-3 (continued) The Concept Matrix: Product/Service and Task Characteristics

6. Correlation Matrix
   - Determine relationship of the Part, Service or Task Characteristic to meet the technical requirements.
     - ● Strong (9)
     - ❍ Moderate (3)
     - ▲ Weak (1)
     - No Correlation (0)

7. Importance Weight
   - Multiply Degree of Importance by numeric value of the correlation.
   - Find summ for each column and record.

   Σ (③ × ⑥) (i.e. the sum of item 3 times item 6 for each column)

8. Relative Weight
   - Convert the Importance Weight to percentages and record.

   $\frac{⑦}{\Sigma ⑦}$ × 100 = Relative Importance Weight for each column.

9. Target Values
   - Establish Value or Tolerance based on specifications, test capabilities, benchmark data or historical information.

10. Sketches
    - As needed to improve communicaiton of how target values should be applied.

This analysis would determine if there are any interactions between various characteristics that should be examined, avoided, or encouraged. This step is usually not as critical at this level of analysis because interactions between the technical requirements have been identified (as in Figure 7-2) and steps should have been taken at that level to identify and coordinate potential interrelations. However, if the team is still concerned about specific interactions between characteristics an analysis can be conducted within a relatively short period of time.

After completion of the Concept Matrix the results should be analyzed and compared (Figure 7-18) to determine priorities for action. The results are then translated into tasks and additional task sheets can be developed and included in the evolving integrated plan as a lower level of detailed activities. Many of the tasks that may be identified will support general activities already developed by the team after completion of the System Matrix, and will create a waterfall effect of activity within that task.

What should result from the Concept Matrix is the further focusing of the team and the project on which parts and mechanisms, service deliverables

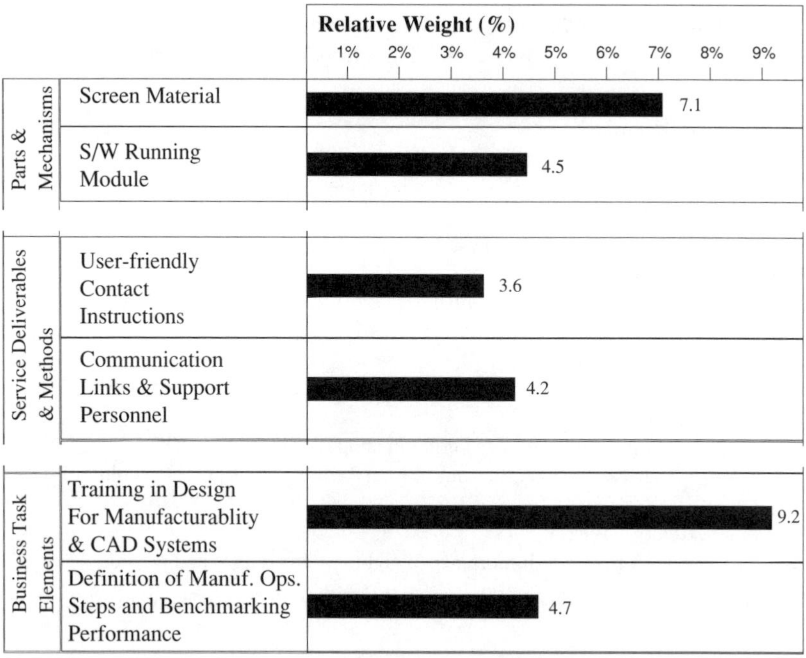

Figure 7-18 The results of the second matrix provide a means to prioritize activities and analysis to support the development of Parts and Mechanisms, Service Deliverables and Methods, and Business Task Elements.

and methods, and business task elements should be the focus of resources to ensure meeting technical requirements that support the critical customer needs. From the priorities shown in Figure 7-18, the team knows that training in design-for-manufacturability and CAD will be critical activities for supporting a host of activities critical to the program. Next would be the development or purchase of appropriate screen material, and any activities associated with that element. In this manner the team retains its focused edge and can move quickly to tackle the pressing issues in development of the interactive video game.

The next step in the Integrated QFD Approach™ is the development of the third matrix: The Process Matrix.

7.1.7 The Process Matrix: manufacturing, service, and business procedures (Step 7)

The Process Matrix is designed to identify the critical manufacturing, service, and business operation procedures to be monitored and controlled. The Target Values for the procedures are developed in order to further define controllable methods of ensuring final compliance with meeting the customer's needs.

Development of the Process Matrix is based upon the same method of analysis for the Core QFD Matrix with a few changes. There are the same number of steps involved as in developing the Concept Matrix, but the results are more detailed-oriented and include more specifics. The Process Matrix begins combining more design and manufacturing information and also brings into play the results of other analysis tools such as Failure Modes and Effects Analysis. Figure 7-19 illustrates the application of the Process Matrix for the CVM game, incorporating the ten elements of the matrix using the six characteristics from the Concept Matrix. The following describes the ten elements of the Process Matrix.

1. **Parts, Service, and Business Task Characteristics** — The first element of the Process Matrix uses the highest priority Parts, Service, and Business Task Characteristics developed in the second matrix. These characteristics are used to form the left side of the matrix, and should be selected as the key characteristics. Selecting the highest priority characteristics usually results in using about 40% of the characteristics developed in the Process Matrix. Because the purpose of QFD is to track the key drivers of a system, no matter the level of detail, it is usually neither beneficial nor practical to use all of the characteristics developed in the Concept Matrix. This is not to say that the characteristics analyzed in the Concept Matrix should be forgotten or abandoned to random chance fulfillment. The characteristics that are not key drivers should still be followed (usually through tasks in the integrated plan) but are not selected for further development in the next stage of the matrix analysis. These characteristics chosen for use in the Process Matrix are selected based upon the results of the Relative Importance Weights developed in Part 8 of Figure 7-17, and for the purposes of illustration, all six of the characteristics developed in Figure 7-17 are used in Figure 7-19 and assumed to be among the highest priority characteristics.
2. **Target Values** — The Targets Values for the Parts, Service, and Business Tasks Characteristics should be included in the Concept Matrix. Characteristic target values provide a measure for support by the procedures that are identified for Manufacturing/Production, Service Procedures, and Business Operations Procedures requirements. At the Process Matrix level of detail, the QFD team is seeking again to satisfy ever-increasing levels of detail. Placing the Target Values on the Process Matrix provides this reminder and aids the QFD team in understanding exactly what is being measured for the characteristics, and how the procedures will help control that measure.
3. **Degree of Importance Ranking** — This ranking is accomplished in the same manner as that done for the requirements in the Concept Matrix (see Item 3 of Section 7.1.6 above). The rankings are based upon the Relative Importance Weight of the Concept Matrix and should reflect the selected characteristics' relative importance to each other using the 1 through 5 designation (5 being the most important).
4. **Manufacturing/Production, Service and Business Operations Procedures** — These are developed using the techniques of brainstorming,

Procedures (4)

(5) Identification Number

(3) Degree of Importance (based on previous weightings)

	(1) Technical Requirements	(2) Targets		Manufacturing Production Requirements: Coating Process = Thickness ± .003 Adhesion = 99% after tests (M1)	S/W I/F demonstration with hdw in loop (M2)
Parts & Mechanisms	Screen Material	Reflectivity = .946 of screen material	5	●	
	S/W Running Module	100% use of object-oriented code, 200 L.O.C.	4		●
Service Deliverables & Methods	User-friendly contact instructions	(5th grade level) Analyze amount of time it takes avg. user to locate and contact. T = 3 min.	3		(6)
	# of Communication Links & Personnel to Respond	3000 to 1 links (users to comlink) 3 people per link	4		
Business Task Elements	Training in DFM & CAD Systems	30% team trained in DFM & CAD with demonstrat. ability	5		
	Definition of Manuf. Ops. & Benchmark Performance	100% Manuf. Ops defined & benchmarked in 6 months	4	❍	●
Importance Weight (7)				334	192
Relative Weight (%) (8)				7.8	4.5
Target Values (9)				10 sec. coat ± 1 sec. No flaking at 180° environ. cycle test	100% of data read across system

Figure 7-19 The Process Matrix: Manufacturing, Service Procedure Operations, and Business Procedure Operations Requirements.

Service Production Requirements		Business Operations Production		Control Points (10)
Field Operators: Respond to 98% callers within 2 min.	Service Locator System – Contact correct dept. 1st try 99.9% time	Training Procedures Std. method for DFM and CAD systems. Validated by exam.	Operational Analysis Procedures: Data collection validates statistical sampling with 95% confidence.	
SP1	SP2	BP1	BP2	
			●	Chemical bond/adhesion test; Env. vibration test; Env. temp tests
			△	Interface "hardware in the loop" demo test phase I
	●		△	Statistical analysis in pilot demo. Monthly spot checks in various demographics.
●				Statistical analysis in pilot demo. Biweekly spot checks in different shifts…
		●	○	Training Cert. Periodic training exams. Efficiency checks…
	○	○	●	Standardization of process plan for data collection and analysis. Calibration checks. …
144	124	228	412	
3.3	2.9	5.3	9.6	
Statistical sampling of efficiency to respond within 2 min.	Pass pilot demonstration of 3000 callers with 99.99% accuracy before launch.	95% Trained Pass standard exam.	Data Collection provides statistical sampling 95% confidence.	

Figure 7-19 (continued).

affinity diagrams, and tree diagrams; however, the development of these procedures should also rely heavily on other analysis such as Failure Modes and Effects Analysis (FMEA), Design of Experiments (DOE), and Statistical Process Control (SPC). These techniques are used to analyze more specific elements of the design and manufacturing processes and are briefly discussed in Section 7.2. The results of these methods of analysis are effectively inserted into the Process Matrix as target values for the procedures, identification of the procedures that are most critical to the success of the characteristics, and as measurements of the detail procedures that will ultimately let the QFD team know if the characteristics are being met and therefore controlled. This level of analysis is often done prematurely without the use of QFD. QFD presents the framework for incorporating the results of these analyses and for allocating requirements to either manufacturing/production procedures, service procedures, or business operations procedures.

5. **Procedure Identification Number** — Just as in the Concept Matrix, the Procedures should be assigned identification numbers for easy reference, tracking, and use in the integrated plan.
6. **Correlation Matrix** — The Correlation Matrix uses the 9-3-1-No Correlation convention for analyzing the relationships between the Parts/Service/Business Operations Characteristics and the Manufacturing/Production, Service, and Business Operations Procedures. The "control" sentence that has been used in previous matrices can once again be employed here, with a slight variation:

   > If I control (manufacturing/production, service, or business operation procedure), then I am meeting my characteristic to provide (what (usually the characteristic target value)).

   The correlation matrix allows the team to know what procedures will lead to meeting the critical characteristics that have been identified as drivers to the success of the project. In Figure 7-19, an example of the use of the "control" sentence can be seen using the "M1" procedure as correlated to the part characteristic of "Screen Material":

   > If I control the "Coating Process" then I am meeting my characteristic to provide "Screen Material" (with a reflectivity = 1.0).

   The "control" sentence aids the team in understanding the strength of the relationship between procedure and characteristic. In the above example, the relationship is a strong one and merits a correlation numerical value of 9 (filled circle). The Correlation Matrix for the

Process Matrix provides the team the ability to analyze cross-functional relationships even at this level. Figure 7-19 once again demonstrates that almost all of the critical procedures have some impact on the ability to meet the Business Task Characteristic of "Definition of Manufacturing Operations and Benchmark Performance" and the associated target value of "100 % Operations Defined and Benchmarked in 6 Months." This makes intuitive sense from the start because if procedures are under control then the ability to meet manufacturing operations and benchmarking procedures will follow. The Correlation Matrix, however, also lets the team visualize the extent and nature of the procedures on the ability to meet this characteristic, and which procedures will be the overall drivers to meeting the remainder of the highest priority characteristics. The Correlation Matrix also looks at the variability in the procedure as an impact to the corresponding characteristic; hence the importance of using other analysis tools to measure variability in the procedure.

7. **Calculation of the Importance Weight** — Calculation of the importance weight is accomplished by multiplying the Degree of Importance (Item 3) for each characteristic by the numerical correlation value of the procedure, and then adding the products for the entire procedure column (see Table 7-4, Item 7).
8. **Relative Weight of the Procedures** — The Relative Weight of the Procedures is calculated by normalizing the Importance Weights (Item 8 in Table 7-4) in the same manner as for the calculation for the Core QFD Matrix in Chapter 4. The Relative Weight can then be used for comparing the importance of the individual manufacturing/production, service, and business operation procedures, and can also be converted into a graphical representation (bar chart).
9. **Target Values for the Procedures** — Target Values for the manufacturing/production, service, and business operation procedures should be developed using established values or tolerances based on a value established through analyses such as FMEA, DOE, or SPC. Targets can also be established through tolerances based upon manufacturing operations, specifications, test capabilities, benchmarked data, and historical information. Just as measurement capabilities were a key consideration in target value selection for the characteristics identified in the Concept Matrix, the importance of measurable targets holds true for the procedures (see Item 9 under Section 7.1.6). An example of the types of measurements that can arise from either SPC or manufacturing operations is given in Figure 7-19 for the "Coating Process." In order for the coating process to achieve a thickness of 0.003 ±0.0002 and adhesion quality of 99% after all environmental cycle tests the process target values are a 10-s coat (±1 s) and no flaking at 180°F for the environmental testing. Measurable target values are also obtained for the service and business operation procedures (see Figure 7-19).

Table 7-4 The Process Matrix: Manufacturing/Production, Service, and, Business Operations Procedures

Inputs:

- Highest priority Parts & Mechanisms, Service Deliverables & Methods and Business Task Elements.
- Target Values and Importance Weights.
- Manufacturing and Production Operations process steps.
- Service and Maintenance process steps.
- Business Operations work flow tasks.
- Failure Modes Analysis of manufacturing, service and business operations if available.

Desired Result:

- Identification of critical manufacturing, service and business operation procedure steps that need to be carefully monitored and controlled.
- Target Values for manufacturing/production, service and business operations procedures.
- Capability to develop company procedures that preserve the customers intent.

1. Parts/Service/Business Operations Characteristics
   - Transfer from the Concept QFD Matrix (items 4 and 8).
   - The critical items are usually within the top 40%.
2. Target Values
   - Transfer from the Concept QFD Matix (item 9).
3. Degree of Importance
   - Assign 1 to 5 based on the ratings established from the Concept QFD Matix (1 being low and 5 being high). (Relates to item 8).
4. Manufacturing/Production/Service/Business Operations Procedures
   - Use affinity and tree diagrams to brainstorm.
   - Use the following phrase: If I control __"how"__, Then I am meeting my customers needs to provide __"what"__.
   - Use additional information from Failure Mode Analysis, Design of Experiments or Statistical Control data.

10. **Control Points** — The Process Matrix contains a new parameter, i.e., the identification of Control Points. Control Points are potential methods that can be used in procedures to validate the characteristics. In Figure 7-19 the Control Points for the "Screen Material" characteristic

Table 7-4 (continued) The Process Matrix: Manufacturing/Production, Service, and, Business Operations Procedures

4. continued.
   – Allocate requirements to either manufacturing/production procedures, service procedures, or business operations procedures.

5. Identification Number
   – Provide an identification number to each element in manufacturing/production, service and business operation procedure for plan monitoring and coordination.

6. Correlation Matrix
   – Determine the impact the procedure will have on the part/service/business characteristics based on the effect of variability in the procedure as an impact to the corresponding characteristic.
   ● Strong (9)
   ❍ Moderate (3)
   ▲ Weak (1)
   No Correlation (0)

7. Importance Weight
   – Multiply Degree of Importance by numeric value of the correlation.
   – Find summ for each column and record.
   Σ (③ × ⑥) (i.e. the sum of item 3 times item 6 for each column)

8. Relative Weight
   – Convert the Importance Weight to percentages and record.

   $\frac{⑦}{\Sigma ⑦} \times 100$ = Relative Importance Weight for each column.

9. Target Values
   – Establish value or tolerance based on manufacturing operations guides, statistical analysis, specifications, test capabilities, benchmark data and historical information.

10. Control Points
   – Identification of potential method that can be used in the procedures to validate the parts, services or business operations characteristics.

are "Chemical Bond/Adhesive Test, Environmental Vibration Tests, Environmental Temperature Tests, etc." Control Points provide opportunities to ensure that the procedures being used will produce the target values necessary in the characteristics. Control Points are often already written into procedures as tests, SPC data points, or other statistical and quantitative measurements, but their direct connection with the procedure supporting the characteristic creates traceability in the Integrated QFD Approach™.

The team can perform an interaction analysis of the procedures (the "roof") if necessary; but as in the Concept Matrix, the critical interfaces between the technical requirements usually identify any potential conflict areas. These are pursued and resolved in the integrated plan through specific action plans, or interfacing requirements communicated to various functional areas as specific tolerance ranges. Still, there is nothing wrong in conducting a brief analysis to determine if any of the procedures conflict with or enhance each other.

The above elements of the Process Matrix are summarized for the reader in Table 7-4. After the completion of this matrix, the procedures should be ranked according to importance in a graphical representation for further analysis (see Figure 7-20).

7.1.8 The Control Matrix: knowing you can get there (Step 8)

The fourth matrix — the Control Matrix — inputs are the highest priority Manufacturing/Production, Service, and Business Operation procedures from the Process Matrix. The purpose of the Control Matrix is to check and control the key production, service, and business operations. The reader experienced in production and manufacturing will recognize many of the individual process control elements that appear in Figures 7-21 and 7-22. In supporting manufacturing, process control charts are often created containing planning requirements, control methods, inspection methods, and test requirements. The Control Matrix takes the highest priority procedures from the Process Matrix and maps them against the items that typically constitute process control charts. By doing this, control is not only maintained, but through the waterfall effect of the Integrated QFD Approach™ the lowest level of tracking in the process control environment can be followed back through the previous matrices to demonstrate control and compliance with the original customer needs.

In Figure 7-21 the Manufacturing/Production Procedure of "Coating Process" for the screen, and the Service Procedure of "Service Locator System" have been chosen as critical procedures to look at from a process viewpoint. These two procedures become the left side of the Control Matrix. The top of the matrix is composed of elements from Manufacturing Process and Procedure (MPP) requirements, Standard Operating Procedures (SOPs) for companies, and from the experience of manufacturing and production control people from within the organization. By mapping manufacturing process operations against specific procedural requirements the QFD team can determine if there are controls in place to meet the target values (Part and Process Characteristic Values in Figure 7-21).

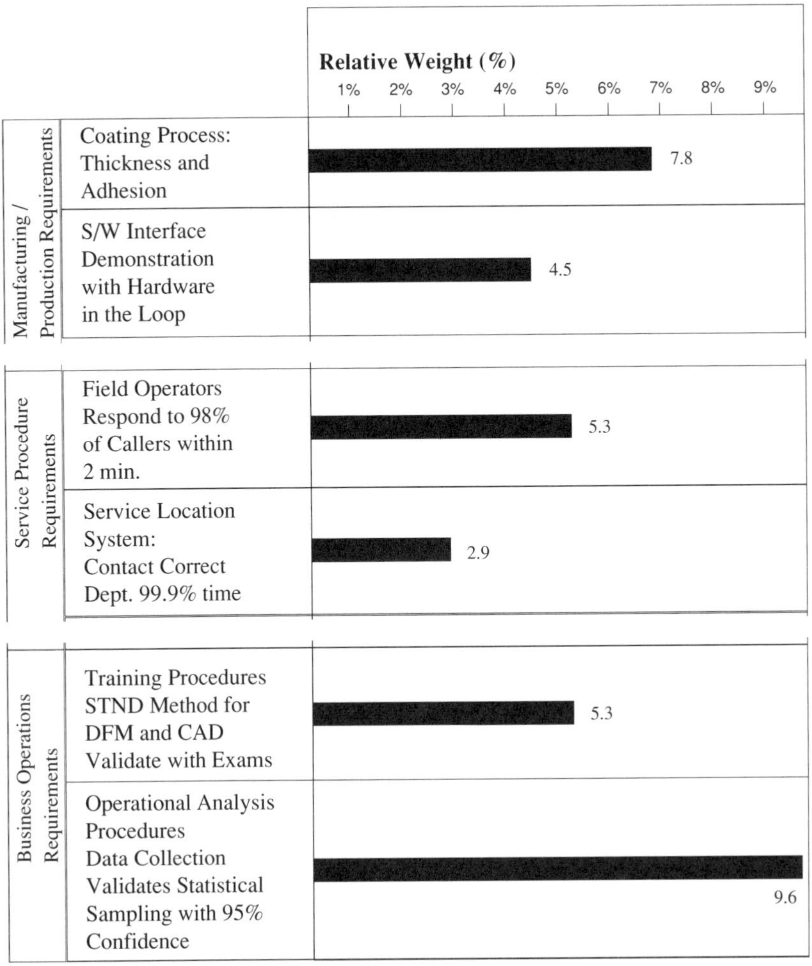

Figure 7-20 The results of the Process Matrix provide a means to prioritize activities and analysis to support the development of manufacturing, production, service, and business operations procedures.

For example, in Figure 7-21 the "Part Characteristic Values" for the Coating Process requires that thickness must be 0.003 ± 0.0002. This requires a series of events to occur in order to insure that the coating thickness meets this requirement. Under "Sample Frequency" the manufacturing team determined that sampling should occur every ten runs for "Thickness Coating," and every 30th item to assure meeting the nonflaking requirement. Under "Data Evaluation" the team determined that process control charts would be kept to measure meeting the requirements. Under "Test/Measurement Tool" the Roller Alignment must be within 0.0001 measures to meet the thickness variation requirements. All of these measures are critical in ensuring proper material thickness, which, in turn, would satisfy the reflectivity characteristic

Service	Product		
8529	PSCR55432	**Part Number**	
SLS	Coating Thickness = .003 in Adhesion/Bond strength	**Part Characteristic Value**	
Service Locator System	Screen System Coating Process	**Process Name**	
SP2	MI	**Routing Sheet Number**	
Correct contact on 1st try 99.9% of time	Temp variation 180° ± 2° Roll alignment variation ±.001 in	**Process Characteristics Value**	
Phone personnel on dept. responsibilities on indiv. to contact	Process Operations Calibrate Alignment Cleaning Techniques	**Training**	**Planning Requirements**
Upkeep phone lists, training program & individual contacts on expectations	Lubrication: Every 2 mo. Ball bearing replacement on schedule	**Maintenance**	**Planning Requirements**
Phone personnel exams	Guard Inspection Temp. Inspection	**Fail Safing**	**Planning Requirements**
Operator control center	Indicated on drawing 0009956	**Location**	**Control Methods**
Every 2 weeks monitor type question/ correct contacts	Thickness coating/ every 10 runs Flaking/every 30th item	**Sample/Frequency**	**Control Methods**
Company Best Practices	ISO	**Standard**	**Control Methods**
Control Charts/ Pareto Diagrams Team meeting resolution	Control Charts Team meeting resolution	**Data Evaluation**	**Control Methods**
Information Access Phone system working condition	Guards Alignment of rollers	**Inspect Item**	**Inspection Methods**
Visual system check	Visual/Each run Calibration	**Method/Frequency**	**Inspection Methods**
Company Best Practices	OSHA requirements	**Standards**	**Inspection Methods**
Tick sheet identifies system not operating/cause	Check sheets Control charts	**Data evaluation**	**Inspection Methods**
Data link indicator Signal/Noise ratio	Roll alignment wrench P3	**Type**	**Test/ Measur. Tool**
Signal to noise = ±3dB	± .0001 in roller alignment	**Calibration requir.**	**Test/ Measur. Tool**
Mable Becker	Pete Jacobs	**Measurement Inspection**	**Person Responsible**
Establish new connections Contact cable company	Recalibrate as necessary	**Follow-up Activities**	**Person Responsible**
None	Freq. of maintenance requires further monitoring and recommendations	**Remarks**	

Figure 7-21 The Control Matrix: Process/Procedure implementation chart for products and services.

Business Operations Procedures

M293	**Dept. Action No.**	
Data collection for coating process provides statistical sampling w/ 95% confidence	**Dept. Action Measurement Values**	
Business Operations Manufacturing Process 298	**Process Name**	
BP2	**Routing Sheet**	
Data collection 95% confidence	**Process Characteristic Values**	
Process Operations overview sessions SPC/Calibration training	**Training**	**Planning Requirements**
Update class/expertise every 3 mo.	**Maintenance**	
Certification of SPC capabilities	**Fail Safing**	
Manufacturing floor 1 / Booth 398	**Location**	**Control Methods**
Critical Process elements including thickness, bond strength, temerature	**Sample/Frequency**	
ISO	**Standard**	
Control charts / Team meeting resolution	**Data evaluation**	
Training materials Field activities of gathering SPC data	**Inspect Item**	**Inspection Methods**
Every 3 mos. Individual discussions / interviews	**Method/Frequency**	
Company Best Practices	**Standards**	
Review comments in team meeting	**Data Evaluation**	
Roller alignment wrench P3	**Type**	**Test/ Meas. Tool**
± .0001 in. roller alignment as spot check	**Calibration requir.**	
John Patterson	**Measurement Inspection**	**Person Responsible**
Perform on the job training and double team verification for accurate measurements	**Follow-up Activities**	
Additional math training is often required to implement SPC training	**Remarks**	

Figure 7-22 The Control Matrix: Process/Procedure Implementation Chart for Business Operations.

for the material selection, which, in turn, meets the Screen Resolution requirement, which satisfies many of the critical Customer Needs. Traceability is, therefore, maintained from the technical requirements to the process and control level.

The Control Matrix allows analysis of whether there are proper controlled processes in place to meet the target values for the procedures that will be employed to meet the Part/Service/Business Operations Characteristics. If the

CVM QFD team could not find a way to control processes to meet the thickness requirements of the screen material, or did not have statistical process control measurements to maintain tolerances, the team would know that the chances of satisfying any of the elements in the other matrices would be slim or merely a guess. Target Values are only important if there are means to meet those targets, and the Control Matrix gives the project a high level of confidence in and methodology to meet the Target Values.

Figure 7-21 also contains a Service Procedure Requirement of "Service Locator System" with a Process Characteristic Value of "Correct Contact on 1st Try 99.9% of Time." For the service process to work effectively, a Target Value of ensuring that the customer reaches the right department 99.9% of the time is critical. To control this the same parameters are used to determine whether the target value can be attained. For "Sample/Frequency" under Control Methods, the team has determined that a measurement can be created to determine if the criteria is being met: "Every 2 weeks Monitor Type Question/Correct Contacts." Methods for measuring "Data Evaluation" and "Calibration Requirement" are also available to determine if the 99.9% accuracy is being achieved. The bottom line is that just as in the manufacturing area control methods are utilized to ensure proper process fulfillment, similar controls can be created in the service area.

Can business operations be controlled in the same manner? Figure 7-22 illustrates how the majority of the Control Parameters used for service and manufacturing controls in Figure 7-21 can be applied to business operations. The critical procedure selected from Figure 7-19 in the Business Operations area is "Data Collection for Statistical Sampling." This process was a critical business operation process in Figure 7-21 and therefore the CVM team determined that the process needed to be controlled in order to ensure success of the video game project. The Process Characteristic Value of "Data Collection to 95% Confidence" level is the target the team needs to hit to ensure meeting the statistical sampling requirements for the coating process. The goal for the team is to ensure that proper data is being taken, and that the data is accurate. The business operation involving data collection is being controlled, which, in turn, assures the team that the parts and process measurements are accurate from a data sampling point of view. For example, is the data sampling for the control charts used in SPC being done properly? Standards for "Inspection" and "Control" methods ("Company Best Practices" and "ISO," respectively) are in place, and training methods are in place to increase adherence to the standards.

By using the Control Matrix in this manner, the QFD team creates standard methods for avoiding problems as well as analysis procedures for solving them. Performance variation is thus minimized.

When building the Control Matrix, the team should consider that this matrix is the most detailed of all the matrices if filled out completely. It requires in-depth understanding of manufacturing, service, and business operation processes, and therefore should be the result of confirmed research

The Control Chart: Process/Procedure Implementation

Inputs:

- Highest priority procedure elements from the manufacturing/production, service and business procedures.
- Target values.
- Process steps to be used in products, services and business operations.
- High priority issues from failure mode analysis concerning manufacturing/production, service and business operations.

Desired Outcome:

- Detailed plans for checking and controlling key production, service and business operations.
- Standard methods to prevent problems.
- Method to minimize performance variation in products, services and business operations.

Figure 7-23 Control Matrix: Implementation monitoring measures for products, services, and business operation procedures.

and experience. Suggested inputs to and probable outcomes from the Control Matrix are contained in Figure 7-23. Communication is a key at this level. Proper communication will reduce a great deal of error and misunderstanding between departments and individuals in charge of carrying out the actual process tasks and measurements. Training should be a key tool in developing proper skills for monitoring and improving daily operations. Figure 7-24 summarizes some of the elements that should be followed when using the Control Matrix.

7.1.9 Refining the integrated plan (Step 9)

After the analysis in the Control Matrix has been completed, the manufacturing/production control, service, and business operation elements of the QFD team should create any necessary subelements to the integrated plan. An example would be the creation of a Process Control Plan for the manufacturing area (an abbreviated example of this plan is given in Figure 7-25). By incorporating the subplans into the integrated plan, the team has visibility into all critical steps of the program, and can best determine if the project is on schedule. Plan integration also continues to help sequence events and operations.

Process Instruction and Standards

Purpose: Communicate operating instructions, specifications, inspection methods and planned production rates.

Content: Standard Operating Procedures
Inspection Instructions
Process Set-up Sheets
Dimensional Standards
Visual Acceptance Standards
Process Route Sheets

Note: Keep them simple, staightforward and revise them as needed.

Orientation and Training

Purpose: Transfer and develop proper skills of employees to perform, monitor and improve daily operations.

Content: Theoretic Knowledge — manuals/lectures/text
Demonstrated Knowledge — hands-on application
Testimonial Case Studies

Process Flow Charts

Purpose: A visual tool to improve communication, planning, reporting and controlling processes and procedures.

Elements: Two symbols
– A rectangle = **Process**
– A diamond = **Decision**

example:

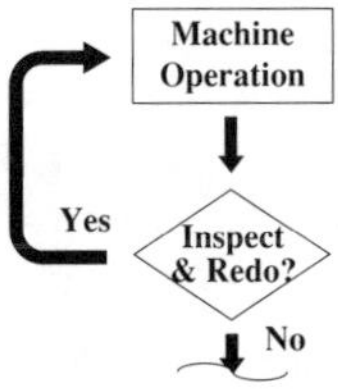

Figure 7-24 Elements to be considered during the development of the Control QFD matrix.

7.1.10 A living document

By the time the team has gotten through Step 9 in the Integrated QFD Approach™ a coherent plan has been developed, responsibilities have been assigned to specific individuals, and the team has a living document to work to at all levels. If redesign is necessary, the team can determine quickly and efficiently all interfaces with any key redesign. If conflicts between requirements arise, the team can determine what impacts any changes will have on overall performance targets.

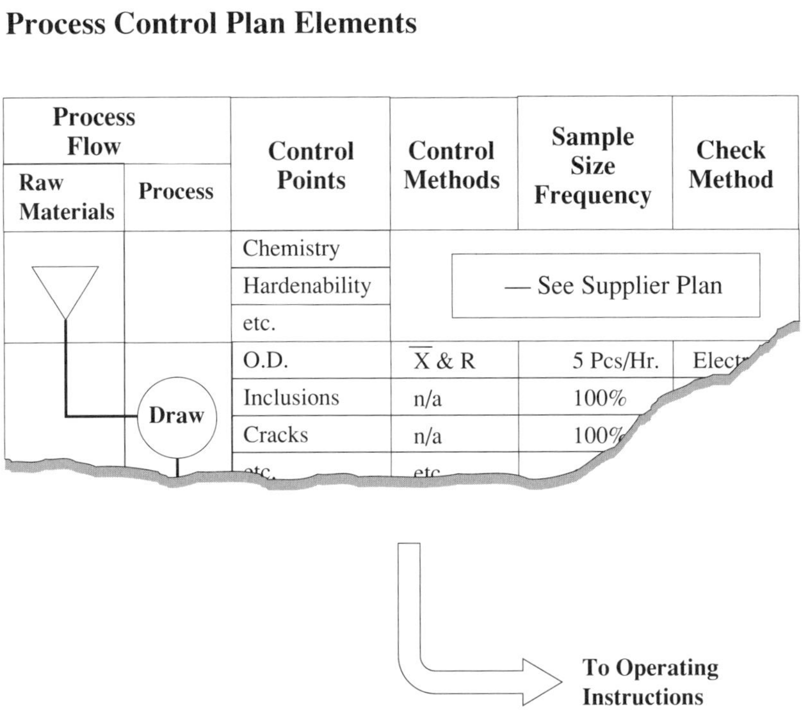

Figure 7-25 The Process Control Plan addresses key procedures and processes, critical characteristics, and control points in the Control QFD Matrix.

The Integrated QFD Approach™ has as its greatest asset the ability to create an organized and methodical approach to requirements development and function integration. It is very well suited to the American business operation in design, service, operations, and manufacturing because of its focused goals and relatively rapid incorporation into existing methodologies. The Integrated QFD Approach™ does not create a huge upheaval in existing operations, but rather brings a sense of order and memorializes useful methods that may already be in place in an operation.

Closely tied to the Integrated QFD Approach™ is the concept of Robust Designs — The use of QFD and other analysis tools to take into account variability as part of the design. The next section discusses how Robust Design compliments QFD, and fits into the Integrated QFD Approach™.

7.2 QFD and Robust Designs

In the environment of systems engineering, QFD is utilized as a major tool in developing Robust Designs. A Robust Design is a design that takes into account variations or variability in the design process, and has also accounted for methodology to control that variation to produce a product that is one of quality. As mentioned earlier in this chapter, just as QFD is essential in developing requirements to meet customer needs, the statistical elements of Robust Design provide the means to measure and ultimately control variation in the process, product, or procedure. Although it is not the purpose of this book to instruct in detail the complexities of using elements such as DOE and SPC (both elements of a Robust Design), the reader should be aware of how QFD and Robust Design integrate, and that requires some understanding of the statistical elements of Robust Design before returning to the interplay of these statistical elements with QFD.

7.2.1 Origins of Robust Design

Robust Design has as its roots statistical experimental design. The beginnings of Robust Design go back to the work of Dr. Genichi Taguchi who was assigned the task of developing high-quality products during Japan's reconstruction after World War II.[30] Taguchi was a recipient of the Deming Award in 1962, and did much of his work for the Electrical Communications Laboratories of Nippon Telephone and Telegraph Company. Taguchi developed mathematical methodology for analyzing the causes of variation and addressing how to make the design less susceptible to that variation, often without actually eliminating the cause of the variation. This was an important value engineering concept because the resources for radical and expensive redesign available to Japanese manufacturing after the war were limited. Thus, inexpensive ways for improving quality were the focus of the industry.

In his book on Robust Design, Phadke states that Robust Design addresses two major concerns faced by all product and process engineers:

- How to reduce economically the variation of a product's function in the customer's environment. (Note that achieving a product's function consistently on target maximizes customer satisfaction).
- How to ensure that decisions found to be optimum during laboratory experiments will prove to be so in manufacturing and in customer environments.[31]

Robust Design allows the engineer to resolve these concerns by using statistical analysis and experimental design. Robust Design has been used extensively in both Japanese and American manufacturing for some time, but the statistical elements of Robust Design are not often used in conjunction with QFD.

7.2.2 Back to the definition of quality

In Chapter 2 the difficulty of defining exactly what people meant when they discussed "quality" was raised, along with the idea that "quality" can mean different concepts depending upon context. The authors tried to paint as broad a brush stroke as possible to define quality in terms of a whole organization in an attempt to incorporate as many of the subdefinitions as possible. When examining Robust Design, the term "quality" has a narrower meaning that takes into account more of a comparison of designed product parameters to actual performance parameters. Taguchi defined "quality" to be a relationship between economics and variation from a target parameter of performance — the farther the variation from the target value, the greater the loss of profit or total loss to society. This concept is based upon the idea that there is no "bright line" distinction between variation from the target value and customer satisfaction. If a fast-food restaurant wants to serve its customers within 3 to 5 min after ordering, the customer would probably not notice a difference between service in 4 min 58 s vs. 5 min 3 s. There is no bright line demarcation that says the customer is satisfied if service takes less than exactly 5 min, and dissatisfied if the service takes longer than exactly 5 min. The parameter is probably more accurately expressed that the longer service takes, the more the customer will become dissatisfied with the quality of the service. This concept is more realistic when gauging what effect variation has on performance parameters than a black-and-white distinction of good/bad quality.

7.2.3 *"Old" vs. "New" philosophies of quality in Robust Design*

In order to determine what a Robust Design is, it is important to understand what is meant by quality with regard to satisfaction of customers expectations and needs, and also what is meant when quality is not achieved, or product failures are encountered.

In the past companies adhered to the old philosophy of quality. That philosophy embraced the concept that there was a lower spec limit (LSL) and an upper spec limit (USL) for any given characteristic, or sets of characteristics, for a product. When taking measurements, a quality product was one that fell within the lower and upper spec limits. Any products that fell outside of those two limits were losses due to either scrap or rework (see Figure 7-26). As a result of this concept of quality, a company's response to a drop in quality products was to increase inspections to insure spec compliance as opposed to re-examining the manufacturing process or re-evaluating the design criteria of the product. However, over the years it has been determined that inspection can only be about 80% reliable. Extensive research has also demonstrated that this concept of quality requires the addition of more manpower to inspect for quality, resulting in a large amount of rework. This can result in scrap costs that are prohibitive or even capable of driving a company out of business. The concept that quality also equals increased inspection is contrary to today's modern concept of being customer-focused and achieving the highest percentage of customer satisfaction.

Over the last 15 years, there has been a shift towards a new philosophy of quality. That philosophy is to measure quality in a product in terms of the total loss to society due to financial variation and harmful side effects. This is the Taguchi concept of quality. Figure 7-27 represents a functional view of product specs with relation to quality. As the reader can see in Figure 7-27,

Old Philosophy of Quality

- Tendency to increase inspections for specification compliance.
- Inspection is only 80% reliable.
- Requires more manpower and results in rework and scrap costs that can be prohibitive.
- Questionable customer satisfaction.

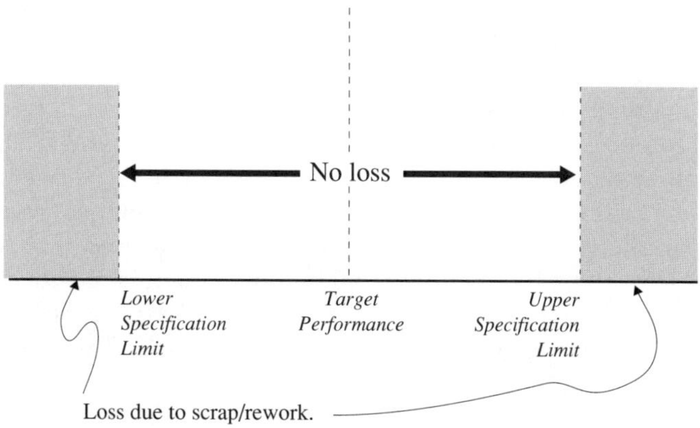

Figure 7-26 Previous definitions of quality focused on specification compliance.

New Philosophy of Quality

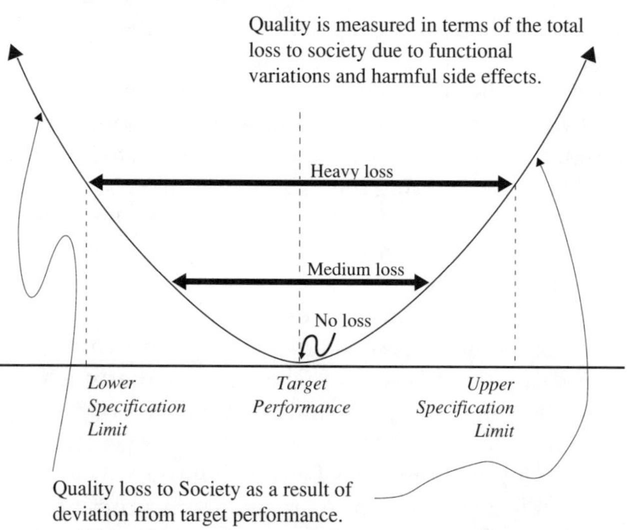

Figure 7-27 The new philosophy of quality as defined by Taguchi.

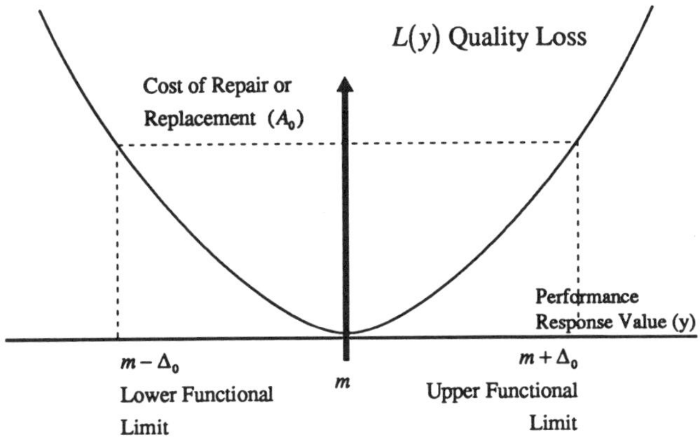

$$L(y) = \frac{A_0}{\Delta_0^2}(y - m)^2$$

y = The quality characteristic or Performance Response Value
m = Target Value for quality characteristic y
A_0 = Cost of repair or replacement of the product
$m \pm \Delta_0$ = Functional limits
(the value of y at which product would fail
in half the applications)

For example:
A power supply circuit used in a video game system

m = target voltage = 110 volts
Δ_0 = ±10 volts
A_0 = \$275 to repair

$$L(y) = \frac{275}{10^2}(y - 110)^2 = 2.75(y - 110)^2$$

If Actual Performance was 105 volts

$$L(y) = 2.75(105 - 110)^2 = \$68.75 \text{ loss of quality}$$

Figure 7-28 Taguchi's Quadratic Loss Function.

a product that lies in the center of defined upper and lower specs creates no loss for the company, but, as product specification *variation* ventures toward either end of the spectrum between upper and lower limits, the loss grows consistent with a curve that can be represented by a quadratic equation.

This quadratic equation has been defined by Taguchi as the Quadratic Loss Function and is represented in Figure 7-28. Taguchi demonstrates that the closer products approach the target value in performance, the less impact, monetarily and socially, they will have on society because of any possible loss of quality. Greater deviation equals greater likelihood that the product will fail or the customer will be dissatisfied because the performance characteristic is closer to the lower and upper specifications limits. Typically, the lower and upper spec limits are defined for the purposes of the equation as the points at which a product would fail in half of the applications. Figure 7-28 illustrates how each of the values for the equation is calculated. Figure 7-28 also contains an example of how the equation could apply to the analysis of a design performance feature of CVM's video game project. By completing the equation using values gathered from the research conducted by CVM in designing its product, CVM could calculate that given the target value for a power-supply circuit — and the functional limits of the value — a monetary value for the loss of "quality" would be $68.75. This information gives quantitative values for the loss of quality that might be experienced as a result of variation in the performance value. It also allows design and manufacturing teams to focus on processes/parts and mechanisms that have a higher potential loss and therefore should be improved the most or receive additional analysis. More detailed analysis of the Quadratic Loss Function and variations of the equation to address situations where the tolerance for variation is not evenly distributed on either side of a target value can be found in Phadke[32] and Kiemele and Schmidt.[33]

The Quadratic Loss Function has been a well-studied analytical technique and is not without its critics. Some individuals feel the equation is too simplistic in its calculation of product value loss and does not give the complete picture when looking quantitatively at quality. The equation is typically used, at least by American companies, in examining the cost of failing to meet the quality target values when the target is a measurable value. Critics complain that the formula does not necessarily create the *best* system, but only allows improvement in quality based upon known parameters. This type of criticism is unwarranted because the Taguchi methodology does not solve all issues associated with product and process improvement. It is a statistical measuring method that can be combined with other types of analysis, such as QFD, for achieving measurable improvement in products and processes.

Taguchi shows us that we can measure the impact or loss that variation can have on a product or process. This is an important statistical measurement. However, having seen what variation can cost, it is more important to understand the causes of variation and the methods that can be used to control variation.

7.2.4 *Understanding variation*

Variation is not a very difficult concept to understand in and of itself. We are all familiar with different types of variation that surround us in our daily lives. The *American Heritage Dictionary* defines "variation" as:

> 1. a. The act, process, or result of varying. b. The state or fact of being varied. 2. The extent or degree to which something varies . . . 4. Something that is slightly different from another of the same type . . . [34]

Variation: A marked difference or deviation from characteristic form, function or structure from another of the same type.

Examples of the impact variation has on society.

Activity / Service or Product	Impact
Golf swing	Ball location off course
Copy machine	Jamming — Loss of time and materials
Self-assembly home furniture	Parts don't fit
Performance appraisals	Unfair promotion practices
Weather	Rain of golf day
Military equipment - Radar detection	Noise — inaccurate target location

Figure 7-29 The loss of quality due to product and process variations.

The first two definitions are not exactly helpful in understanding variation, but the fourth definition is the one that we are most concerned with when dealing with variation in a design or manufacturing environment. From a product standpoint, variation can be defined as a marked difference or deviation from a characteristic form, function, or structure from another of the same type.

The most common form of variation that we have all experienced occurs almost every day of the week. When driving to the same destination every day, whether it be work, school, or the shopping center, we know that the distance is the same and that there is a target value representing the approximate amount of time that it takes to get from home to that location. However, we also know that the time that it takes to get to that destination can vary as a result of traffic, speed, weather, or other forces which can impact (sometimes literally in the form of an accident!) the movement of the car from home to the destination. Depending on the type of variation, for example, fog or ice on the freeway, the drive time can be significantly increased resulting in a substantial delay. Figure 7-29 lists various types of variation and the impact of that variation. Most of the time, we see variation as a negative element in any process that impacts our ability to reach a target value. For example, variation in a golf swing, however seemingly small, can greatly affect the outcome of the golf score by causing the golf ball to veer off its target path. Anyone who has looked at golf on television, or who has played the game for that matter, knows that the function of the golf swing appears to be straightforward and simple: place the golf club head behind the ball, adjust the stance, back swing, follow through. However, anyone who has actually tried

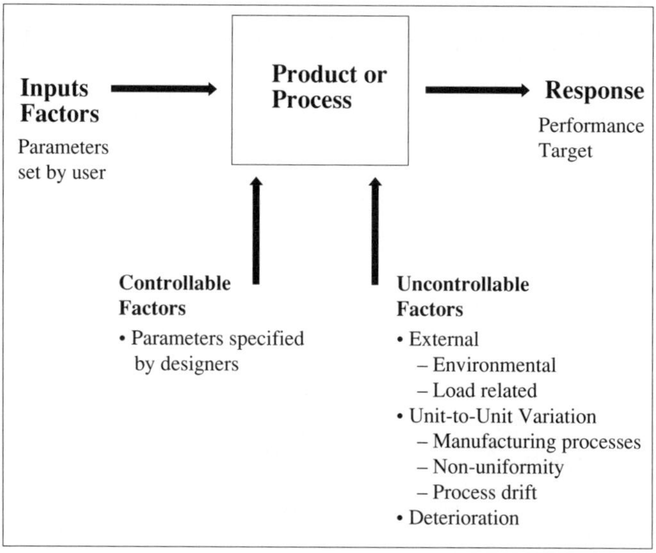

Figure 7-30 The loss of quality most often results from uncontrollable variation sources.

to play the game also knows that any slight variation in the speed of the swing, turn of the shoulder, grip on the club, and a myriad of other small variables can affect the trajectory of the ball. In other words, it is very difficult to keep that ball in the fairway because of the effect of small variations in the golf swing process. Those who have played the game also know how difficult it is to repeat the exact golf swing that gives you a 300-yard drive down the center of the fairway, or provides pinpoint accuracy when chipping onto the green (there are also those who insist that the only way to control variation in golf is to be "one" with Celestial Forces!).

With regard to a product, too much variation ultimately degrades the quality of the product and causes a company to lose profits. Variation can cause specification noncompliance, require extra resources to repair items under warranty, and finally can drive customers away to a more reliable product.

Although it would be ideal to eliminate all variation from a product and hit the target every time, there is no such thing as a process without some degree of variation, however minute. But what are the actual sources of variation?

Figure 7-30 illustrates both controllable and uncontrollable factors on a product or process and the three major sources of uncontrolled variation: external, unit-to-unit variation, and deterioration. Most of the time any loss in quality or effect on actual performance vs. designed performance can be traced to uncontrollable variation. Therefore, the primary goal in achieving a quality process or product is to either eliminate uncontrolled variation, or to find methods to place the factor under control. This concept is in perfect

	Type of Variation			
	Controllable Factors	Uncontrollable Factors		
		External	Unit-to-Unit	Deterioration
Refrigerator	• Amount of refrigerant available • Delta pressure drop during cool down cycle • Diameter and length of pipes • Volume of refrigerator space • Pipe material • Refrigerant type	• No. of times door opens & closes • Amount of time door left open • Initial temp. of food • Supply voltage	• Tightness of door closure • Amount of refrigerant used	• Leakage of refrigerant • Mechanical wear of compressor parts
Automobile Braking System	• Brake material • Pad pressure • Pad surface area	• Wet or dry road • Concrete or asphalt • No. of passengers in car	• Friction coefficient of the brake pads and drums • Amount of brake fluid	• Wear of drums and break pads • Leakage of brake fluid

Figure 7-31 Examples of controllable and noncontrollable factors that influence the performance of a Refrigerator System and an Automobile Braking System.

alignment with the goals behind the Integrated QFD Approach™ which is to provide traceability, and therefore control.

Figure 7-31 gives examples of types of variations, controllable factors, and uncontrollable factors using a refrigerator and an automobile brake system. External variation, which can also be called "noise," is often the most familiar type of variation to see, and sometimes is the most troublesome for the very reason that it cannot be directly controlled, or is more difficult to control. Examples of external variation for a refrigerator would be the number of times the door is opened and closed, and the amount of time that the door is left open. For an automobile, whether the road is wet or dry is a type of variation that must be anticipated, but cannot necessarily be controlled. Unit-to-unit variation would be the tightness of door closure for a refrigerator, and the friction coefficient of the brake pads in the drums for a car. Deterioration is another type of variation that must be anticipated, but is sometimes very difficult to control in product development and usage. In Figure 7-31 deterioration could be exemplified as leakage of refrigerant from a refrigerator compressor or the wear of the brake drums and the brake pads for a car.

With variation there must also be control factors. Figure 7-31 also lists control factors to be anticipated in the two examples. For a refrigerator, the amount of refrigerant available would be a control factor that would anticipate all three of the types of variation. With the automobile example, the construction of the brake material would be a control factor. As one can see, control factors are those elements that are selected in the design or manufacturing process in an attempt to control or mitigate variation.

But once you have identified variation and understand some of its sources, what is the next step?

7.2.5 Methods to reduce variation

Figure 7-32 graphically demonstrates the importance of hitting the target in terms of measuring the quality of a product. If a large number of products are being produced that vary greatly from the target value but may still marginally lie between the lower and upper specification limits, it is said that process is only "barely capable" because of the high potential for failure. This concept is consistent with the Taguchi equation. The more a process is under control and the narrower the band of variation, the higher the likelihood for

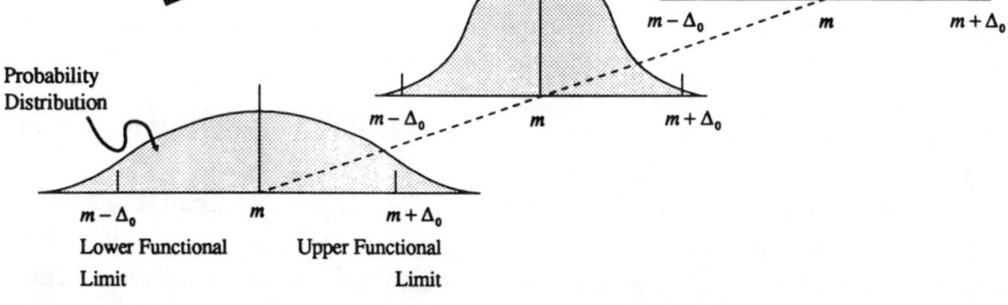

Figure 7-32 A reduction in the system variation increases the probability of delivering a system that performs to the specified target parameters.

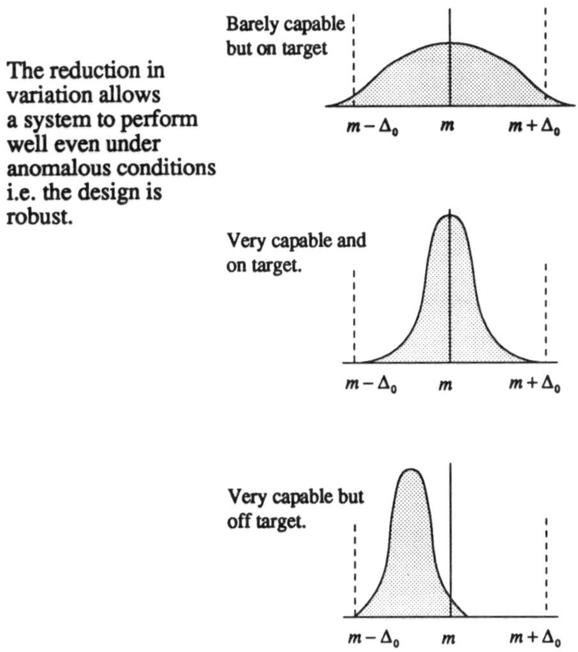

Figure 7-33 The reduction of system variation increases the robustness.

customer satisfaction because the probability of delivering a product that performs to the specifications is greater.

There is another reason why a narrower band of variation creates a more Robust Design. In Figure 7-33 three graphs are displayed showing three types of variation. In the first graph, the product or process is on target with equal distribution of variation, but it is only "barely capable" because of the wide divergence of the variation. The third chart shows a product or process that is "capable" but is missing the target. There are few or no products coming off the line that hit the target exactly, but there's very little variation in terms of the actual specification. Therefore, even if the product is not exactly on target, each varies little and remains within the upper and lower tolerances.

Ideally, a quality product would not only be capable in the sense that the variation is very slight, but also be on target placing it directly between the lower and upper specification limits, as shown in the second graph.

If controlling and reducing variation is the key to ensuring product quality when developing a robust product, what are the methods used to reduce the variation and thereby increase the quality? One of the major methods has already been discussed and that is QFD. QFD is essential in creating the original specifications or Target Values and is the first step in reducing variation. By proper identification of the Target Values, methods can be developed to control design and processes to help attain those Target

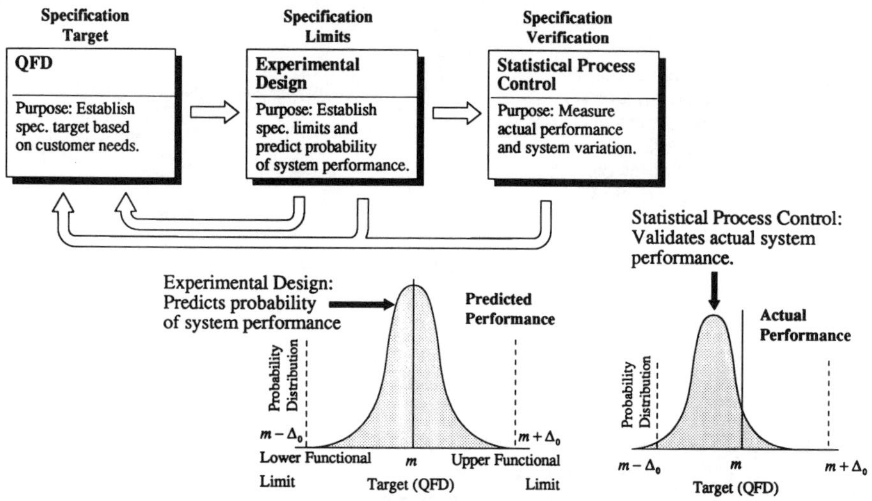

Figure 7-34 The interaction of Robust Design techniques with QFD.

Values. QFD ultimately lays the framework for application of tools such as Functional Analysis (which we have seen previously), Design of Experiments (DOE), and Statistical Process Control (SPC). We have examined some of the effects of variation, and why a Robust Design takes into account variation and tries to minimize it. The next step is to look at how DOE is used to analyze the sources of variation among design parameters.

7.2.5.1 Design of Experiments: voice of the engineer

If QFD employs the concept of using the voice of the customer to develop new products and services, then DOE incorporates the concept of "listening to the voice of the engineer." If there is one element that an engineer, or technical specialist, wants or desires when developing new designs for products, that element is quantitative measurable data. An engineer wants to be able to see and hold in his hands information that allows him to make a reasoned decision. DOE is the tool used to established quantitative data through a measured experiment. Figure 7-34 shows the relationship of QFD to DOE and SPC (discussed in Section 7.2.7). The purpose of QFD is to establish the Specification Target that is based on a customer need. In the Integrated QFD Approach™ this specification can be established at any of the various levels of detail. DOE established the specification limits and predicts the probability of the system performing to the required performance criteria. SPC finally measures the actual performance and system variation. Figure 7-34 illustrates how QFD, DOE, and SPC relate to the probability distribution of variability of a system.

7.2.5.2 What is a designed experiment?

A designed experiment is not merely conducting a single experiment to gather data, but a methodical and reasoned approach to developing the maximum amount of useful data using the minimum amount of time and resources. Too often engineers construct experiments without focus and

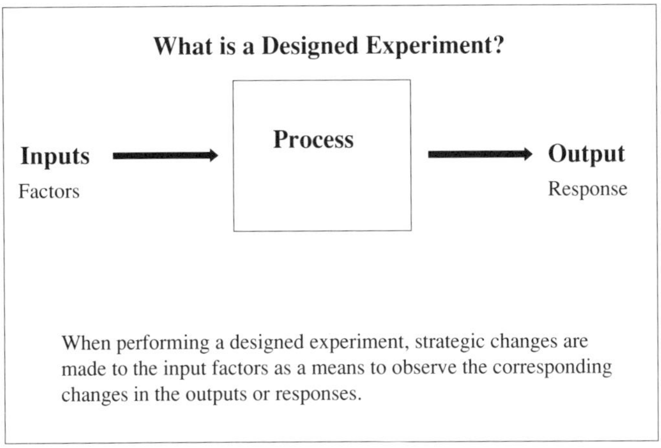

Figure 7-35 Experimental Design is an analysis method to evaluate the impact of variation on system performance.

great quantities of data are produced, but it is later discovered that most of that data does not assist the engineers in making qualitative decisions about design. DOE is a specialty in itself and a great deal of effort and time can be put into designed experiments that produce the desired data. It is not the intent of this section to provide the consummate discourse on DOE, but rather to give the reader sufficient information about DOE and how it correlates with QFD in producing a Robust Design. For those readers who have an understanding of DOE, this section is probably remedial, but the authors also show the correlation of DOE to QFD.

To begin with, a designed experiment starts with the concept that there are inputs to a process that yield outputs (Figure 7-35). A designed experiment then is composed of purposeful changes of the inputs (factors) to a process in order to observe corresponding changes in the output (response). The objectives of a designed experiment are as follows:

- Obtain maximum amount of information using a minimum amount of resources;
- Determine which factors shift the average response, which factors shift the variability, and which factors have no affect;
- Find the factor setting that optimizes the response and minimizes the response; and
- Build empirical models relating the response of interest to the input factors.

Using DOE one can identify how a factor affects system performance. Figure 7-36 shows how variation in four different types of factors can affect the system performance as a measure of variability related to the target value. Factor A variation affects the *average* system performance, moving the center of the curve away from the Target Value. Factor B affects *system variability*,

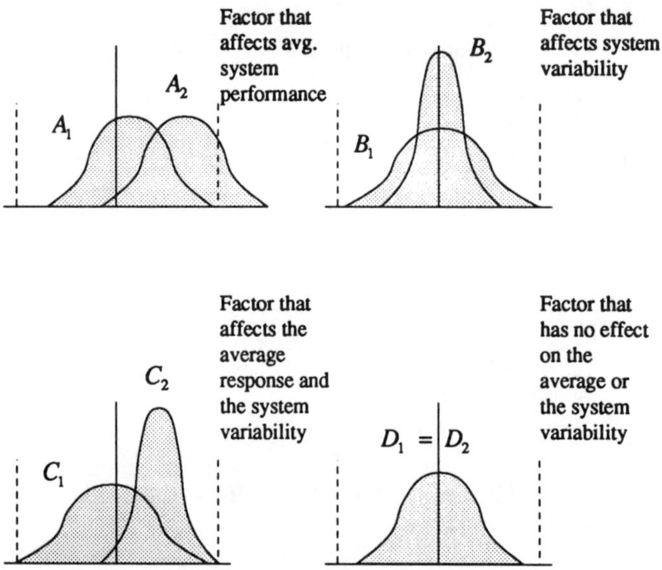

Figure 7-36 Experimental Design identifies the impact of each factor on system response.

causing more variability but evenly distributing the variability on either side of the Target Value. Factor C in Figure 7-36 effects the *average response* and the *system variability*. Factor D has no affect on the *average* nor on the *system variability* because changing the input does not affect the variability of the system at all.

So how do you do a designed experiment?

DOE uses **orthogonal arrays** to combine factors for experiments to determine how the factors affect the measured response. Taguchi methods also use orthogonal arrays to analyze the effects of factors, but Taguchi methods are sometimes criticized as being too simplistic in examining the system and interrelationships between the factors. What is important here is that whatever DOE school of philosophy is chosen it can complement QFD. To give the reader a brief understanding of DOE, the following looks at how DOE can apply to candy-making, taking into consideration only two factors: time and temperature.

Step 1. The first step is to select the factors that can affect the response. This selection process is where QFD is first employed because it can give the critically important factors to examine in terms of customer needs. For our candy-making example, the two factors are:

- Factor A = Temperature
- Factor B = Time
- Response (y) = Candy consistency (Hardness rating 1 to 10 with 10 being very hard.)

(See Figure 7-37.)

Step 2. Next is to select the test settings. Test settings can come from engineering or process knowledge, technical limitations, preestablished target figures or other sources of information that apply to the factors. In Figure 7-37, the test settings are divided into low and high settings. For Factor A they are 200°C (low) and 220°C (high), and for Factor B they are 6 min (low) and 10 min (high). The first question that arises is what about taking tests at intervals? The purpose of DOE is to use only a few experimental runs to determine the boundaries from which intermediate settings can be arrived at through computation. Therefore, factors are examined using, for example, lower and upper specification limits, if those limits are available. These limits are represented by two symbols: low = (–), and high = (+), which are shorthand for +1 and –1.

Step 3. The third step is to set up the orthogonal array. The array is composed of a series of test runs representing the possible combinations of Factors A and B. Run 1 would be at a temperature of 200°C (–) for 6 min (–). The measured response (y_1) would then be recorded. In order to understand the possible interaction between Factors A and B, a third column contains the mathematical representation of A × B, or as in Run 1, –1 × –1 = +1. The array contains at the bottom the measured response averages for runs with low settings and runs with high settings, and the average for "low" and "high" combinations in the interaction of A and B (Column (3).

Step 4. After the array has been set up, the next step is simply to run the experiment and collect the data.

Step 5. The fifth step is to analyze the results. Figure 7-38 shows hardness results for the four experimental runs for the candy-making process. As the array shows, the average hardness for all the runs was 7. The average run for "low" and "high" settings for each factor and the interaction is also given in Figure 7-38. The next step is to fill in the constants in the Prediction Equation. The prediction equation uses the change in A, B, and A × B factors to yield a predicted response value for any value of A and B. In the example, the equation for predicting the hardness is simplified to: $y = 7 + A + 2B$, where A and B are the scalar numbers between –1 and +1 for each factor A and B, and the factor for A × B drops out because the change is 0. The graph at the bottom of Figure 7-38 shows the range of response values for the average low and high settings for the factors, and the vector for each represents the value from –1 to +1. For example, if the prediction equation was used to find the values for a response hardness of 7, then A and B must be 0 and the actual factor values would be 210°C for A and 8 min for B (see Figure 7-38).

After the runs have been made, the next step might be to make a run at the predicted values for the factors to corroborate the prediction equation and to ensure that there are indeed no unpredicted variations caused by factor interactions. If the prediction run has a result that does not conform

Step 1: Select Factors

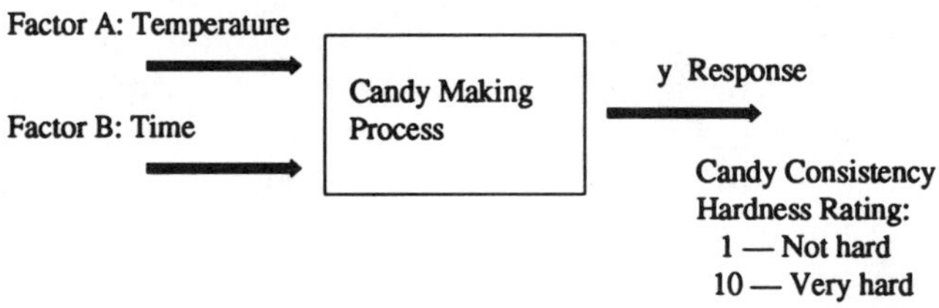

Step 2: Select Test Settings

	High Setting (+)	Low Setting (–)
Factor A: Temperature	220°C	200°C
Factor B: Time	10 min.	6 min.

Step 3: Set Up An Orthoganal Array

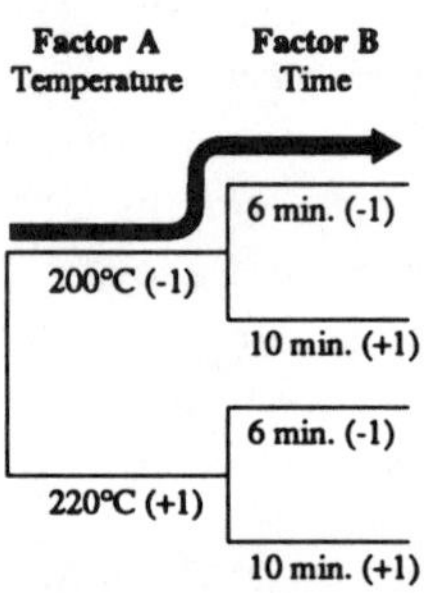

Full Factorial Design

Orthogonal Array (L_4)

	Run	Factor A Setting	Factor B Setting	Interaction A & B	Measured Response (y)
	1	–	–	+	y_1
	2	–	+	–	y_2
	3	+	–	–	y_3
	4	+	+	+	y_4
Average Response	Low Setting (–)	$\frac{y_1+y_2}{2}$	$\frac{y_1+y_3}{2}$	$\frac{y_2+y_3}{2}$	
	High Setting (+)	$\frac{y_3+y_4}{2}$	$\frac{y_2+y_4}{2}$	$\frac{y_1+y_4}{2}$	
	Δ	ΔA	ΔB	Δ = (avg. +) – (avg. –)	

Figure 7-37 Experimental Design example for a candy-making process: Steps 1 through 3.

with the equation, then the interaction must be examined in greater detail and a new orthogonal array should be created and tests run to account for the interactions that may be cumulative or nonlinear.

Taguchi uses a "signal-to-noise ratio" in designing experiments utilizing logarithmic transformations that consider both mean values and variability. Taguchi equations are also designed to consider situations when the variation may not be positive and negative relative to a target value, such as in the case of radiation emitted, when the target value is no radiation emitted.

Step 4: Run Tests

	Run	Test Setting Factors A Temp °C	Test Setting Factors B Time (min)	Interaction A & B	Measured Response (y)
	1	200	6	+	4
	2	200	10	−	8
	3	220	6	−	6
	4	220	10	+	10
Average Response	Low Setting (−)	6	5	7	$\bar{y} = 7$
Average Response	High Setting (+)	8	9	7	
	Δ	+2	+4	0	

The Prediction Equation:

$$\hat{y} = \bar{y} + \left(\frac{\Delta A}{2}\right)A + \left(\frac{\Delta B}{2}\right)B + \left(\frac{\Delta A \cdot B}{2}\right)A \cdot B$$

$$= 7 + \frac{2}{2}A + \frac{4}{2}B + 0(A \cdot B)$$

$$= 7 + A + 2B$$

The Average Equation:

$$\bar{y} = \frac{y_1 + y_2 + y_3 + y_4}{4}$$

Step 5: Analyze Results

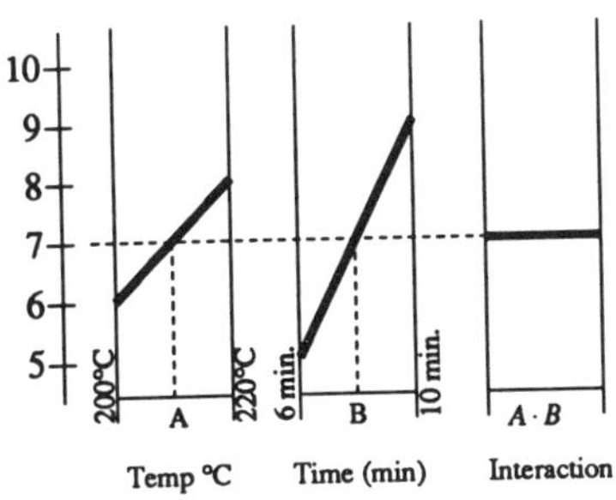

For maximum hardness
A = 220°C and B = 10 min

For minimum hardness
A = 200°C and B = 6 min.

For a hardness of 7 ($\hat{y} = 7$)
we set B at 8 min.
A at 210°C and substitute in
Prediction Equation:
$\hat{y} = 7 + 0 + 2(0) = 7$

Figure 7-38 Experimental Design example for a candy-making process: Steps 4 through 5.

7.2.6 *Use of QFD and DOE in a case study*

To demonstrate how DOE can apply to a more complicated situation, Figures 7-39 through 7-47 represent data taken from a designed experiment conducted by Martin Marietta Astronautics Group for the Satellite Servicing System (SSS).[35] The SSS was essentially a service station in space to conduct repair and maintenance of satellites. The SSS involved a mechanical arm that would conduct some of the maintenance on the satellites and was also equipped to select tools from a tool package. The scope of the project was to optimize manipulator dexterity using a work station and robotics simulation

Customer Objectives

Design Configuration			Degree of Importance	Current Company Rating	Planned Level	Improvement Ratio	Sales Point	Importance Weight	Relative Weight
	Low Risk	Low qualification risk	2	4	4			2	1.7
		Ease of integration	4	4	4		○	4.8	4
		Max. use of existing technology	4	5	5		○	4.8	4
	Low Cost	Max. use of existing hdw. and facilities	5	3	4	1.3		6.7	5.5
		Cost effectiveness of NSTS Integration	3	4	4			3	2.5
		Cost effectiveness of SSS I/Fs	4	3	4	1.3		5.3	4.4
		Minimum number of flights	4	3	3			4	3.3
	Safety, Rel. Q & A	Fail Ops./Fail safe	4	4	4			4	3.3
		Optimize man/machine I/F	4	3	4	1.3		5.3	4.4
		High success probability factor in design	2	4	4			2	1.7
		Engineering excellence in design/dev. to meet requirements	3	2	3	1.5		4.5	3.7
		Manuf. excellence & productivity	3	2	2			3	[illegible]
	Systems Engin. & Integr.	Integrated Operations	5	3	4	[illegible]			
		Evolutionary growth of SSS	3	[illegible]					
		System verification							

Developed at Martin Marietta Astronautics Group, Denver, CO

Figure 7-39 Satellite Servicing System QFD/Experimental Design case study: Definition of Customer Needs.

software package. The dexterity was defined as including the angle of orientation of the tool plate averaged over multiple work space points.

The first step in the analysis was the application of QFD to determine the customer needs and the key requirements. Figure 7-39 defines the customer needs and shows the relative importance weights taking into consideration the sales points and company benchmarking. Figure 7-40 graphically represents the prioritization of the customer needs, showing that two key needs were "Integrated Operations" and "Supervised Autonomy Control." Figure 7-41 is the next step in the QFD, illustrating how the Flight Telerobotic System (one of the systems in the SSS) requirements would correlate to the customer needs. Figure 7-42 illustrates the prioritization of the relative weights of the technical requirements for the Flight Telerobotic System, showing that the "Reach/Mobility" requirement was the most critical in terms of meeting the customer needs. (A sensitivity analysis was conducted excluding the Sales Points and Improvement Ratios with regard to the Flight Telerobotic System elements as shown in Figure 7-43. The results still indicated that "Reach/Mobility" was the key requirement.)

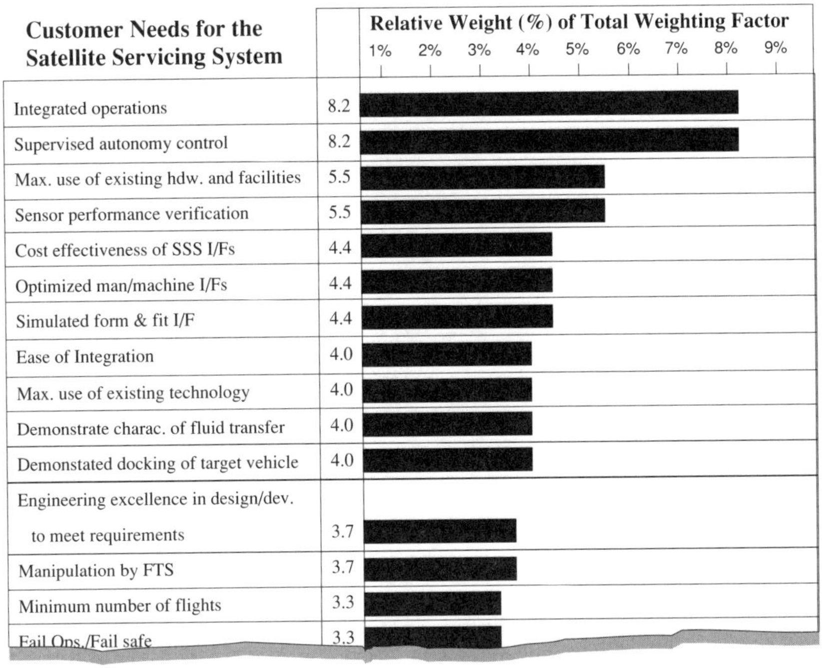

Developed at Martin Marietta Astronautics Group, Denver, CO

Figure 7-40 Satellite Servicing System QFD/Experimental Design case study: Prioritization of Customer Needs.

Based upon the results of the QFD, the design team agreed that an experimental design would be used to optimize the manipulator dexterity ("Reach/Mobility"). This was done using an IRIS workstation and CimStation Robotic Simulation Software. Figure 7-44 illustrates the scope of the DOE and the inputs and results for the Flight Telerobotic Service Manipulator.

The purpose of the experiment was to select trial runs utilizing various parameters for each of the inputs and to measure the noise factor, which happened to be the shoulder roll position. Figure 7-45 illustrates a marginal mean graphical analysis for the dexterity. Based on this analysis, the factors that affect the shift average were the "Joint Travel Limits" and the "Link Length Ratio." Figures 7-46 and 7-47 show graphical representations of analysis using standard deviation for the two key factors.

The benefits of doing the designed experiment on the SSS were the establishment of quantitative data for dexterity improvements, and a focused and streamlined process to do the trade studies. Ultimately the program achieved the same results as a similar flight telerobotics servicer study with only 25% of the actual effort and resources expended. Some of the lessons learned in development of this DOE were: (1) differences in terminology will result in different end results; (2) there must be enough time allowed for brainstorming the experimental array; (3) understanding statistical analysis

Flight Telerobotic System

				Flight Telerobotic System											
				Utility I/Fs			Camera Light Mount Assembly		Manipulators: No. of arms			Arm Pos. Mech	FTS Proc. Req.		
			Relative Weight	Power	Data	Video	CMA's Positioning Mechanism	Cameras: Number and Location	Joint Order & DOF	Exchangable End Effectors	Standard Tools	Reach and Mobility	Positioning	Interfaces	Telemetry/Data
Design Configuration	Low Risk	Low qualification risk	1.7			▲						❍	▲	▲	▲
		Ease of integration	4	❍	❍	❍	▲	▲				❍	▲	▲	▲
		Max. use of existing technology	4	▲	▲	❍	▲		❍			▲			
	Low Cost	Max. use of existing hdw. and facilities	5.5	▲	▲	▲	❍	▲	▲	▲		❍	▲	▲	▲
		Cost effectiveness of NSTS Integration	2.5	▲	▲	▲							▲	▲	▲
		Cost effectiveness of SSS I/Fs	4.4	❍	❍	❍	❍	▲	❍	▲	▲	●	❍	❍	❍
		Minimum number of flights	3.3												
	Safety, Rel Q & A	Fail Ops./Fail safe	3.3	❍	❍	❍	▲	▲		▲	▲	❍	❍	❍	❍
		Optimize man/machine I/F	4.4	▲	▲	▲		▲	▲	▲	▲	❍	▲	▲	▲
		High success probability factor in design	1.7			▲	▲	▲	❍	❍	❍	❍	▲	▲	▲
		Engineering excellence in design/dev.	3.7	▲	▲	❍	▲	❍	❍	❍	▲	●	❍	❍	❍
		to meet requirements	2.5												
		Manuf. excellence & productivity	8.2	▲	▲	▲	●	▲	▲	▲	▲	●			
	Systems Engin. & Integr.	Integrated Operations	2.5	▲	▲	▲	❍	❍	▲	▲	▲		●	●	●
		Evolutionary growth of SSS					▲	▲							
		System verification													

Figure 7-41 Satellite Servicing System QFD/Experimental Design case study: Correlation of System Product Requirements and Customer Needs for the Flight Telerobotic System.

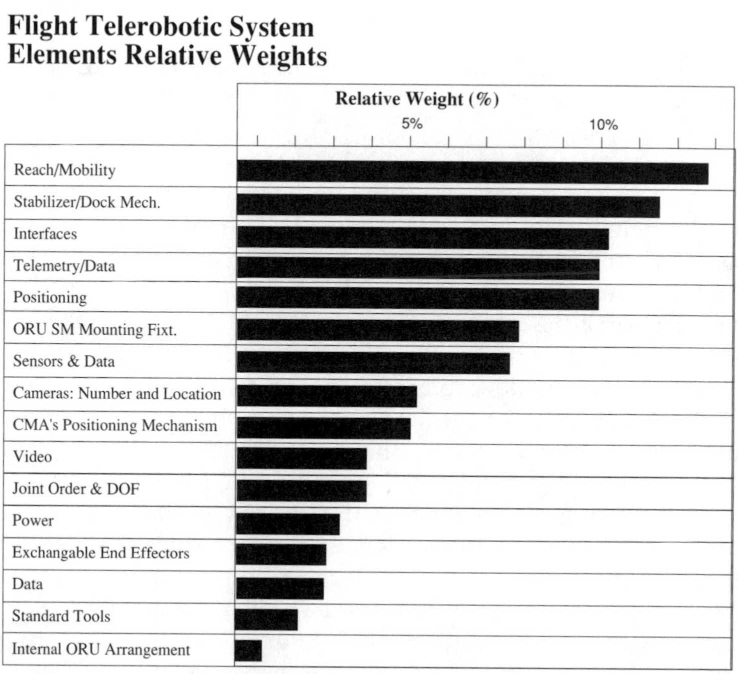

Figure 7-42 Satellite Servicing System QFD/Experimental Design case study: Prioritization of System Product Requirements for the Flight Telerobotic System. (Developed at Martin Marietta Astronautics Group, Denver, CO.)

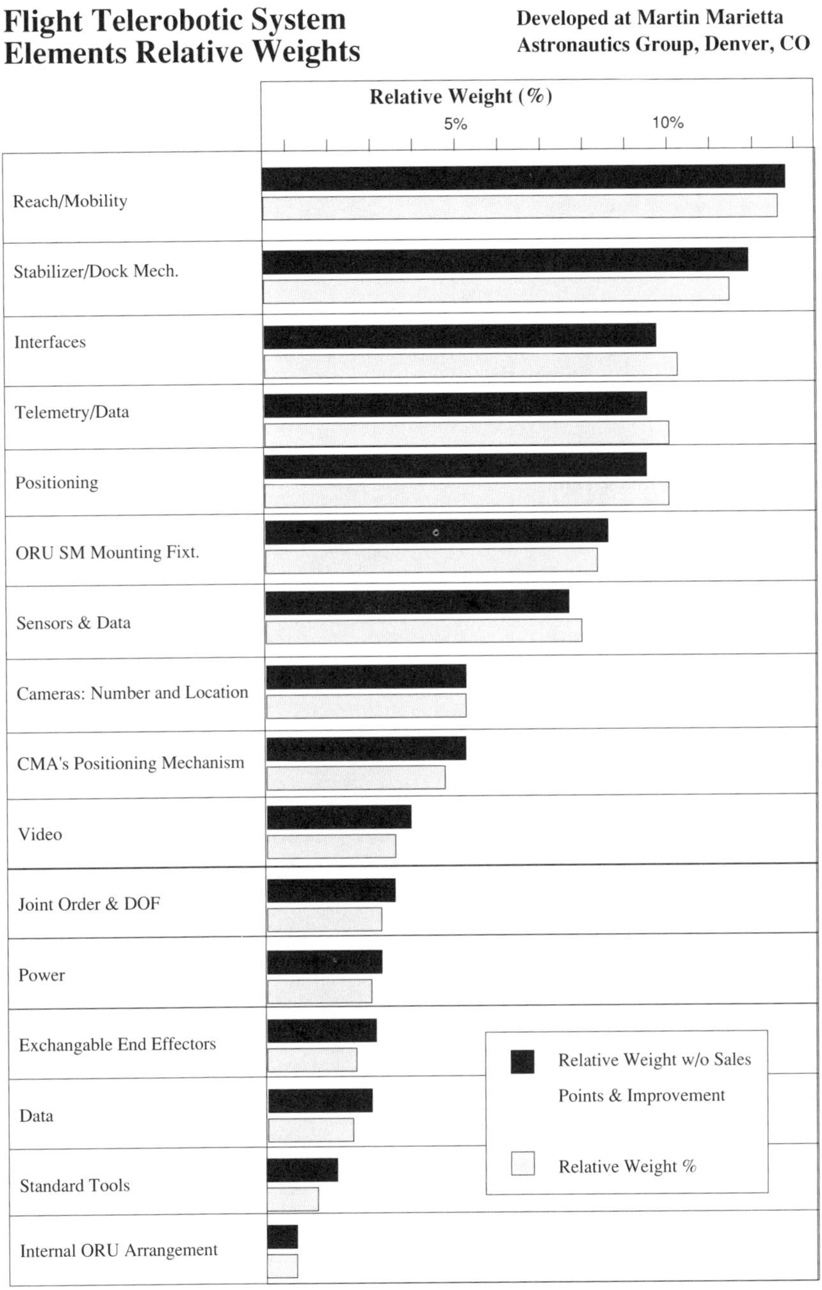

Figure 7-43 Satellite Servicing System QFD/Experimental Design case study: Sensitivity Analysis of Excluding Sales Points and Improvement Ratios.

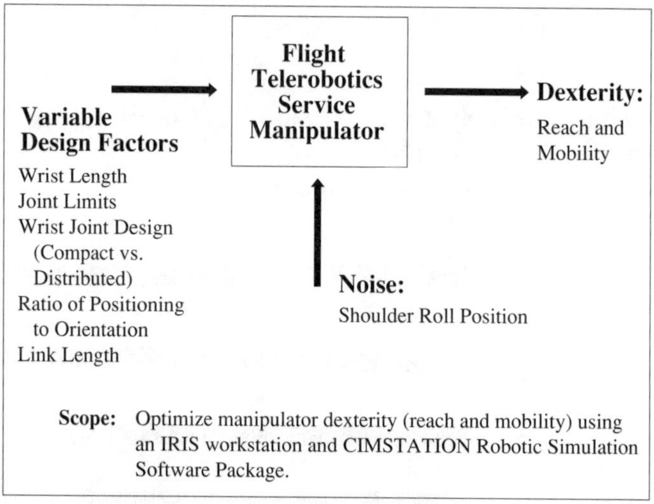

Figure 7-44 Satellite Servicing System QFD/Experimental Design Case Study: Scope of the Experimental Design.

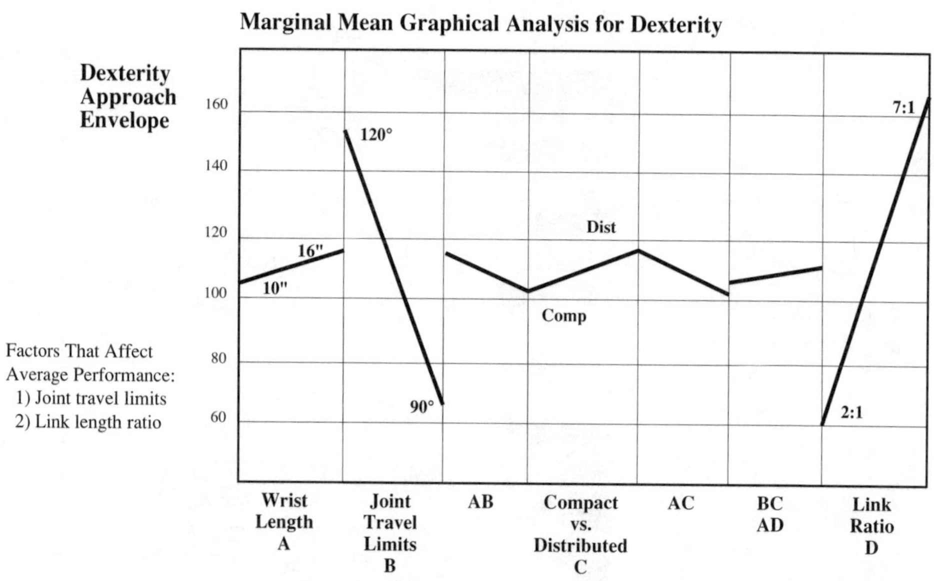

Figure 7-45 Experimental Design Marginal Mean Graphical Analysis of Dexterity (Reach and Mobility) for the Flight Telerobotic System.

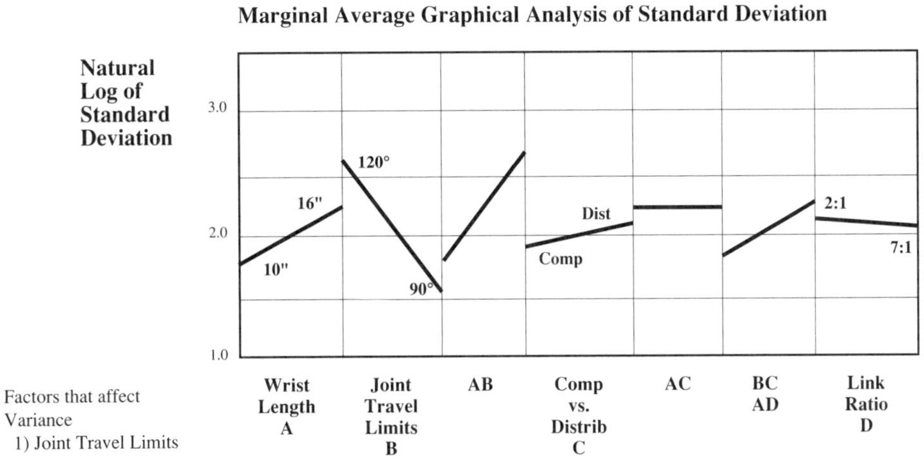

Figure 7-46 Experimental Design Marginal Average Graphical Analysis of Standard Deviation for the Flight Telerobotic System.

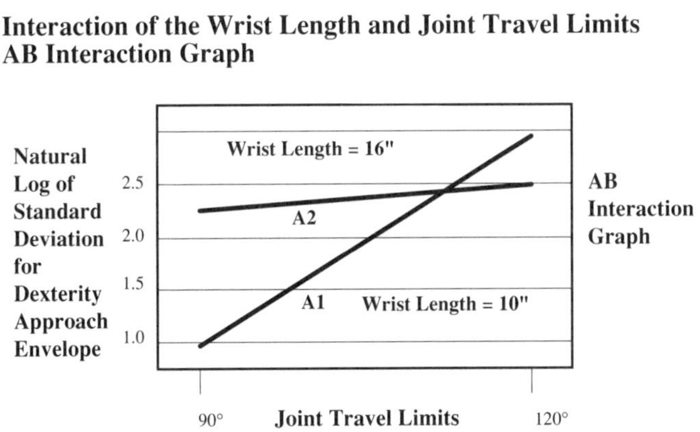

Figure 7-47 Experimental Design Interaction Results of the Wrist Length and Joint Travel Limits for the Flight Telerobotic System.

will accelerate the learning process; and (4) customer involvement assisted in a focused approach to the DOE.

The combination of using QFD to rapidly get to the critical technical requirements, and DOE to establish design criteria, created a robust design in the SSS.

7.2.7 *Statistical Process Control: listening to the voice of the process*

If DOE is listening to the voice of the engineer, then SPC is listening to the voice of the process. By this the authors mean that any time a process is

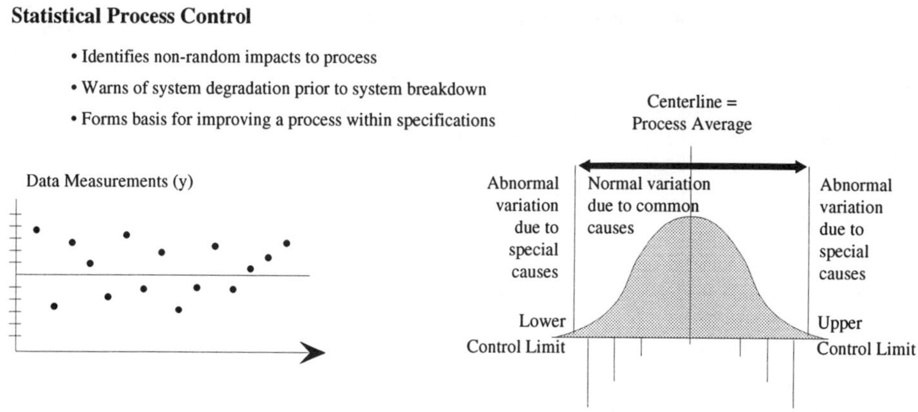

Figure 7-48 Components of Control Charts.

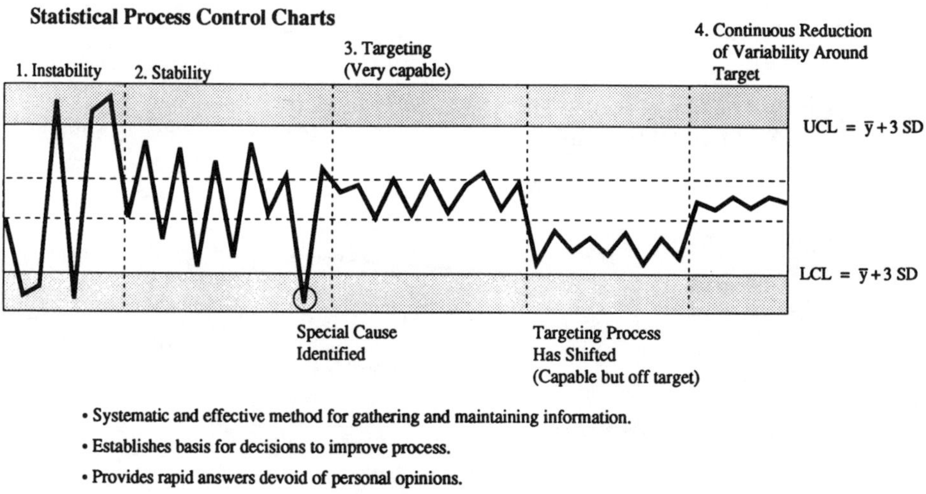

Figure 7-49 Stages of SPC improvements.

decomposed and fully understood, the "voice" of the process comes through, which is really the information necessary to control that process to yield the best methodology for reducing variance and increasing the ability to hit the target parameter. SPC naturally follows DOE when integrating these elements into a Robust Design. Just as QFD gave the target and DOE gave the best combination of factors to achieve the target, SPC creates a methodology for hitting the target consistently. SPC is a systematic and efficient method for gathering and maintaining information: it establishes the basis for decisions to improve a process, and it provides rapid answers devoid of personal opinions with regard to quality measures. In its simplest form SPC measures a process over time. It can lead to the improvement of a process, as well as identify when a process begins to vary to the extent that upper and lower specification limits are in danger of being exceeded.

Figure 7-48 shows two graphs that are critical to understanding SPC. The graph on the left side shows that SPC is a sampling of the process products compared to the target value. SPC recognizes that every process has some variation and classifies these variations in two ways: (1) common cause variation, and (2) special cause variation. Common cause variation is variation that cannot be reduced without a process change. Special cause variation is variation that is unexpected and the result of a unique factor. Therefore, special cause variation can be isolated and controlled or eliminated. The second graph in Figure 7-48 illustrates the proposition that if samples of product are randomly taken in the process, the mean of the samples will equal the process mean, and therefore a normal distribution will occur and 99.73% of the samples taken will be within three standard deviations of the overall mean.

Based on this statistical information, the gathering of information to establish the upper and lower control limits based on three standard deviations yields the process capabilities. Once the process capabilities are established, future data collection can tell if the process needs improvement, or is operating efficiently. Figure 7-49 shows a standard sample process control chart, and the improvement of a process. In (1) on the chart the process is unstable and needs to be analyzed to establish control limits. In (2) the control limits have been established and the process is stable because it is operating within those limits (note the aberrant reading where a special cause has been identified). In (3) the process is very capable because it is narrowing in on the target value. However, a process can be very capable, and shift off the target value. This means that there is very little variance in the process, but it is not centered on the target value. Finally, in (4) the process has centered on the target and is showing little variation. This is the goal of process improvement using SPC.

7.3 Some final comments on the Integrated QFD Approach™ and Robust Design

Robust Design tools have yielded substantial improvements in quality and cost reduction in American manufacturing and processing. But as competition increases, and the more obvious improvements have been made using these statistical tools, the ability to improve has become more difficult. By using Robust Design tools as elements of the Integrated QFD Approach™ limited resources can be focused on the key areas of improvement that remain, and a framework for improvement can be maintained. The authors contend that in the future, Robust Design cannot be done without the requirements, prioritization and planning attributes of QFD.

7.4 Chapter summary

7.4.1 The Integrated QFD Approach™

1. The Integrated QFD Approach™ breaks the customers' needs into four distinct matrices: the System Matrix, Concept Matrix, Process Matrix, and the Control Matrix. However, these matrices include product, service, and business operation related requirements.
2. The Integrated QFD Approach™ offers advantages beyond the other methods in that it combines not only an entire organizational view, but it also simplifies the process, stimulates team creativity for developing new products, services and operations, and provides the road map to create an integrated plan for action.
3. There are nine steps to the Integrated QFD Approach:

   Step 1. Develop Systems Matrix
   Step 2. Develop Initial Integrated Plan
   Step 3. Develop Functional Flows
   Step 4. Develop System Alternatives
   Step 5. Evaluate System Alternatives
   Step 6. Develop Concept Matrix
   Step 7. Develop Process Matrix
   Step 8. Develop Control Matrix
   Step 9. Refine Integrated Plan

4. Functional Analysis plays a key role in concept development and understanding methods for design in conjunction with QFD.
5. Functional Analysis helps to define alternative concepts for analysis and evaluation against the QFD technical requirements.
6. The Integrated Plan becomes a "living document" in the design and manufacturing process and is updated as additional information is discovered in the QFD analysis.

7.4.2 Robust Designs

1. Robust Design has its roots in statistical experimental design. Dr. Taguchi developed a method of using experimental design techniques to analyze the cause of variation in a design.
2. Robust Design addresses reducing the economic impact of variation in a design, and ensuring that decisions found to be optimum during laboratory experiments will be proven in manufacturing and use.
3. The "old" philosophy of quality said that any characteristic that measures between the lower specification limit and the upper specification limit was acceptable. The new philosophy measures quality in terms of a loss to society due to financial variation and harmful side effects, using a quadratic equation to show the greater loss as the product characteristic is farther away from the "ideal" specification.

4. Variation includes both controlled and uncontrolled parameters. It is important when analyzing variation that the source of the variation is understood.
5. Methods to reduce variation include Design of Experiments (DOE) and Statistical Process Control (SPC).
6. Robust Design tools are included in the Integrated QFD Approach™. QFD is critical in prioritizing requirements and focusing resources on those elements critical to design success.

chapter eight

QFD: the basis for Systems Engineering and Concurrent Engineering

Synergy means behavior of whole systems unpredicted
by the behavior of their parts.
R. Buckminster Fuller, from
What I Have Learned, How Little I Know.

8.0 Synergy and engineering

Buckminster Fuller is well known for more than his avant-garde architecture. He is perhaps best described as a true American visionary, whose fascination with the world drew him into projects as diverse as his interests and only limited by his imagination. He is often quoted, and, therefore, it should come as no surprise that a bit of his wisdom should appear in a book on QFD. In fact, Fuller was well know for his ability to take the "parts" of different disciplines and meld them into what he liked to call "synergy." His idea that the behavior of systems could not always be predicted by the behavior of their parts is both reality and an anathema to an engineer trying to balance a design for a new product. The goal of engineering is to predict behavior based on the application of scientific and engineering applications. But unseen interactions often provide unpleasant surprises in the engineering, and business, worlds. Systems and Concurrent Engineering can be viewed as methods for creating "synergy," but in a more predictable way.

This chapter will examine two engineering concepts — System Engineering and Concurrent Engineering — and how QFD can be used in these concepts to define requirements and focus resources. The reader who may be familiar with these two engineering concepts may notice similarities between the Integrated QFD Approach™ and elements of Systems Engineering, and the resemblance is purely intentional. Systems Engineering and Concurrent Engineering have been around for some time; they have been used extensively in the aerospace industry for identifying requirements and approaching engineering challenges from a whole or "systems" point of view. However,

QFD is not always associated with nor used in these concepts, and the authors have tried to point out how QFD can compliment these approaches. There are several excellent texts on Systems Engineering and the authors' intent is not to supplant these teachings but to help the reader understand the operation of QFD using the terminology associated with Systems Engineering. Systems Engineering will be discussed first because understanding how QFD operates in Systems Engineering will provide the foundation of how Concurrent Engineering builds upon this foundation.

However, before examining these concepts in detail it is first important to understand their synergistic relationship. Systems Engineering and Concurrent Engineering are often confused, and their relationship to each other is still a topic of debate. The authors believe that Systems Engineering is concerned more with applying methodology to integrate systems efficiently to produce and service a product, whereas Concurrent Engineering is concerned with the architecture in the company that supports the application of Systems Engineering. Figure 8-1 uses the RSM™ flow chart from Chapter 2 to illustrate where Systems Engineering and Concurrent Engineering fit into a company profile. As the reader can see in Figure 8-1, the hierarchy is as follows:

- RSM™ — Encompasses the entire corporation as a "system" with all activities measured against the company mission goals.
  - Concurrent Engineering — Uses QFD to improve engineering processes to provide architecture for Systems Engineering.
    - Systems Engineering — Focuses on the development of new products.

With this general hierarchy in mind, the next step is to investigate the two subjects in more detail, and determine if we can *predict* some of the synergy between them.

8.1 Systems Engineering and QFD

8.1.1 Defining Systems Engineering

Defining Systems Engineering[36] is not a simple task, and perhaps the easiest way to understand the concept is to start with the definition of the term "system" itself. *The American Heritage Dictionary* defines system as follows:

> 1. A group of interacting, interrelated, or interdependent elements forming a complex whole. 2. A functionally related group of elements, esp.: . . . c. A group of interacting mechanical or electrical components.[37]

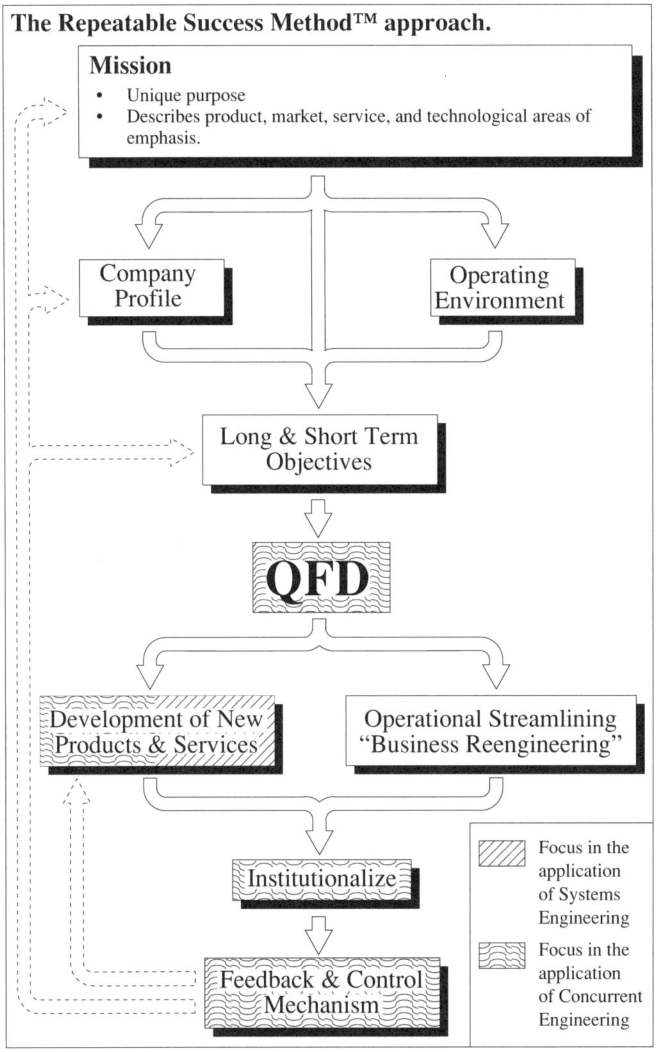

Figure 8-1 Integration of Systems Engineering and Concurrent Engineering in RSM™.

This common theme that a "system" is interaction among or between elements to form a whole is consistent with the word's etymology. Blanchard reminds us that "system" comes from the Greek word *systema*, which means an organized whole. Chapman, Bahill, and Wymore take a more functional approach to defining a system: "[A]ny process or product that accepts inputs and delivers outputs."[38] (See Figure 8-2.) The Department of Defense, which has long had a standard for conducting Systems Engineering, defines a "system" as follows:

System Definition:

- Stems from the Greek word "systēma" meaning an organized whole.
- Any process or product that accepts input and delivers output.
  – Chapman, W.L., Bahill, A.T., Wymore, A.W.,
  "Engineering Modeling and Design" pg. 10
- "A set of interrelated components working together with the common objective of fulfilling some designated need."
  – Blanchard, B.S.,
  "Systems Engineering Management" pg. 10

System Characteristics:

1) Constitutes a complex combination of resources

2) Contained within some form of hierarchy
   - Broken into subsystems and components that interact

3) Has a purpose
   - Functions to respond to some identified need, with the goal of achieving the overall objective in a cost effective manner

Type of Systems:

- Natural vs. Man-made
- Physical vs. Conceptual
- Static vs. Dynamic
- Closed Loop vs. Open Loop

Figure 8-2 System definition and related characteristics.

> A system is a composite of people, products, and processes that provide a capability to satisfy stated needs. A complete system includes the facilities, equipment (hardware and software), material, services, data, skilled personnel, and techniques required to achieve, provide and sustain system effectiveness.[39]

This definition takes into consideration nearly all of the elements required to develop a product, including the support functions. The military and aerospace industries have long been advocates for some form of Systems Engineering for decades. Much of what is accepted today as hierarchical systems design has been developed or modified by the military and aerospace industries as a result of the need to analyze complex electronic and materials engineering, and the complexities involved in interaction of individually complex subsystems.

Systems have similar characteristics, such as hierarchy, organization and subsystems. There are also many different types of systems. Since the time of the Greek philosophers, who often mused about "natural" systems, humans have attempted to order elements into categories, and categories into systems. Systems are often organized into competing concepts, such as natural vs. man-made, or static vs. dynamic. Figure 8-2 lists a few definitions of a system, some system characteristics, and types of systems.

Systems Engineering operates in concept as any type of system: It combines interrelated individual elements to form a whole. In describing what Systems Engineering is, Blanchard states that it is not an "engineering discipline" in itself:

> In summary, system engineering per se is not considered an engineering discipline in the same context as electrical engineering, mechanical engineering, reliability engineering, or any other design specialty area. Actually, system engineering involves the efforts pertaining to the overall design process employed in the evolution of a system from the point when a need is first identified, through production and/or construction and the ultimate distribution of the system for consumer use. The objective is to meet the requirements of the consumer in an effective and efficient manner.[40]

This description of Systems Engineering incorporates the concept central to QFD — meeting the requirements of the consumer or customer. QFD may be considered a system as well because its objective is to provide a hierarchy within which requirements are identified, prioritized and incorporated into the design. QFD, which the authors have shown can apply to a variety of disciplines, is a specific methodology that is used to maximize the system to which it is applied. Systems Engineering is a type-specific methodology used to foster development of products, and is a **complete requirements and design methodology.** But where Systems Engineering emphasizes completeness, QFD emphasizes *focus.*

Systems Engineering has its own terminology, although that terminology can vary depending upon its application, such as aerospace vs. commercial industry. Systems Engineering incorporates many of the same tools that are used in the Integrated QFD Approach™, but because it is focused heavily on incorporating product design, software, hardware, and process/production it is a specialty application. The rest of Section 8.1 will focus on the conventions of Systems Engineering and the interplay between it and the Integrated QFD Approach™ in particular and QFD in general.

8.1.2 An overview of the Systems Engineering Life Cycle

The Systems Engineering Life Cycle begins with Concept Exploration and ends with Operations and Maintenance. In between is a series of planned activities, design reviews, and specification development milestones. Systems Engineering creates a hierarchy of the design process. The program is divided into several phases, and each phase has specific activities and review points. This systematic approach to design and production strives to provide measurable criteria by which the design can be evaluated at each phase of the process. The idea behind Systems Engineering is that there are certain activities that must occur prior to requirements development, and this is true independent of the type of product or service being developed. Figure 8-3 illustrates the activities, design reviews, and specification developments for each of the five program phases. The reader can see that many of the activities associated with the program phases are familiar components. For example, in the Concept Exploration phase of Systems Engineering, Marketing Analysis is conducted to begin to analyze the requirements for the product. Although Figure 8-3 contains a substantial amount of activities and information for discussion, the following provides a conceptual understanding of what is involved at each phase in order to analyze QFD interfaces with Systems Engineering:

Concept Exploration — Concept Exploration can best be described as the phase when the requirements for the program are first developed. The design team makes use of marketing, early concept design, technology evaluation, and customer information to develop the initial requirements for the program. During this phase all efforts are aimed at the System Requirements Review (SRR) (see Figure 8-4), which is the milestone that must be met prior to moving into the next phase of the program. The System Specification that is developed during this phase is often called the "A Spec," or the top-level specification of requirements (see Figure 8-5). The A Spec contains the technical, performance, operational, and support characteristics for the product or service, and defines the system interfaces. The initial Functional Analysis in a top-level systems analysis is conducted during this phase to provide understanding of the operational elements of the program. An addition to this phase is the "Integrated Program Plan With Definition of Configuration Control." The inclusion of a plan is not always part of the Systems Engineering Life Cycle as defined by some sources because the planning function is often considered separate from engineering. The planning function is often described as the "Systems Engineering Management Plan" (SEMP), which includes a program management guide for meeting milestones. However, the authors believe that a more detailed plan, such as the integrated planning document developed as part of the Integrated QFD

Program Phases	Concept	Preliminary System Design	Detailed System Design	Production/ Construction	Operation & Maintenance
Activity	• Marketing Analysis • System Feasibility Analysis & Risk Assessment • Conceptual Design • Operational Requirements • Maintenance Concept • Evaluation of Technology • Begin Functional Analysis • Integrated Program Plan w/ Definition of Configuration Control	• Functional Analysis • Requirements Allocation • Design Synthesis • Evaluation of Concepts • Preliminary Design • Test & Evaluation of Design Concepts (bread boards) • Interface Definitions • System Modeling • Detailed Program Plan	• Detailed Design of Subsystems & Components • Trade-off Analysis • Prototyping Models • Test & Evaluation of Components • Interface Evaluation • Change Evaluation/ Control • System Modeling • Production Planning	• Production and/or Construction • Supplier Production • Distribution • System Operational Use • Maintenance & Support • Collection of Data • Readiness Reviews	• System Operational Use • Logistic Support • Operational Test and Evaluation • Data Collection & Analysis • System/Subsystem Modifications • Customer Service
Design Reviews	Conceptual Design Review/ System Requirements Review (SRR) ▲	System Design/System SW Reviews SDR ◄——► SSR Preliminary Design Review (PDR) ▲	Equipment/SW ◄——► TRR ▲ ▲ Critical Design Review (approved design for fabrication)	Engineering Change Reviews FCA ▲ PCA ▲ FQR ▲	
Specification Development	System "A" Specification ▲	Prelim. Development "B" ▲ Prelim. Process "C" ▲ Prelim. Product "D" ▲ Prelim. Material "E" ▲ Prelim. "ICDs" ▲	Development "B" ▲ Process "C" ▲ Product "D" ▲ Material "E" ▲ "ICDs" ▲		

Figure 8-3 Relationship of Specification Development with System Engineering life cycle.

SRR *System Requirements Review*	• Ascertain adequacy of system requirements. • Conducted when systems level functional analysis is complete.
SDR *System Design Review*	• Evaluate the optimization, correlation, completeness, and risks associated with the allocation of technical requirements. • Summary review of engineering process and engineering plan for next phase. • Conducted when system characteristics are defined and the configuration items identified.
SSR *Software System Review*	• Review of Computer Software Configuration Item (CSCI) requirements and operational concepts. • Conducted when CSCI is sufficiently defined to evaluate the system, segment, or major level requirements.
PDR *Preliminary Design Review*	• Evaluate the progress, technical adequacy and risk resolution of selected design approaches (hardware & software). • Evaluate degree of definition and assess technical risk associated with the selected manufacturing methods/processes. • Establish the existence and compatibility of the physical and functional interfaces. • Evaluate compatibility between software requirements and preliminary design.
CDR *Critical Design Review*	• Determine detail of design satisfies physical and functional requirements. • Establish detailed design compatibility between hardware elements, software systems, equipment, facilities and personnel. • Assess Risk. • Review Specifications.
TRR *Test Readiness Review*	• Conducted for each computer software configuration item to determine whether the software test procedures are complete. • Evaluate software test procedures for compliance with software test plans and descriptions, and adequacy in validating test requirements.
FCA *Functional Configuration Audit*	• A formal audit to validate that the development of a configuration item has been completed satisfactorily and that the item has achieved the performance and functional characteristics specified.
PCA *Physical Configuration Audit*	• Confirmation that the "as built" item conforms to the technical documentation.
FQR *Formal Qualification Review*	• Test, inspection or analytical process by which system elements are verified to have met specification requirements.

Figure 8-4 Major design reviews held during a project life cycle.

System Specification *(Type A)*	• System technical, performance, operational and support characteristics. • Defines allocation of requirements to functional areas and interfaces. • Defines system elements and interfaces.
Development Specification *(Type B)*	• Subsystem technical performance, operational and support characteristics. – equipment, assembly, computer programs, facility, critical item support, etc.
Product Specification *(Type C)*	• Subsystem technical requirements that can be: – procured "off the shelf" – standard system components, software programs, and spare parts.
Process Specification *(Type D)*	• Subsystem technical requirements that cover a service performed on a system component. – machining, marking, packing, and processing.
Material Specification *(Type E)*	• Susbsystem technical requirements that pertain to raw materials, mixtures, and semi-fabricated materials used in the fabrication of a product. – e.g. paints, electrical cable, piping.
Interface Control Document *(ICD)*	• System and subsystem interfaces including technical, performance and support characteristics. • Describes both sides of interface: – information data exchange such as file transfer formats, equipment, type of data and method of transfer.

Figure 8-5 Types and content of system specifications.

Approach™ is necessary to provide the level of program insight necessary for success. Once the A Spec has been developed and the SRR has been held, the team is ready to move into the next phase of the project.

Preliminary System Design — In the Preliminary System Design phase the activities are more focused on the system elements and their interaction. (see Figure 8-3). The Functional Analysis is continued at a lower level of detail in order to conduct concept development and evaluation. The preliminary design is created based on the results of the concept analyses, testing, and evaluation of the design concepts. Electronic assembly "bread boards" are developed to analyze the functional design of the electronic systems, and modeling is conducted in order to understand the interfaces between the system elements, such as electronic, mechanical, and subsystems. This is also the phase at which program resource allocations are made for the remainder of the project. The System Design Review (SDR) and System Software Review (SSR) are conducted during this phase and can often comprise several stages of analysis depending upon the complexity of these elements in the overall project (see Figure 8-4). At the end of the phase, the Preliminary Design Review (PDR) is held wherein the preliminary design is approved prior to continuation of the project. The PDR considers risk, progress of the project, initial selection of manufacturing processes, and software requirements/interfaces. During this phase several elements of specification development are defined (see Figure 8-5). The preliminary development (B Spec), process (C Spec), product (D Spec), and material (E Spec) specifications are defined as described in Figure 8-5. These specifications typically define target values and components for procurement. In addition to these specifications, the Integrated Control Documents (ICDs) are initiated. An ICD is a specification document that describes system and subsystem interfaces including technical, performance, and support characteristics. This document provides guidance to the different design and manufacturing disciplines of the program during the design phase and into the production phase. For example, ICDs are critical in defining software interfaces between complicated electronic subsystems. The program plan is further defined in this phase based upon the increased level of detail and knowledge. Once the PDR has been held and the design approved, the next phase is entered.

Detail System Design — The Detail System Design phase is in many ways a continuation of the activities in the Preliminary System Design phase. Further decomposition of subsystems is conducted and trade-off analyses are conducted on specific materials, processes, and designs. Prototype models are developed during this stage and the system modeling includes more detailed test analysis (such as FMEA and DOE). Further definition of the specifications is conducted during this phase, with the goal being the creation of the complete specification document to be used during production and operation and maintenance. The goal for the Detail System Design phase is to complete the analysis necessary to conduct the Critical Design Review (CDR). The CDR determines that the detail design satisfies physical and functional requirements for the project. It should also determine that specifications are complete and that producibility has been properly assessed (see Figure 8-4). All ICDs should also be completed by the end of this phase.

Production/Construction — Production and construction of the project elements and products is the phase that should not be entered into until the final design has been approved (more on why this is critical is given in the section on Concurrent Engineering). The system is tested and data is taken on production units to verify performance criteria. Any modifications to process or design is made typically during this phase, although this is the most costly time for redesign or change. During this phase the maintenance and support activities are brought to the forefront. Four reviews are conducted during this phase of the Systems Engineering Life Cycle: (1) Test Readiness Review (TRR); (2) Functional Configuration Audit (FCA); (3) Physical Configuration Audit (PCA); and (4) Formal Qualification Review (FQR) (see Figure 8-4.) The TRR is conducted for each computer software configuration item to determine whether the software test procedures are complete. The FCA is a complete review of the functional attributes of the system or product in order to determine if the functional characteristics meet the performance characteristics. The PCA confirms that the "as built" item conforms to the technical documentation. This is especially critical not only for sale to the customer, but also if the product must comply with other specification requirements such as ISO 9000 for the European Community. Lastly, the FQR is a formal test, inspection, or analytical process confirming that the system elements are verified against all specification requirements.

Operations and Maintenance — This last phase of the life cycle is concerned with the use of the system or product. It deals with the logistics of supporting the system in the field or in the customer's facility, customer service, and data collection on the performance of the product or system for any modification or redesign. This phase of the Systems Engineering Life Cycle is often overlooked because of less "engineering-related" activity, but is an important phase for customer satisfaction and verification of the design in actual use. Engineering is most often involved in this stage in any change reviews.

A sixth phase of Systems Engineering not yet fully recognized as part of the Life Cycle in most companies, but becoming increasingly important, could be called the **Recycling/Disposal** phase. The increased focus on the environmental impact of disposing of products is beginning to take more of an economic shape each year as companies try to reduce the waste at their own facilities, as well as provide for how their products will eventually be disposed. Governmental regulation is also being imposed on companies to plan for the recycling or disposal for their products. As a result of the decreasing or nonexistent landfill space in Europe, regulators there have begun to assess disposal fees on such items as medical products if the products cannot be recycled. The obvious consequence of this will be an increased emphasis on engineering to not only provide for more recyclable materials in products, but to even assist in the design of after-market disposal or recycling procedures.

8.1.3 Difficulties encountered in the Systems Engineering Approach

As ordered and methodical as Systems Engineering appears, it can be fraught with difficulty in application in a real production scenario, especially when state-of-the-art technology is involved. Within each of the first three phases of the process there can be significant iterations of the design and specifications. Without focus, Systems Engineering can be a nightmare of redesign as concepts developed in the Concept Exploration phase are found to be unworkable or too costly in the Preliminary System Design or Detail System Design phases. As a result of design rework, A Specs are often not fully completed until well into the detail design or even in production. Timetables are often compressed to stay on schedule and the ordered approach to design becomes chaotic as review phases are delayed but activities continue to progress into the next phase. The following summarizes some of the criticisms of Systems Engineering as a methodology:

- Lack of ability to prioritize requirements early in the design to avoid later problems;
- No mechanism to create visibility into the interrelation of the competing subsystems;
- Process is cumbersome — difficulties in dealing with individuals and departments in coordinating information thereby creating confusion;
- Iterations are often a *repeat* vs. *refinement*;
- Existing business processes heavily relied upon without being brought into the design process. Therefore, if design and production processes are not in control and streamlined, Systems Engineering can only operate to the current operational level of those processes;
- Development of specifications often done in a "vacuum" without insight into marketing knowledge and its potential impact on design;
- Technology transfer occurs through the exchange of data packages as opposed to direct communication and discussion;
- The "System engineer" becomes responsible for the design and, therefore, "team" accountability does not exist; and
- Systems Engineering focuses on product development that is limited by current design methods, manufacturing capabilities, and business operations.

Although Systems Engineering can eventually provide completeness of the design and manufacturing, the reality of shortened concept-to-market windows and increased competitiveness can make the application of Systems Engineering bulky without a focusing mechanism. QFD as part of the Integrated QFD Approach™ provides that focus.

8.1.4 Systems Engineering and the Integrated QFD Approach™

How and when does QFD interact with the phases of the System Engineering Life Cycle?

The reader can already see where QFD may apply to the System Engineering Life Cycle from the realization that a primary function of QFD is to define requirements. An obvious first application of QFD is in the initial exploration of the system concept and definition of customer needs. But for a complete understanding of the optimum use of QFD in Systems Engineering, a comparison of the nine steps of the Integrated QFD Approach™ (see Figure 7-1 for a review of the nine steps) and the Systems Engineering Life Cycle phases is required.

8.1.4.1 Definition of need and concept exploration

Figure 8-6 shows the interrelation of the Integrated QFD Approach™ and the Systems Engineering Process. In the figure, the Concept Exploration phase is actually preceded by a Definition of Need step, which looks at applied research to determine if a need even exists for the product. This prevents the creation of "technology-driven" products vs. market-driven, products, although the authors realize that just because a product may be technology driven does not preclude its success. Rather, this step allows for an understanding of the marketplace need, and if there is no need, the realization of the fact that the product may be one driven by technology. This step is accomplished through the use of the System QFD Matrix comprising Step 1 of the Integrated QFD Approach™, and by the creation of the customer needs for the matrix.

Concept Exploration involves the completion of the System Matrix and the inclusion of the product, service, and business operations requirements which are the initial *system* requirements for the project. What is added by having QFD in this phase of the project is the prioritization of the driving subsystems for the project, and a way to discover the interrelations between those systems. Systems Engineering will provide the detail for all of the requirements but QFD focuses the resources of the organization on the important or critical requirements.

The Concept Exploration phase should also include the creation of the initial Integrated Plan which is Step 2 of the Integrated QFD Approach™. The Integrated Plan becomes the connecting thread throughout the Systems Engineering process and helps to define interfaces between subsystems and departments. This creates avenues of communication and discussion for real-time problem solving.

8.1.4.2 Preliminary System Design

The authors have divided the Preliminary System Design phase into three activity subphases: (1) System Functional Analysis; (2) System Analysis Trade-off Studies and Optimization; and (3) System Synthesis and Definition. (see Figure 8-6). The System Functional Analysis applies the activities in

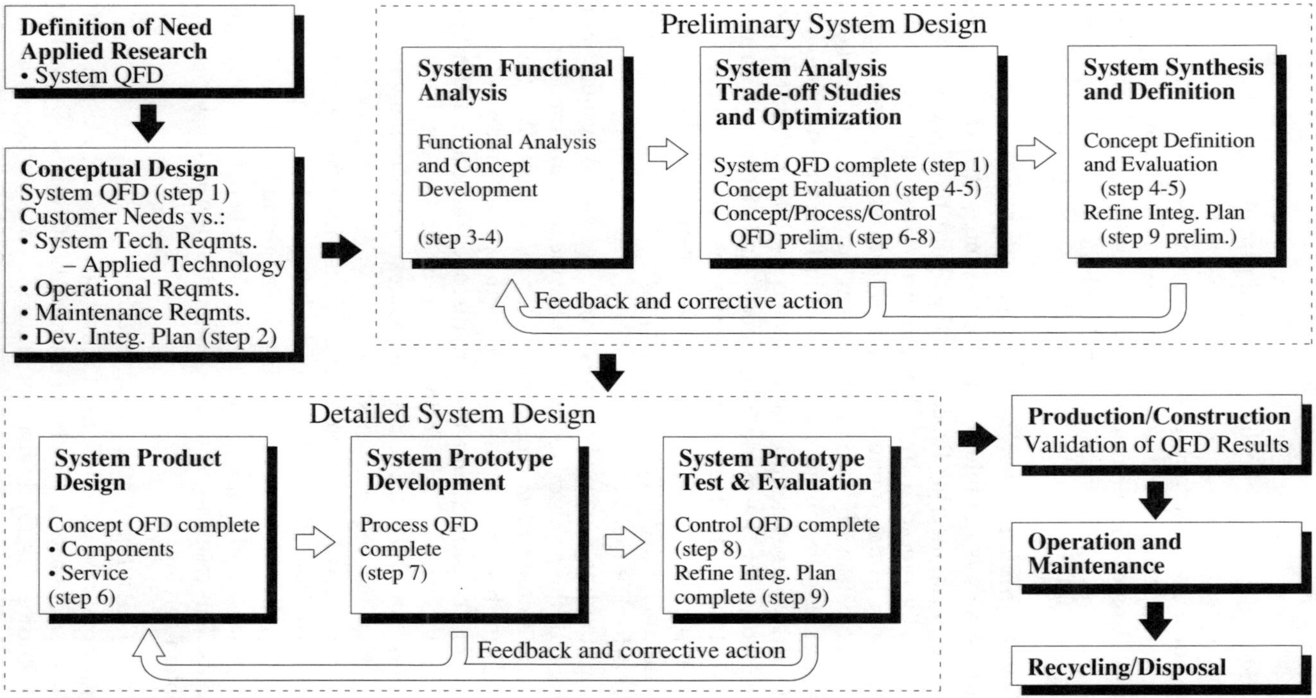

Figure 8-6 Application of the Integrated QFD Approach™ to the System Engineering Process. Steps correlate to the Integrated QFD Approach™.

Steps 3 and 4 of the Integrated QFD Approach™, which include developing the functional flow and the system alternatives. Functional analysis helps to further define the extent of the design in terms of alternative concepts. The design concepts are tested and evaluated and the interfaces are further defined in the System Trade-off Studies and Optimization, involving Steps 4 and 5 of the Integrated QFD Approach™. The third activity subphase — System Synthesis and Definition — also utilizes Steps 4 and 5 of the Approach, but focuses on the interaction between the subsystem interfaces.

The Preliminary System Design illustrated in Figure 8-6 contains a "Feedback and Corrective Action" loop indicating the iterative nature of the process. It also references the preliminary development of the next three matrices of the Integrated QFD Approach™: the Concept (Step 6), Process (Step 7), and Control (Step 8) matrices. By initially completing these three matrices the design team identifies the key elements for each decomposed level of detail and can expend resources in these areas early in the design phase to resolve technical problems and interactions. The feedback recognizes the fact that changes will be necessary, but within the confines of the preliminary design. Thus, changes should be refinement and not repetition. Not shown in Figure 8-6 is Step 9, which is the refinement of the Integrated Plan, which continues during the Systems Engineering process. Completion of the preliminary matrices allows further definition of the plan and helps to maintain a framework for the entire process.

8.1.4.3 Detail System Design

Like the Preliminary System Design phase, the Detail System Design phase contains three subphases and a feedback loop (see Figure 8-6). The three subphases contain the product design, the prototype development, and the testing and evaluation of the prototype:

- **System Product Design** — This subphase involves development of the complete product, service, and business operations QFD matrix, which is the Concept Matrix (Step 6 of the Integrated QFD Approach™). Components are finalized and services defined prior to proceeding to the prototype development.
- **System Prototype Development** — The Step 7 Process Matrix is employed in this subphase of prototype development. Processes and procedures are refined as the prototype is developed.
- **System Prototype Testing and Evaluation** — The Control Matrix (Step 8 is finalized during the final testing and evaluation of the prototype to determine if the design is meeting the target specifications.

The final system design is completed during this phase and the feedback loop represents a further refinement of the design, not a "back-to-the-drawing-board" approach that can sometimes characterize poor systems engineering when concepts are changed or abandoned.

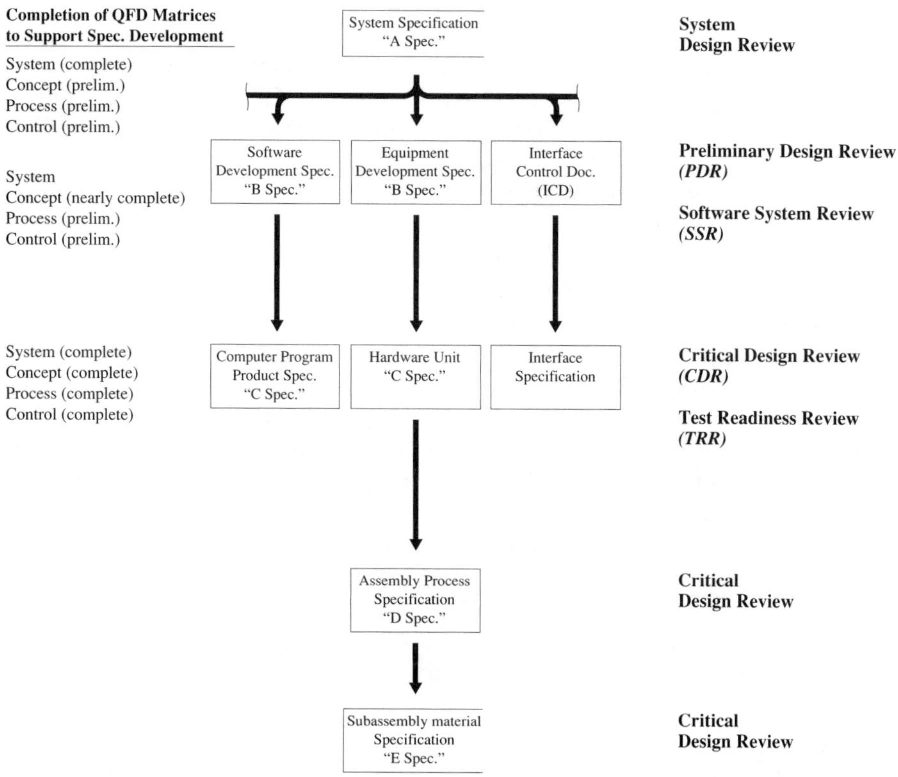

Figure 8-7 Relationship of QFD matrices to major project reviews and specification development.

8.1.4.4 Production/construction and operation and maintenance

If the QFD has been properly performed, no design changes are required during the Production/Construction phase of Systems Engineering. During production analytical techniques such as SPC can be used to determine if the product is conforming with the design specifications. This is where process control, collection of manufacturing data, supplier control, and distribution feedback are all taken into consideration in determining whether or not the as-built product is being produced within the target parameters defined during the QFD matrices development. The traceability aspects of the Integrated QFD Approach™ is a key element in the overall Systems Engineering process to provide insight into any problems and the solution of those problems during the previous phases.

For the Operation and Maintenance phase, this step provides additional insight and feedback into the product performance. In the event that changes are necessary, the Integrated QFD Approach™ has provided the documentation for understanding the impact to the overall design based upon the integration of the subsystems and the interrelationships between any of the design elements.

8.1.4.5 Relationship of QFD matrices to major project reviews and specification development

One question that seems to come up often when Systems Engineering is combined with QFD is, "When should the various QFD matrices be completed in relationship to the specification development and design reviews?" Although there is no cut-and-dried answer to this question that would apply to all situations, some generalizations can be made.

Figure 8-7 illustrates the relationship of QFD matrices in the Integrated QFD Approach™ to design specification completeness and program reviews. For example, in order to support the A Spec or System Specification, the System Matrix should be complete *and* the Concept, Process, and Control matrices should be preliminarily completed. Completion of these matrices to these levels will support the SDR. Figure 8-7 shows that in order to support the CDR and the remaining specifications, all four of the QFD matrices should be completed.

8.1.5 Benefits of using QFD and Systems Engineering together

Many of the complaints about Systems Engineering disappear when the Integrated QFD Approach™ is used as an integral part of the process. The increased communication and interfacing provided by the multifunctional design team eliminates "communication by paper," which is sometimes not communication at all. The prioritizing ability of QFD focuses the systems process and helps to speed it to conclusion. Development of the Integrated Plan helps keep the program from getting off track and identifies the functional and systems interfaces that are key to the success of Systems Engineering. QFD also creates iteration by refinement rather than iteration as repetition.

As a footnote, the associated organization, managerial, and team building benefits that are associated with the Integrated QFD Approach™ and RSM™ apply with equal force to Systems Engineering.

8.2 Concurrent Engineering and QFD

Concurrent Engineering (also called Integrated Product Development) is a term that is familiar in the aerospace and defense industries but is not as recognized in the commercial world. It has often been described as "good" Systems Engineering, but there is more to understanding exactly what it is than trying to use a positive modifier on an already complex subject! The purpose of this section is not to describe the application of a method without application to the commercial environment, but to give the reader some background in understanding the relationship between Concurrent Engineering and Systems Engineering applications, and the role of QFD in Concurrent Engineering.

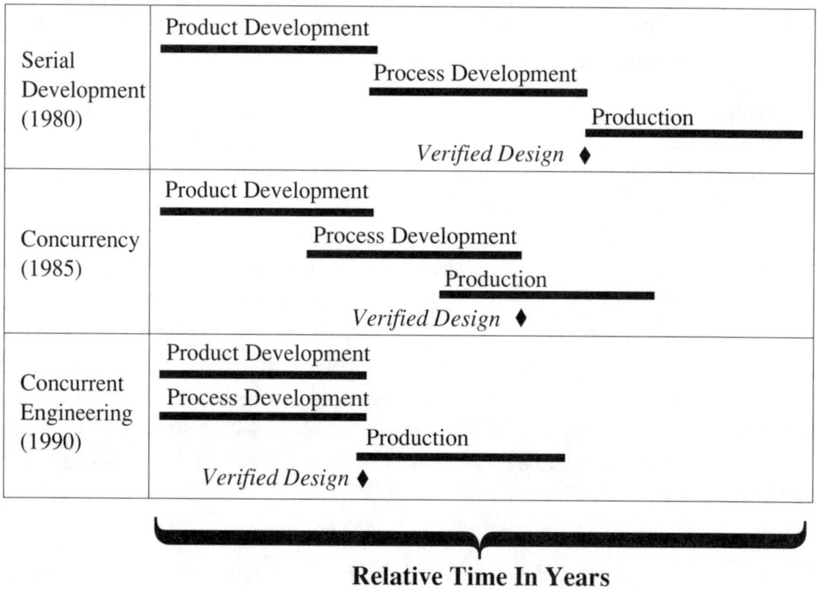

Figure 8-8 Efficiency Achieved through simultaneous product and process definition.

8.2.1 *What is Concurrent Engineering?*

The term *Concurrent Engineering* implies that several things are being done together in an engineering discipline. This is not a facetious statement. The basic premise behind Concurrent Engineering is the simultaneous development of the elements of a system or products to reduce expending resources and time. The Institute for Defense Analysis issued a report in 1988 on the role of Concurrent Engineering in weapons system acquisition, and included the following definition of Concurrent Engineering:

> Concurrent engineering is a systematic approach to the integrated, concurrent design of products and their related processes, including manufacture and support. This approach is intended to cause the developers, from the outset, to consider all elements of the life cycle from conception through disposal, including quality, cost, schedule, and use requirements.[41]

Although this definition of Concurrent Engineering comes from a defense-oriented source, the concept encompasses the broader application of Concurrent Engineering to commercial applications. And even if the above definition sounds similar to the definition for Systems Engineering, Concurrent Engineering is really the architecture supporting the application of Systems Engineering. Shina's definition of Concurrent Engineering is more specific in referencing specific program disciplines and incorporates three critical factors — cost, quality, and time:

> *Concurrent engineering* (CE) is defined as the earliest possible integration of the overall company's knowledge, resources, and experience in design, development, marketing, manufacturing, and sales into creating successful new products, with high quality and low cost, while meeting customer expectations. The most important result of applying CE is the shortening of the product concept, design, and development process from a serial to a parallel one.[42]

Shina invokes the idea that Concurrent Engineering operates to shorten design time, reduce cost, and result in high quality by creating a "parallel" approach to design and manufacturing. This concept of a parallel approach is graphically represented in Figure 8-8. Figure 8-8 shows that the basic approach to design and manufacturing prior to 1980 was really "Serial Development." Serial Development operated by first completing the *product* development, which included concept and design, and then conducting *process* development, which would yield the *verified design* prior to engaging in production. This process reduced engineering changes during production but was cumbersome and time-consuming. As a result of the stepped-up competition of the 1980s, and the shortened life cycle of many high-technology products, products fell victim to Concurrency — the overlapping of product development, process development, and production. This approach created significant costs because production was being entered into prior to the establishment of a Verified Design.

Concurrent Engineering is really the parallel development of product and processes to produce the verified design faster, but prior to production (see Figure 8-8). This approach requires heavy interfacing between engineering and manufacturing in order to simultaneously develop the design that is also using the best manufacturing processes — often referred to as Design for Manufacturing (DFM).

Because Concurrent Engineering is the architecture that supports the activities in Systems Engineering, Concurrent Engineering is sometimes confused as just another *type* of Systems Engineering. This is an erroneous assumption. Systems Engineering can exist without Concurrent Engineering because it is a methodology that uses scientific and engineering efforts to transform an operational need into a defined system configuration. Concurrent Engineering, however, allows the methodology of Systems Engineering to operate efficiently by allowing the transmission of information within a well-established system so that the elements of Systems Engineering can be concurrent. As pointed out earlier in this chapter, Concurrent Engineering

Meet Critical Objectives

- Reduce Cost
- Reduce Time
- Increase Quality

how

Change in Engineering Activities with Regard to Timing, Process and Philosophy

- Provide early and complete understanding of customer requirements
- and priorities.
  Translate requirements concurrently and in an integrated fashion into optimal products and manufacturing and support processes.
- Provide continuous review and improvement of product, process and support characteristics.

how

Establish Capabilities to Support Engineering, Production, and Support Processes

- Data capture (leasons learned) on comparable products, processes and support.
- Define and capture data from new systems development.
- Synthesize requirements into design of product, process and support.
- Validate design of product, process and support.
- Manage product, process and support data.
- Disseminate product, process and support data.
- Deliver product or data for manufacturing and support product.
- Perform rapid prototyping.
- Incorporate process robustness into system.
- Use intelligent oversight for impact assessment of changes & effective change control.
- Provide current system & design information to all project members simultaneously & be proactive with design iterations.

how

Technical Building Blocks

- Establish common information architecture for data processing and data structures.
- Provide Information Framework: Procedures for establishing, storing, exchanging, and evolving information-based policies and tools.
- Use tools and models: i.e. process models, assembly and cost models.
- Implement manufacturing systems — improve integration of design with manufacturing.
- Define and control design processes.

Figure 8-9 Concurrent Engineering framework.

focuses on the design and manufacturing process improvements that allow Systems Engineering to move faster and more efficiently.

But what is the architecture of Concurrent Engineering that permits the concurrent development of product and process? In Section 8.1.3, one of the drawbacks seen in Systems Engineering was that it focuses on product development which is limited by current design methods, manufacturing capabilities, and business operations. For example, if a company's internal design change process was cumbersome, paper-oriented, and required a series of approvals from various project functions before incorporation, Systems Engineering would be hindered by the limitations of that *process* of change approval even if the company was practicing Systems Engineering from the viewpoint of phase activity. Concurrent Engineering would look at the engineering change process itself and incorporate functional analysis, coupled with automation, to improve the process to serve the needs of engineering better.

This concept of improving the system that does Systems Engineering is expressed as a framework of four major concept areas: (1) Meet critical objectives; (2) Change engineering activities with regard to timing, process and philosophy; (3) Establish capabilities to support engineering, production, and support processes; and (4) Develop technical building blocks. Figure 8-9 illustrates the detailed elements of these concept areas.

The first concept area involves meeting the critical objectives of Concurrent Engineering, which are reduced cost, reduced time, and increased quality. Concurrent Engineering's main purpose is to get a design accomplished faster and in production sooner at a lower cost to the company. All of the other concept areas of Concurrent Engineering are focused to accomplish this purpose. Concurrent Engineering does not necessarily focus on the product itself, but on the support architecture to design and produce the product.

The second concept area involves change in the way that engineering may be done in a company. In order to achieve the critical objectives, Figure 8-9 shows that *early* and complete understanding of the customer requirements and priorities are key to the success of the effort and *how* the objectives are to be met. Translation of requirements concurrently and continuous review and improvement of product, process, and support characteristics are the other methods to meet the Concurrent Engineering objectives.

The third concept area contains the methods for meeting the elements of changing engineering activities. Figure 8-9 lists 11 capabilities that will be critical in the support of engineering and production. The 11 concepts describe important information interfaces, and focus on the rapid and accurate transmission of data and requirements. This is accomplished by establishing the technical building blocks described in the fourth concept area. The technical building blocks involve common information architecture for data processing, procedures for information transfer, modeling, manufacturing systems (such as SPC), and focus on design processes.

Concurrent Engineering Objectives Through Demonstration/Validation Phase

			…	Degree of importance	Current Company Rating	…	Planned Level	Improvement Ratio	Sales Point	Important Weight	Relative Weight
External Customer	Clear Description of What to Build	Working relationship with customer		3	3		5	1.67		5.0	1.5
		Well defined contractual requirements		3	2		3	1.50		4.5	1.3
		Well defined product requirements		5	1		3	3.00	❍	18.0	5.3
Efficient, Effective Management	Company Management	Leadership		3	2		3	1.50		4.5	1.3
		Administration		1	2		3	1.50		1.5	0.4
	Program Management	Leadership		5	2		3	2.00		10.0	2.9
		Administration		4	3		4	1.33		5.3	1.6
	Productive Use of Automation	Tools that support program process		3	1		4	3.00		9.0	2.7
		Collection, control and distribution of information		4	1		3	3.00		12.0	
High Quality Product Design (Hardware and Software) that are Testable and Producible	Manufacturability	Ensure manufacturability in all program phases		5	2		3	2.00			
		Cost effective integrated manufacturing process		3	2		4	2.00			
	Subcontractors and Materiel Defined	Credible selection process									
		Support design, manufacturing and test									
		Management of vendors									
		Management of subcontractors									
	Integrated Test Program Defined	Ensure testability in all program phases									
		Cost effective integrated testing plan									
	System Design	Coordinated product documentation									

Figure 8-10 Development of a Concurrent Engineering Road Map using QFD: Definition of Concurrent Engineering Objectives.

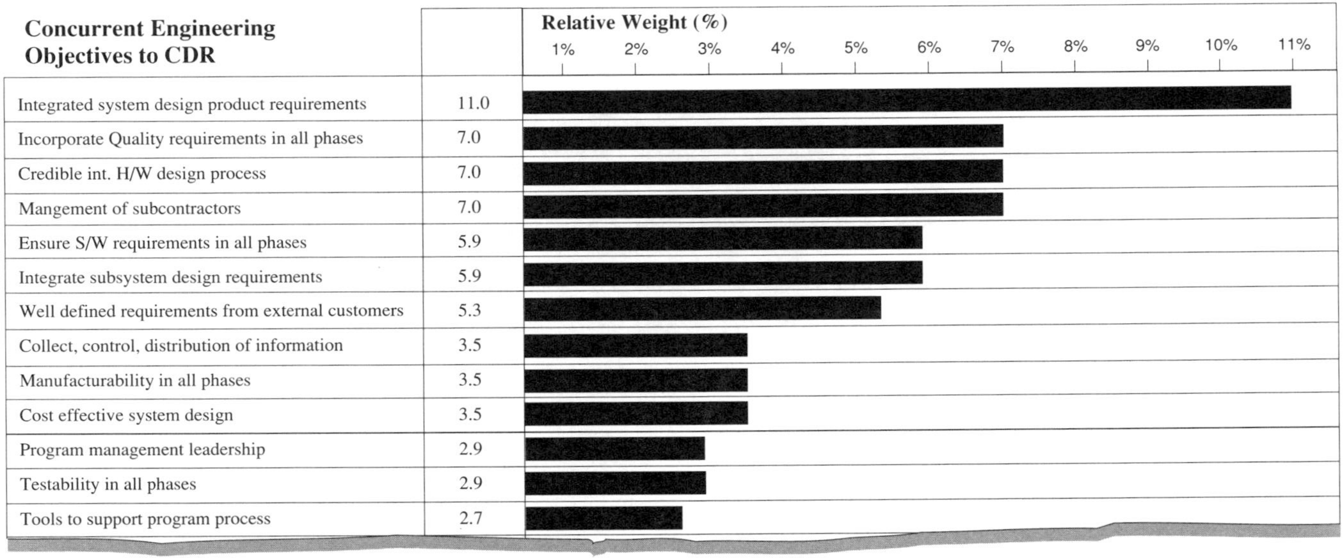

Figure 8-11 Development of a concurrent engineering road map using QFD: Prioritization of concurrent engineering objectives.

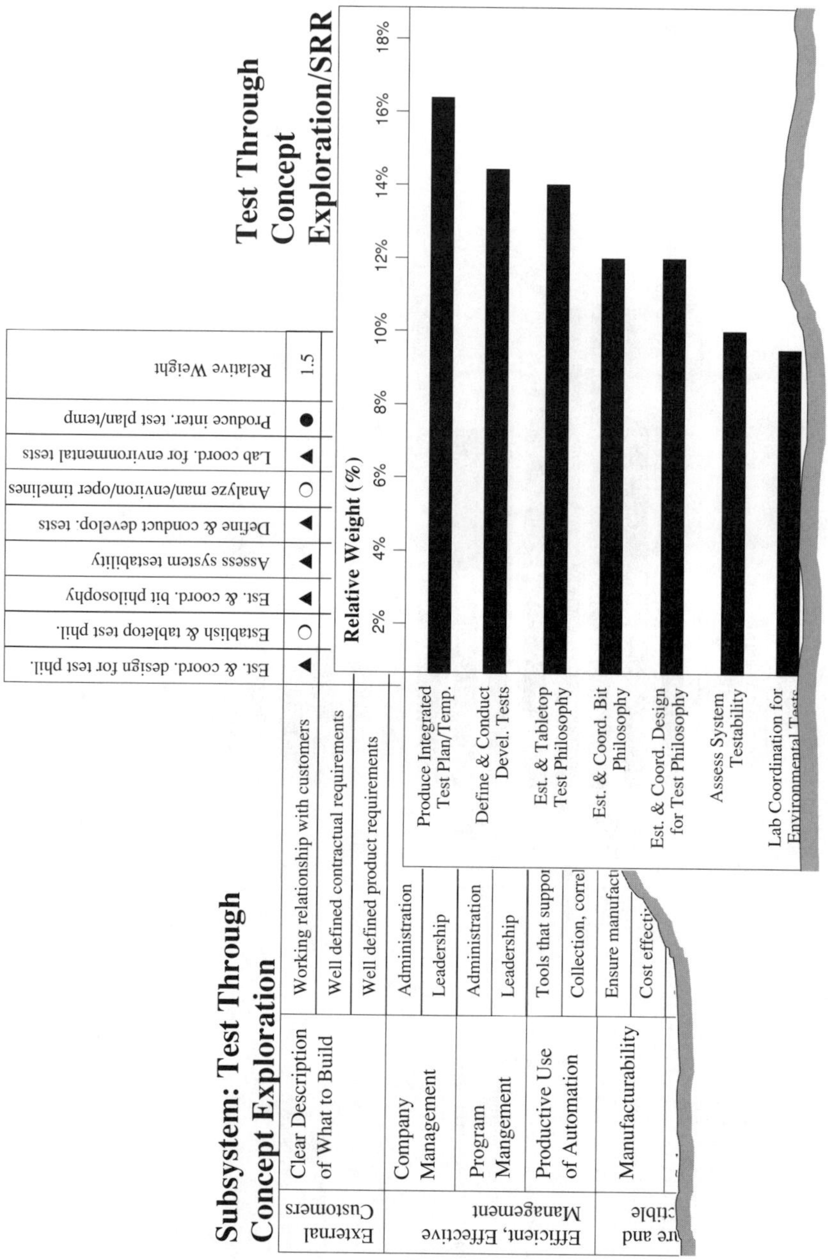

Figure 8-12 Development of a concurrent engineering road map using QFD: Prioritization of the system testing activities during the concept exploration phase.

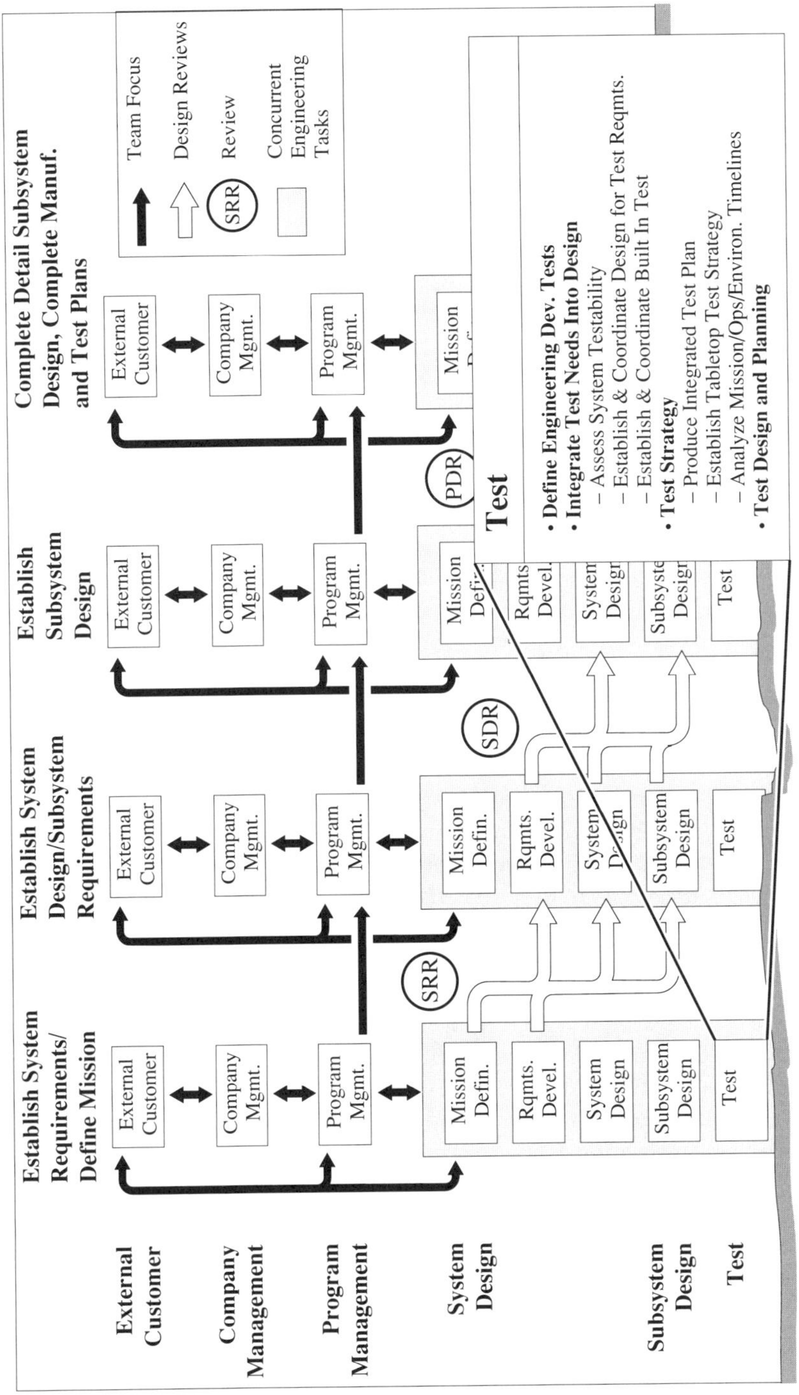

Figure 8-13 Translation of the QFD results into a concurrent engineering environment.

8.2.2 Concurrent Engineering and QFD

Because Concurrent Engineering focuses on *how* to do engineering and production, rather than *what* is being engineered and produced, many companies may be practicing elements of Concurrent Engineering already. Concurrent Engineering uses techniques like SPC, DOE, Functional Analysis, CAD/CAM, and QFD. In fact, the authors contend that Concurrent Engineering cannot be functionally accomplished without the inclusion of QFD as an integral part of the design and production process. QFD analysis focuses the Systems Engineering effort on the customer and satisfies the objectives of Concurrent Engineering to control cost, time, and quality.

But if Systems Engineering involves QFD already as shown in Section 8.1, and Concurrent Engineering encompasses Systems Engineering, is there an additional role for QFD in Concurrent Engineering?

The answer to this question is an unequivocal "yes." The reader has seen where QFD fits into the Systems Engineering paradigm as a method for focusing the efforts of the project team and as part of the Integrated QFD Approach™. QFD also plays a role in Concurrent Engineering in focusing the company on the engineering and production elements supporting the project. One of the best ways to see how QFD can be involved in this effort is to examine how QFD was used to focus Concurrent Engineering in the creation of the Spacecraft Technology Center (SPC) at the Martin Marietta Aerospace Group.

In 1991 the Martin Marietta Aerospace Group sought to develop an automated spacecraft design center using Concurrent Engineering to bring down the cost and time associated with spacecraft design. The first step was to develop a Concurrent Engineering "road map" for implementation. That was accomplished by taking the Concurrent Engineering objectives and using those as the customer needs for the initial QFD matrix. Figure 8-10 shows part of the QFD matrix that developed the prioritized Concurrent Engineering objectives for a program through the demonstration and validation phase, which is equivalent to CDR in the Systems Engineering world. These objectives were then ranked according to relative weight so the SPC team knew which Concurrent Engineering elements would give the "most bang for the buck" in terms of a architecture for design and production in meeting the requirements through CDR (see Figure 8-11). This process was repeated for subsystems and also mapped to each review milestone in the Systems Engineering phases. For example, Figure 8-12 shows a sample matrix for Test through Concept Exploration, and the resulting prioritized activities for Test through SRR.

After completion of the QFD matrices, the next step was to translate the results into the Concurrent Engineering architecture to support design and production. Figure 8-13 shows the responsibilities of the Concurrent Engineering team, and the activities for each of the Systems Engineering functions, in relationship to the review milestones of the Systems Engineering phased approach. What the Concurrent Engineering analysis did was to define what activities had to be accomplished prior to the move to the next phase of the Systems Engineering process. It also assigned responsibility to individuals on the Concurrent Engineering team, rather than making a "systems engineer" responsible for ensuring the completion of the activities for review readiness and acceptance.

The next step in the process was to create the now-familiar Integrated Plan incorporating the activities that needed to be accomplished to support the Concurrent Engineering initiatives. Figure 8-14 shows a sample task sheet for integrating test needs into the design. The task sheets helped to define the activity to support the project and gave the inputs and outputs as well as defined risk for that activity. After the Integrated Plan was developed, the next step was to create an architecture for information flow through the program phases. Figure 8-15 is a partial picture of how information was tracked between the functions in relationship to review milestones and time. A numbered coding system was adopted to simplify the information flow.

Each department, or engineering/production function, then created a checklist for those activities that had to be completed prior to completion of a phase. Figure 8-16 shows the "Go/No Go" checklist for Test through SRR. This checklist shows that these activities needed to be completed, and a support mechanism had to be in place for each, prior to the end of the SRR phase of the systems approach.

Once the checklist was completed, process improvement techniques were applied to improve the design process and the individual procedures involved. Figure 8-17 shows the development of test metrics for improving the process to integrate test needs on the design. This kind of analysis allowed the Concurrent Engineering team to develop the best approaches to minimize time, cost, and resources while ensuring quality in the test process.

Using QFD as part of the Concurrent Engineering approach to the STC allowed Martin Marietta to create an integrated center that cut the design time for spacecraft by 70%, which represented a substantial cost savings as well. In addition, because communication interfaces were not only created for the efficient exchange of information, but also automated as part of the design process, design changes could be incorporated real time into the process making the iterative nature of Systems Engineering truly one of refinement and not repetition.

8.3 *Summary*

The interrelationship of Systems Engineering and Concurrent Engineering is best represented in Figure 8-1, showing how these concepts fit into the overall management concept of the organization and RSM™. RSM™ encompasses the corporation and treats it as a single system with all activities measured against corporate goals. Concurrent Engineering focuses on the design and manufacturing processes themselves as opposed to the product. Systems Engineering focuses on the product systems and is a methodical approach to the design and manufacture of a product.

Task Number:	**TS02**
Task Name:	**Integrate test needs into design (level 2)**
Task Description:	Define, in terms of guidelines and design practices, how the design needs to accommodate testing. Based on the process of testing defined, how is the design of the system actually going to facilitate the process. For SRR this includes: • Assess System Testability (TS0201). • Establish & Coordinate Design-For-Test Rqmts. (TS0202). • Establish & Coordinate Built-In Test Rqmts. (TS0203).
Inputs:	System Design Concepts, Top-Level Test Strategies.
Outputs:	Impacts and Requirements for the Design of the System to Assure that the Testing Program is Facilitated by the Design. DOCUMENTS: Formal/Informal Guidelines and Design Practices (ERPM or EPDR)
Controls:	
Outputs:	DOD 4245.7-M Templates: Design for Testing • Built-In Test • Integrated Test.
Process Measurements:	Test SRR FGo/No-go Checklist — **Tools/Models:**

	Test Engineer Test Labs Engineer	**Duration (months)**	**% Effort**
#1	Test Engineer		
#2	Test Labs Engineer		
#3	Concurrent Engineer. Team		
#4			
#5			
#6			
#7			
#8			

Program Phase:	**Start/Stop Dates:**	
Start - SRR	1/1/90	1/31/90

Risk:

– Test article may not be effectively testable
– May not be able to accomplish requi[...]
– May cause additional risk [...]
to unnecessary ope[...]

Figure 8-14 Translation of QFD results into concurrent engineering activities.

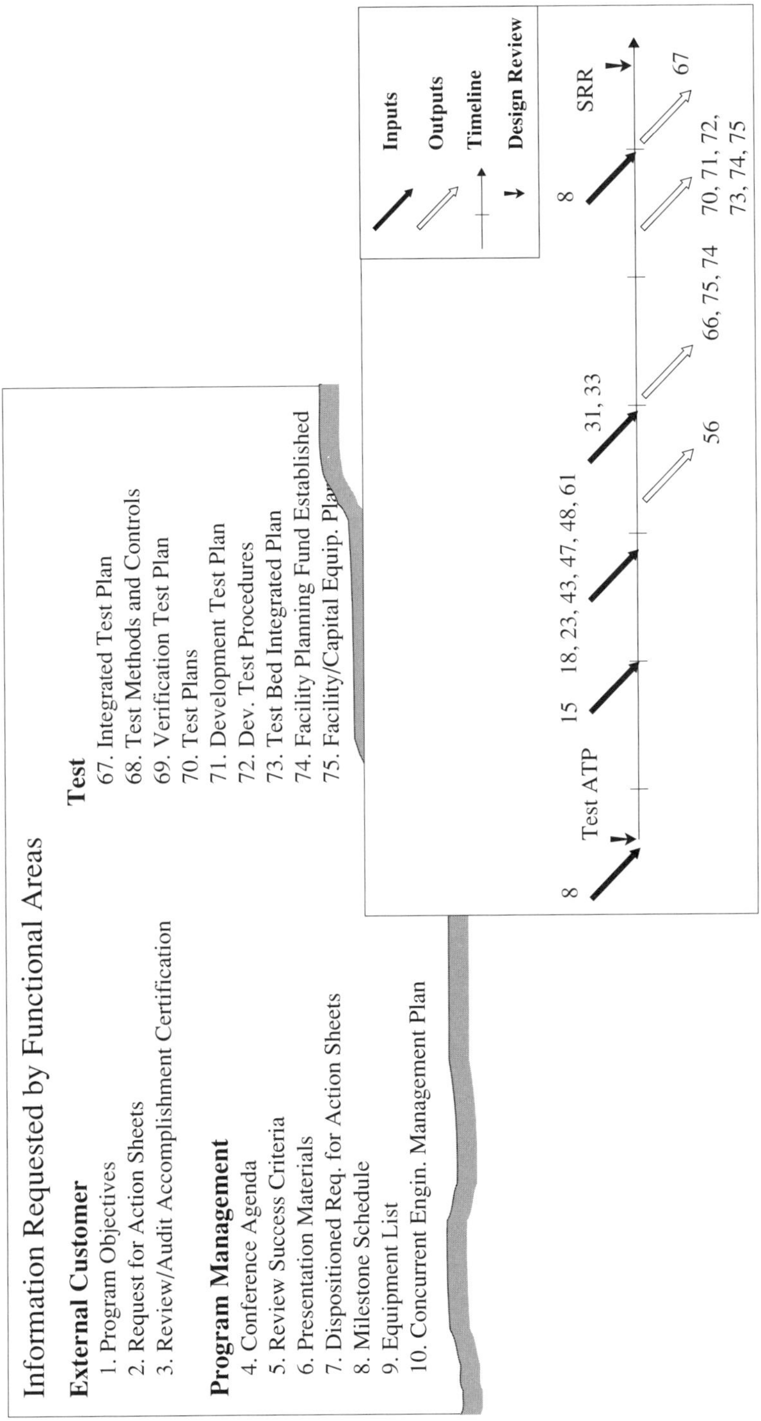

Figure 8-15 Analyzing the information flow for program phases in a concurrent engineering environment.

Define & Conduct Engineering Development Tests

- Tests cover critical technology, interfaces and high risk items.
- Development test program exists.
- Timing of development test program relative to designs it affects.

Integrate Test Needs Into Design

- Use of design-for-test/built-in-test considered troubleshooting and remove-and-replace operations.
- All handling and movement operations can be safely and easily performed.
- Test points support data requirements and efficient troubleshooting.
- Pass/fail requirements are quantitative and appropriate.

Test Strategy

- All levels and test organizations are included.
- Tests are integrated, efficient and not repetitive.
- Coverage of mission and contingency operations.
- Customer coordination and agreement of stratefy/approach.
- Reliability testing is included (trending, etc.)
- Trade studies have been performed to optimize the test strategy/approach.
- Complete, no TBDs.
- Appropriate manning for test early on.

Test Design and Planning

- Test schedules are correct and adequate (duration, integration, networked).
- Analysis, inspection, test and demonstration efforts are integrated.
- All test instrumentation has a process capability > 2.
- Risk assessments performed on the test flows.
- Covers all test requirements efficiently.
- Facilities identified and defined.
- Test plans cover all lab requirements.

Test Conduct and Management

- Planning for management of test conduct in work.
- Trending requirements and coordination underway.

Figure 8-16 An example of checklists and completion criteria for concurrent engineering activities for testing through the System Requirements Review (SRR).

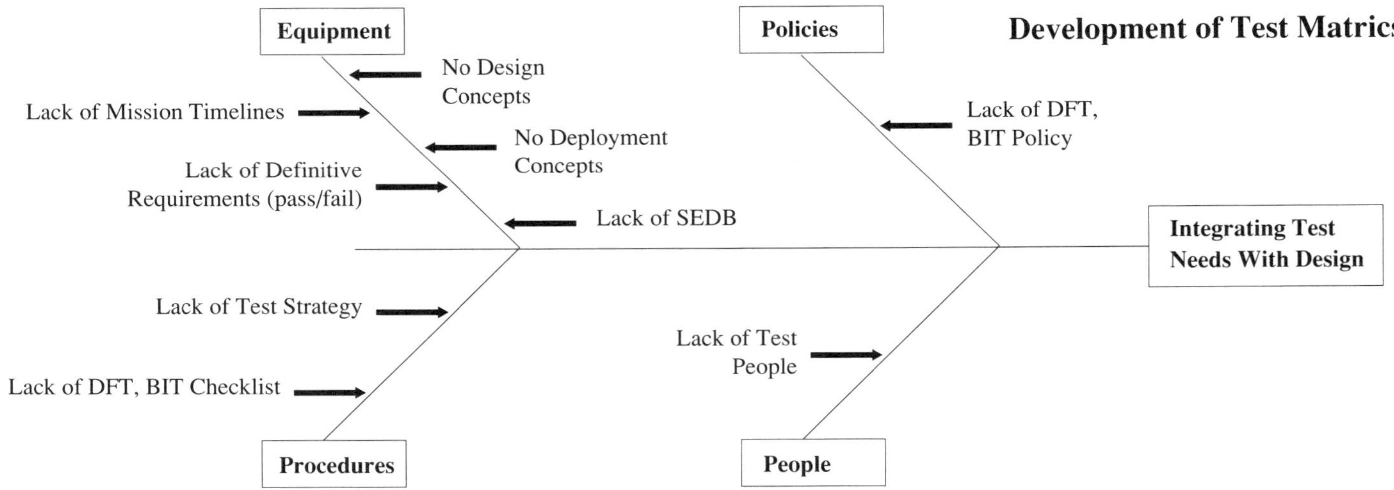

Goals

- Minimize the amount of time required to troubleshoot in test.
- All systems and subsystems identified in line replacement units.
- Minimize number of major movements during manufacturing and test.
- Eliminate extra dissasembly required during remove and replace.

Measures

- Number of major movements.
- Troubleshooting can be isolated to an LRU.
- Percentage of 2nd level functions that can be self-tested.

Figure 8-17 Application of Process Improvement techniques to improve the Design Process and Establish Test Process metrics.

Concurrent Engineering:

Is Not	Is
A magic formula for success. The best system cannot compensate for lack of talent.	An approach for improving the efficiency of good people who work hard.
Arbitrary elimination of a phase of the existing, sequential, feed-forward engineering process, or the elimination of any engineering functions.	Co-design of downstream processes toward a more all-encompassing cost-effective optimum design.
Simultaneous or overlapped design and production.	Simultaneous design of the product and of the downstream processes.
Simultaneous design of the product and execution of the production process.	Completion of all design efforts prior to the production initiation.
Just design for producibility or for reliability or for maintainability.	All of these plus design optimization to integrate these domains within a cost-engineering process.
Conservative design (derated parts, redundancy, close tolerances, etc.)	Achieving robustness by optimizing over a larger set of processes and by determining how to achieve target values with the lowest cost parts.
Conservative with respect to incorporation of new technologies.	Disciplined approach to integrate new technologies with minimum technical, cost and schedule risk.
Conservative testing strategy (refining the design through test-and-fix cycles).	One Pass Design in place of test-and-fix cycles.
Conservative inspection.	Manufacturing repeatability through product robustness and by designing a manufacturing process that includes means for monitoring and controlling itself (manually or automatically).

Figure 8-18 Concurrent engineering summary. (From Institute for Defense Analysis Report — R-358, 1990.)

Because Systems Engineering is usually better understood than Concurrent Engineering, and Concurrent Engineering is still confused with Systems Engineering by many people, the authors have included Figure 8-18, which summarizes what Concurrent Engineering is and is not.

8.4 Chapter summary

1. Systems Engineering and Concurrent Engineering are often confused. They are not different names for the same engineering methodology, but intimately associated with each other on a hierarchical basis.
2. Systems Engineering is concerned more with applying methodology to integrate systems efficiently to produce and service a product, whereas Concurrent Engineering is concerned with the architecture in the company that supports the application of Systems Engineering.
3. Both System Engineering and Concurrent Engineering are part of the overall RSM™ methodology.
4. QFD is a specific methodology that is used to maximize the system to which it is applied. Systems Engineering is a type-specific methodology used to foster development of products, and is a *complete requirements and design methodology*. System Engineering emphasizes completeness — QFD emphasizes *focus*.
5. Systems Engineering utilizes specifications and design reviews to organize the design and production cycle. QFD provides focus in that cycle to maximize resources and help provide traceability of the requirements in each phase. The Integrated QFD Approach™ also identifies the functional and systems interfaces that are key to the success of Systems Engineering.
6. QFD creates iteration by refinement, rather than iteration as repetition.
7. Concurrent Engineering is defined as the earliest possible integration of the company's knowledge, resources, and experience in design, development, marketing, manufacturing, and sales into creating successful new products, with high quality and low cost that meet customer needs.
8. Concurrent Engineering focuses on *how* to do engineering and production, rather than on *what* is being engineered and produced. Because of this, Concurrent Engineering encompasses System Engineering.
9. QFD focuses the company on the engineering and production elements supporting the project and streamlines the engineering/production process.

chapter nine

Process improvement, business re-engineering, and QFD

There are more things in heaven and earth, Horatio,
Than are dreamt of in your philosophy.
Hamlet, Act I, Scene V.

9.0 QFD for process improvement?

These words of admonition from William Shakespeare's tragic Dane, Hamlet, to his compatriot, Horatio may seem an unlikely introduction to a chapter on process improvement and business re-engineering. Although Shakespeare's works (especially *Hamlet*) are often subject to widely disparate interpretations, we can probably conclude that the Bard was not thinking of QFD and its relationship to process improvement and business re-engineering when he penned his famous tragedy nearly 300 years ago.

However, Hamlet's reference to a larger universe than that known to Horatio might be a rather literary parallel to what QFD is to process improvement and business re-engineering. Just as QFD is the centerpiece of Concurrent Engineering, it is also the key ingredient to what has been loosely termed process improvement or process re-engineering. Process improvement, in the authors' experience, is usually that function in an organization whereby processes themselves (as opposed to product or hardware designs) are the focus of improvement techniques and prioritization. Process improvement is typically a term used more frequently in service-oriented companies as opposed to manufacturing companies. Although service organizations often provide some "product" with the services they provide (for example, a software manufacturing company typically will provide service for its software as well as the actual software itself) their activities are extremely process-intensive. Therefore, improvements in how processes are done result in improvements to the service/product they provide and yield the largest gains in both resource utilization and profitability.

Even though the terminology is somewhat different, the concept of RSM™ still applies when dealing with process improvement, or what is sometimes referred to as "process re-engineering." It also applies to "business re-engineering" which is used by Michael Hammer and James Champy to describe a process of redesigning a company's operations completely. Hammer and Champy describe "re-engineering" as "starting over." They say that incremental changes are not part of business re-engineering but tossing out "old systems" is. They formally describe business re-engineering as:

> (T)he fundamental rethinking and radical redesign of business processes to achieve dramatic improvements in critical, contemporary measures of performance, such as cost, quality, service, and speed.[43]

Business re-engineering is looked at as a "radical" and "dramatic" improvement of business processes — a sort of Process Improvement on Steroids. Companies have achieved dramatic success with business re-engineering, but its radical nature is also synonymous with high cost, and the system is not without critics. Companies are not always ready to take the plunge of business re-engineering all at once, nor can they afford to; but all companies can benefit by some form of process improvement, whether radical or more gradual.

This chapter is focused on how to use QFD in a process improvement environment, whether it is business re-engineering on a grand scale or focused pilot improvements. QFD and process improvement use the language inherent to processes, emphasizing those elements which have a larger impact in process improvement, as opposed to overall product development or manufacturing. The reader will note that in essence QFD is used the same way, but there are additional factors for consideration, including a shift from project-oriented concepts to organizational concepts geared towards continuous process improvement.

9.1 *What is a process?*

Process is one of those words that everyone understands but no two people define the same. It's a word like "thing" — it is used every day to describe a multitude of activities, none of which are alike. This is why "process" improvement is often a difficult concept because it can encompass so much as to become almost meaningless. In dealing with process improvement it first becomes necessary to define exactly what we're talking about when we say "process."

The dictionary is always a good place to look when trying to define a term. The *American Heritage Dictionary* defines *process*, the noun, as follows:

> 1. A system of operations in the production of something. 2. A series of actions, changes, or functions that bring about an end or result. 3. Course or passage of time. 4. Ongoing movement; progression.[44]

These four related definitions gives us at least an idea of what a process is, and is helpful in understanding what a process is to a company. The point

Process Elements

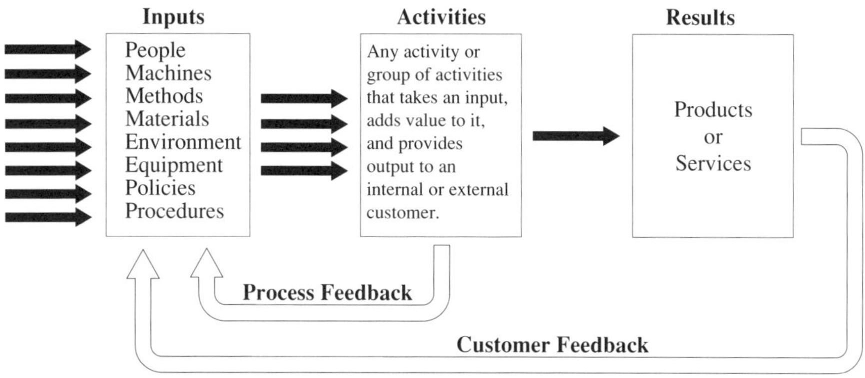

Figure 9-1 The goals of Process Re-engineering are to optimize daily activities to obtain results based on customer needs and satisfaction and to remove causes of problems as a means to improve productivity, efficiency, and quality in the resulting products and services.

of defining the term is to be able to know what you are focusing on in an organization in order to improve it. The authors believe that the above definition contains an important point regarding the definition of concepts: a process is "some thing" that brings about a *result*. This, above anything else, should be paramount to an organization when talking about processes.

The authors have defined a process as having three parts: (1) Inputs, (2) Activities, and (3) Results. Figure 9-1 illustrates how these three elements combine to make a process. But the authors have found that often what people think of as a process is more a description of what they do everyday, rather than the activities or groups of activities that are focused on taking inputs and producing results. Figure 9-1 gives a definition of a process that is functional and allows an organization to focus on what goes into a process and what are the results. The **inputs** are "things" that most of us are familiar with as part of a process: people, machines, methods, etc. The **results** are usually understood as well: products or services (even internal products and services). It is the **activities** that go into the process that are sometimes the most difficult to understand and categorize.

For example, if waxing your car is a process, then it may consist of buying wax and cloths, buying a polisher, parking your car in the garage, applying wax, removing wax, polishing the finish, and taking the car out of the garage. It can be broken down into the three elements as follows:

- **Inputs** — You, the car, materials, polisher
- **Activities** — Buying the materials, buying the polisher, moving the car, applying the wax, removing the wax, polishing the finish
- **Result** — A car with a wax finish

This is a straightforward example. But what if you start to add some other activities? The phone rings while you wax the car and you answer it. It is your cousin across the street. He is calling to tell you that you wax your car too much. In fact, it just so happens that *every* time you wax your car the phone rings and *every* time, it is your cousin telling you that you wax your car too much. The phone call from your cousin delays the result, it upsets you, you miss some of the dried wax, causing you to redo some of your work. You answer the phone every time you wax the car and its always your cousin — Does this mean answering the phone is part of the process of waxing your car? No. Even though you do it every time you wax your car, your cousin's critiques do not add value to the wax job. Perhaps this is an activity that should be removed from the process to bring about improvement. Do not answer the phone; move; get your cousin to move; close the garage door when you wax. There is a multitude of ways to improve the process by taking out the activity that delays the result and does not add value to the product or service.

The goals of process improvement are to take those daily activities that are part of the processes and obtain results based on customer needs. Removing problems, or nonvalue-added activities, in the process will result in improved productivity, efficiency, and quality in the results — products and services. This is why QFD can play an important role in process improvement because the benefits QFD bestows on the product environment also result from the application of QFD to process improvement.

When looking at improving processes, it is also important to understand the process hierarchy from a large process down to small processes and finally down to tasks. Most organizations have internal hierarchies for their processes, whether they recognize them or not. Figure 9-2 shows process hierarchy: Macro-processes are broken down to Sub-processes, which are broken down to Activities, which are broken down to Tasks. Companies can be viewed in this manner to understand what it is that they do on a daily basis to get a product or service out the door.

This can be illustrated by looking at the game of golf (a process many of us are continually trying to improve!). If a round of golf on an 18-hole golf course is looked at as a macro-process, one subprocess might be playing hole number one. The activities involved in playing hole number one would be driving, hitting fairway shots, chipping, and putting. Tasks might be the decomposition of "putting" into its elements of lining up the putt, back swing, and follow through. The result of golf may be entertainment (HA!) or a lower score. We can look at this analogy again from a value-added standpoint. Although you may eat lunch every day and, as part of the golf round, you take time out between the ninth and the tenth hole to have lunch; having lunch is not part of the process. It is an activity that you do every day no matter where you are, but it does not add value to the process of playing golf. When people start to break down processes and examine them more closely, they discover that they do a lot of activities during the day that may not add any value to the process.

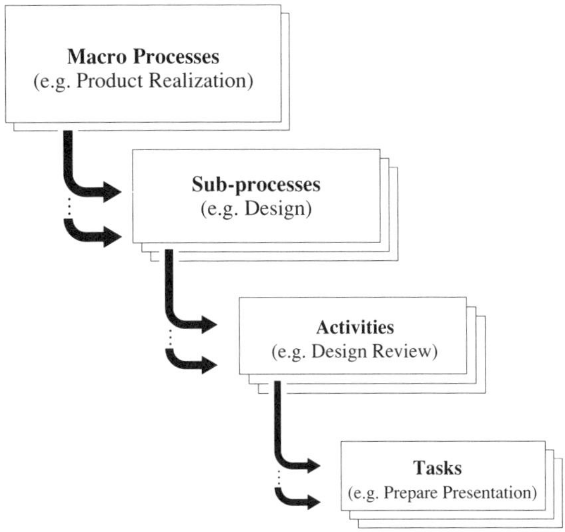

Figure 9-2 An example of process hierarchy.

This is all part of understanding the process; and defining and analyzing the process can be a key component to process improvement.

9.2 The steps to process improvement

The steps taken for process improvement are similar to the steps taken to improve the quality of a product or service. Just as product quality improvement requires planning, prioritization, and execution, so does process improvement require the analysis of the business framework and identification of the critical processes prior to taking action to improve or change the way business is done. The focus remains on the customer because any process or activity should somehow relate to a result that is geared to respond to a customer need.

Process improvement can be broken into five steps or tasks:

Task 1.0 — Business Framework and Establishment of Critical Processes for Improvement
Task 2.0 — Empower Working Teams
Task 3.0 — Identify, Analyze, and Develop Process Alternatives
Task 4.0 — Improved Process Recommendation
Task 5.0 — Implementation and Continuous Improvement

These five tasks describe a plan of action for process improvement and can be applied to any company or organization, no matter the size.

But how does QFD fit into process improvement? Given the fact that the above tasks talk about "critical processes," "analyze and develop process alternatives," and "improved process recommendation," the reader

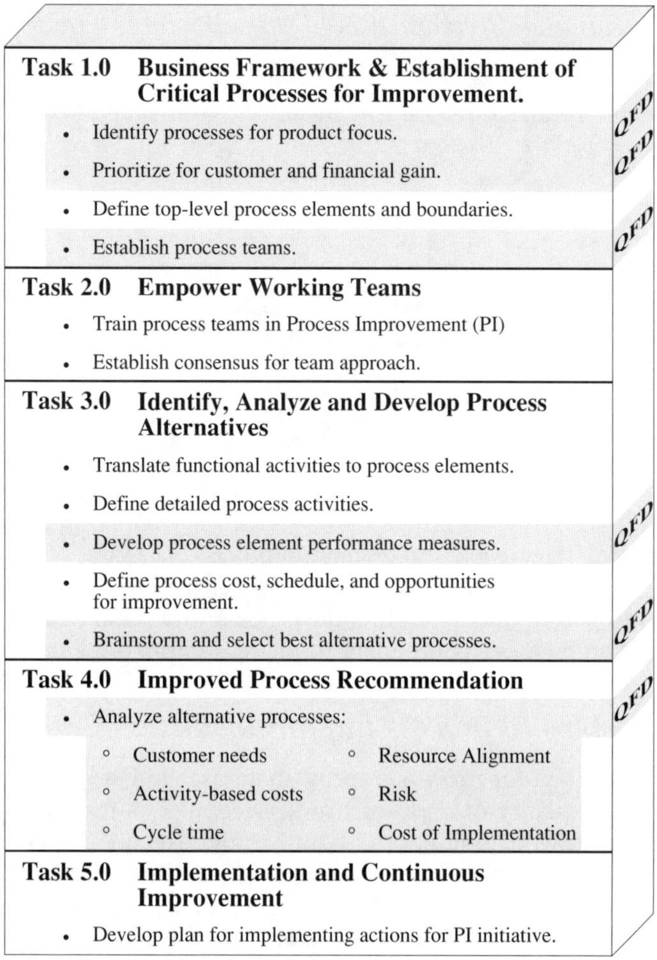

Figure 9-3 The interactions of QFD to support the implementation steps for Process Re-engineering. Highlighted areas indicate where QFD can be effectively applied.

has probably surmised that process improvement relies on prioritization and correlating process elements — meaning that the Core QFD Matrix can play a role in making process improvement work. Figure 9-3 illustrates the activities that comprise the five tasks of process improvement, and defines where QFD comes into play to support the tasks. QFD is used in several areas to brainstorm alternatives and prioritize alternative processes, and components of processes, to achieve the best businesses operations within the company.

In order to understand how process improvement can work in a service-oriented environment, the authors will use a financial services company to illustrate the five tasks.

9.2.1 *Vision and direction: Task 1.0 — Business Framework and Establishment of Critical Processes for Improvement*

When an organization does process improvement it needs to first establish its goals for the process improvement. Will this be a company-wide implementation effort, much like the radical approach advocated by Hammer and Champy, or will it be directed at a few of the key processes of the company — the ones that could have the most impact on the bottom line and therefore should be tackled first? The leadership of the company needs to decide the scale of the process improvement (although once a company begins to look at key processes and begins implementing change, the authors find that the effort snowballs, and leadership usually wants to take the next step and incorporate process improvement company-wide).

After deciding the scope of the implementation, the next action is to identify the business framework for improvement and to identify the company's Support Processes, Planning Processes, and Core Processes. At this level of process definition, the Support Processes are usually functional areas such as accounting, human resources, administrative services, etc. Planning Processes can be the new products and services area, strategic planning, and business planning. Core Processes are usually the products the company produces. These products can be services. Figure 9-4 demonstrates how these top-level core, planning, and support processes interact with the company leadership.

The first place that a form of the Core QFD Matrix can be used in Task 1.0 is in identifying, prioritizing, and aligning company processes, services, and products with customer needs — what QFD does best! Just as in developing a product to meet customer needs, the process planning matrix is the result of brainstorming the customer needs first, and then identifying the planning, support, and core processes that address those customer needs. An example of a matrix for the hypothetical financial services company is given in Figure 9-5. In defining these customer needs and processes a team is usually assembled from the company leadership or upper management. Representatives from marketing, accounting, product planning, human resources, and the other functional and product areas of the company usually comprise the team. The authors have even done this step with the executive management of the company including the president. Because this step involves the company leadership it accomplishes a variety of goals:

- Brainstorming the key customer needs for the company;
- Defining the key processes;
- Educating senior management about QFD and its relationship to prioritization and process improvement;
- Ranking the key customer needs;
- Providing a competitive benchmarking for customer needs and top-level processes; and
- Determining which are the company's key processes that most affect customer needs.

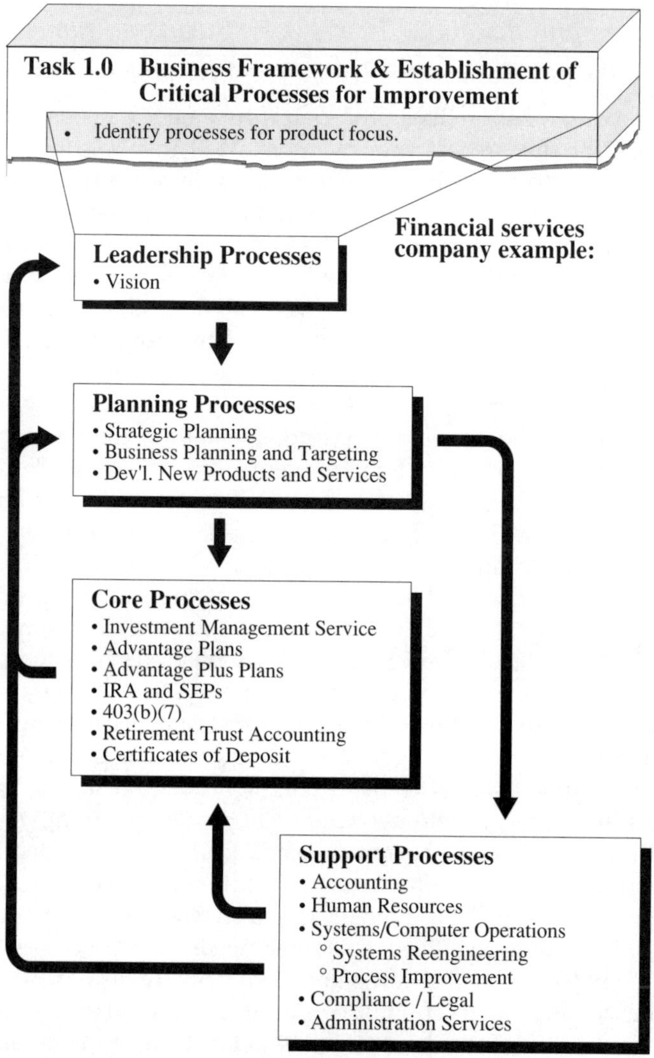

Figure 9-4 An example of the interaction of top-level processes for a financial company.

The above process sensitizes upper management to what truly drives the company and prepares them for what the company will do to improve processes. The steps for completing the process matrix are the same as those described in Chapter 4 for Sections 1 to 7 of the matrix. For example, brainstorming the customer needs relies on information gathered about what the customer wants from the company — marketing surveys, focus groups, etc. Management can also set goals (targets) for the identified processes. (See also Chapter 11 for facilitation guidance.) Having defined the key planning, support, and core processes for the company, management can then make

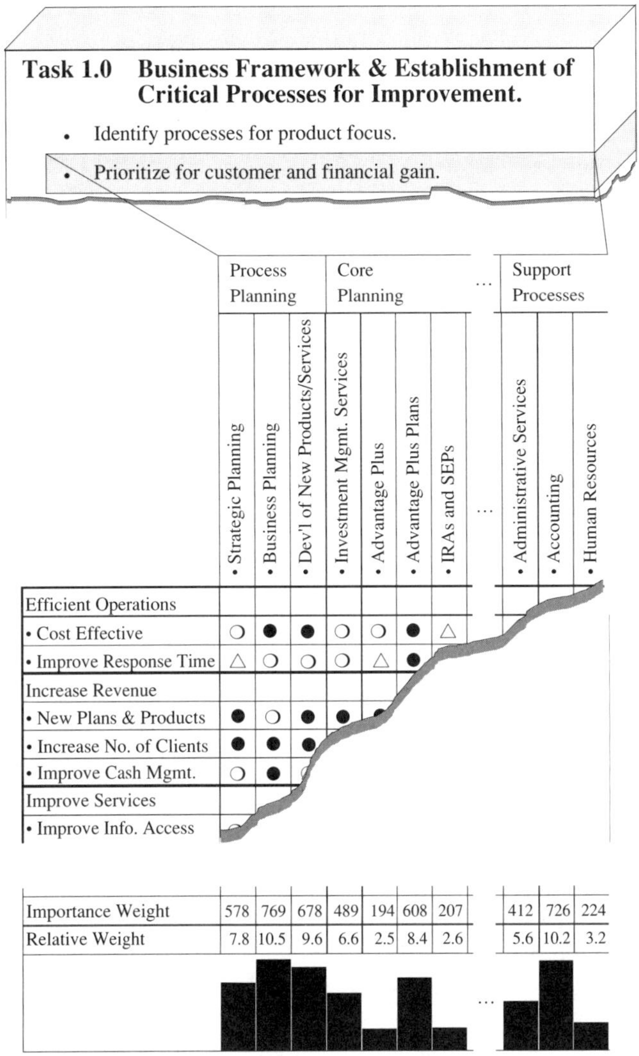

Figure 9-5 QFD provides a framework for prioritizing and aligning company processes, services, and products toward meeting the customer needs.

informed decisions about how to next proceed. Should two of the processes be selected for pilot improvement programs? If only three or four of the processes have the greatest impact on customer needs, should they be the focus of the improvement? Is the company as a whole going to tackle the effort?

Once management has answered these questions, the next step is to define the top-level process elements and boundaries. To do this the team should use the criteria for identifying the three parts of a process defined in Figure 9-1. The inputs, activities, and results (based on customer needs)

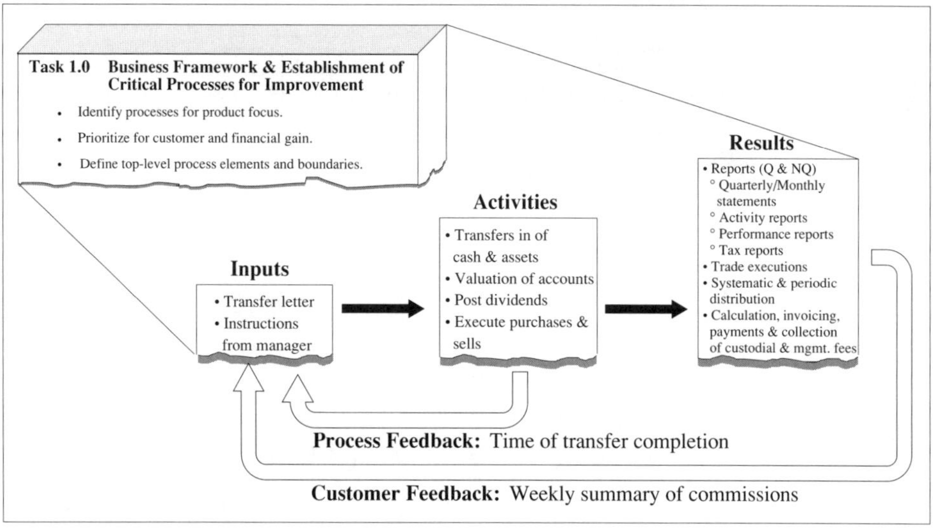

Figure 9-6 An illustration of a Qualified Plan Process Elements for a financial service company. In this illustration, customers include money managers, account owners, and registered representatives.

should be defined for all of the processes that are going to be examined for improvement. Figure 9-6 shows what a process breakdown may look like for the financial services company using Figure 9-1 as a guide. For each process (such as Investment Management Services) the inputs can be defined, the activities listed, and the results of the activities shown. Definition of the process to this extent allows the selection of the process improvement team that will actually begin to look at all of the inputs, activities, and results at a detailed level for improvement.

Now that the process has been defined to this extent, the team selection can be done using a modified Core QFD Matrix. The key needed attributes of the team can be developed with input from human resources, senior management, and management for the identified process area. Based on this criteria the individuals who will participate in the process improvement activities can be compared with the criteria to ensure adequate knowledge and skill to carry out the improvement activities. Figure 9-7 is an example matrix showing how potential team members compare with capability requirements. This is not done to "rank" individuals against each other, but to ensure that the team capabilities are there to address the task.

Approaching process improvement from this direction gives management traceability to the company processes from the strategic-plan level down to the individual activities of each employee. Attacking process improvement from the top down results in a structured improvement, much like using RSM™ as described in Chapter 2. It removes guesswork about which processes to tackle first and focuses the improvement effort on satisfying

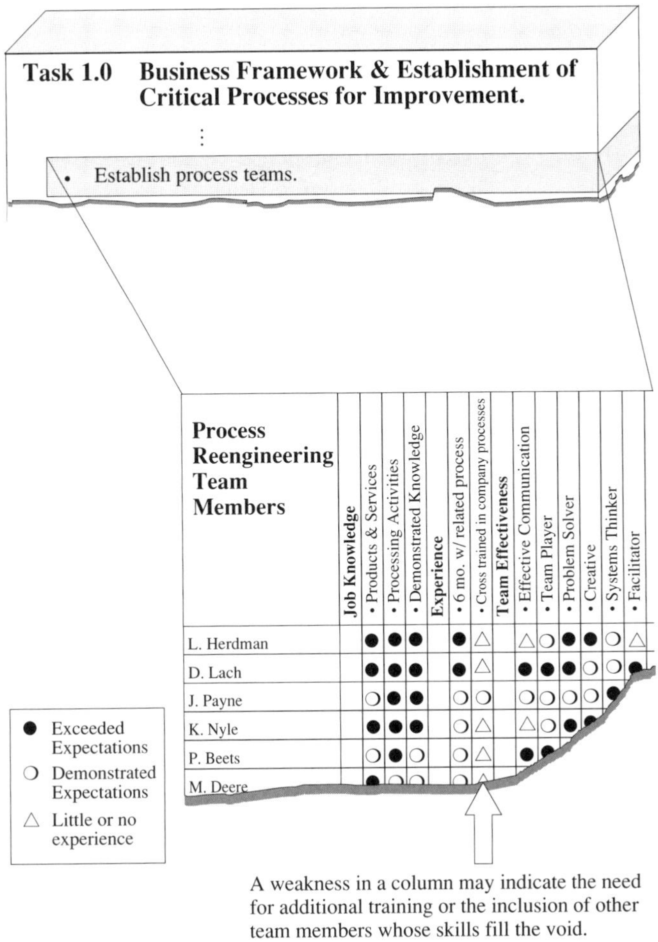

Figure 9-7 QFD can be used to assess individual and team capabilities.

customer needs. Process improvement is not an activity done for its own philosophical reasons, but because the company is trying to be more competitive, cut costs, or improve profit.

Once the process improvement working team or teams have been selected, the next step is to empower the teams.

9.2.2 Assuring success: Task 2.0 — Empowering the Working Teams

The first step in empowering the selected "working teams" is to bring them into the process improvement plan. This involves letting them know the goal of the process improvement, including what management is planning on doing with the results of their efforts. If the goal of the company is truly to downsize to accommodate either a decrease in revenues or to ensure profitability, then the working teams — and the rest of the company — should know this is one of the reasons for the process improvement. If the goals of the process improvement effort are not clearly explained to the teams, not only will they struggle to embrace the tasks, but also rumor and innuendo about team selection will sabotage the process. The working teams should know that their skills are necessary to the survival of the company, and no one will be singled out for layoff as a result of the process improvement. Team members must feel they are empowered to help the company, not to cut fellow workers.

Hopefully, most companies will be using process improvement for increased growth and not downsizing. But even in a more "positive" environment such as growth, the working team members must understand the goals of process improvement. The authors recommend presenting the teams with the findings of management in key process selection. This will bring them into the activities. The working teams should also thoroughly understand how QFD works, either through training or facilitation through the rest of the process improvement. Training in process improvement includes training in functional analysis, as discussed in Chapter 7, and also in techniques such as brainstorming, tree diagramming, affinity diagramming, and any other analytical tools needed for the specific jobs within the defined process.

After the teams have been trained in the basics of process improvement, the next step is for each team to come together and establish a consensus for the team approach. This does not mean how they are going to design their own process improvement approach, but how the team will interact. The teams must know the parameters for involvement and should consider the following criteria:

1. How often will they meet;
2. What is the scope of their involvement (half-time, full-time);
3. What formal communication methods will they use to keep all members fully informed;
4. What will be the timetable for execution of their assigned task;
5. How will they interface with other members of the company; and
6. Do they know the limits of their charter as defined by management.

These are important questions and, just as in any quality effort, the team must know what they are being asked to do in order to be truly empowered. The working teams should have some autonomy over their activities and in making some decisions about the process improvement or they are not truly empowered. However, along with decision-making authority comes responsibility and the working teams should also know what management is expecting in terms of results. If, for example, the process working teams discover ways to cut time for a process or transaction, and save money in the process, they should not be given an arbitrary goal of 30% improvement. Target values are good when developed as part of the QFD analysis because they are backed by data, experience, and knowledge. Mandatory goals without support can easily cause failure based on unrealistic goals. The team should be part of the target and goal creation process, with management's ultimate approval.

The final element of Task 2.0 should be an approval by management of the teams' approach to the process improvement effort. The teams should "own" the effort and make regular reports to executive management so they can get ongoing feedback and feel that management is supporting their efforts.

9.2.3 Tackling the Job: Task 3.0 — Identify, Analyze, and Develop Process Alternatives

Task 3 involves a systematic decomposition of the processes down to the task levels, and puts into action the process depicted in Figure 9-2 above. The working teams will require the most time to complete this task. The first step is to translate the functional activities to process elements. This means taking the process that was selected for improvement, such as Investment Management Services in our financial services example, and decomposing all of the functional activities involved in that process into process elements — input, activity, result (see Figure 9-6).

The next step is to prioritize the activities using the Core QFD Matrix to compare them against customer needs that were developed using the brainstorming technique described earlier. This step, illustrated by example in Figure 9-8, shows how the core process (Investment Management Services) is further decomposed into its subprocesses (reports, trade executions, etc.) and then ranked against customer needs to determine which subprocesses are critical. Competitive benchmarking as described in Chapter 4 is used and target values are developed. At the end of this task the team needs to develop operational improvement performance measures or targets (Section 6 of the Core QFD Matrix). These measures will become the target values for the process improvement. For example, one of the key activities under the Investment Management Services is Reports, and under Reports is the Number of Data Collection Stops. The target value corresponding to the Number of Data Collections Stops is "Reduce by 50%." This is the goal the team has selected and the process improvement techniques will be structured to achieve a 50% fifty percent reduction in data stops for the reports.

After prioritizing the activities associated with the Investment Management Services, the next step would be to decompose these activities further into constituent parts. This includes breaking processes into subprocesses, subprocesses into activities, and activities into tasks. This is really defining the processes completely and understanding how the company functions to complete a top-level process, i.e., provide a service to the customer. All of this is part of defining workflow through the company and identifying the activities that touch the process. The team does this through conversations with company departments to verify accuracy and gather information on the daily

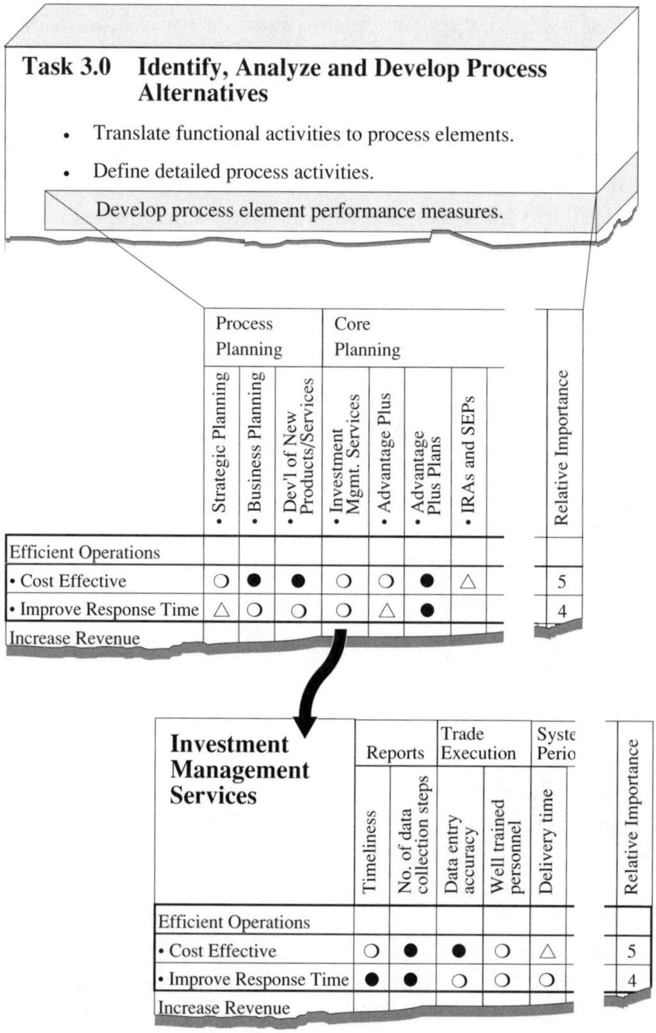

Figure 9-8 Systematic decomposition of corporate processes to tasks using the QFD methodology provides a framework for establishing operational improvement performance measures.

activities for each person coming into contact with the Investment Management Services. Flow charting and fault-tree analysis are the tools used in this stage to define how the current processes operate.

The next step requires setting target goals for each of the elements of the process as illustrated in Figure 9-9. Targets goals represent measureable values for each of the elements such as "Trade Execution." After the target values have been set, process activities need to be brainstormed in order to develop concept alternatives to accomplish the target goals established in Figure 9-9.

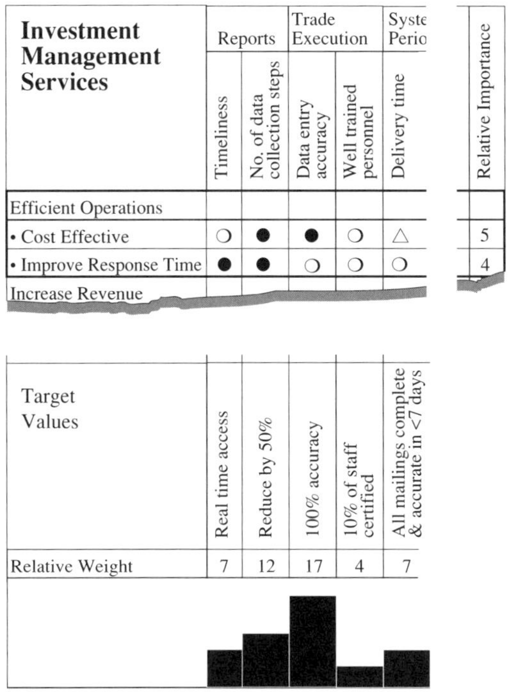

Figure 9-9 Setting target values and determining relative weights of tasks within a process for later evaluation of process concepts.

The process alternatives are developed by using tools such as Functional Analysis to develop the best approach to meet the customer needs. Using the Investment Management Services example, Figure 9-10 illustrates the QFD matrix that prioritizes what processes will have the greatest impact on meeting the target values previously developed. Prioritizing of these targets and process alternatives is critical in order to know what most impacts the core process and what will have the greatest effect on meeting customer needs. Although all processes can be analyzed and improved, organizations need to know which ones to work on first; i.e., which improvements will provide the "biggest bang for the buck." Out of this ranking come the process system alternatives represented by "Option A" and "Option B."

Figure 9-11 illustrates the development of process concept alternatives and ranks each process "option" against the desired targets that have been developed. The team should focus on the option that most fully satisfies the targets (and, therefore, the customer needs), but the team should also consider if the best elements of any of the options can be selected to make a "best practices" solution.

At the end of Task 3.0, the working teams have established the best alternatives to recommend to management and are ready to move to the next stage of process re-engineering.

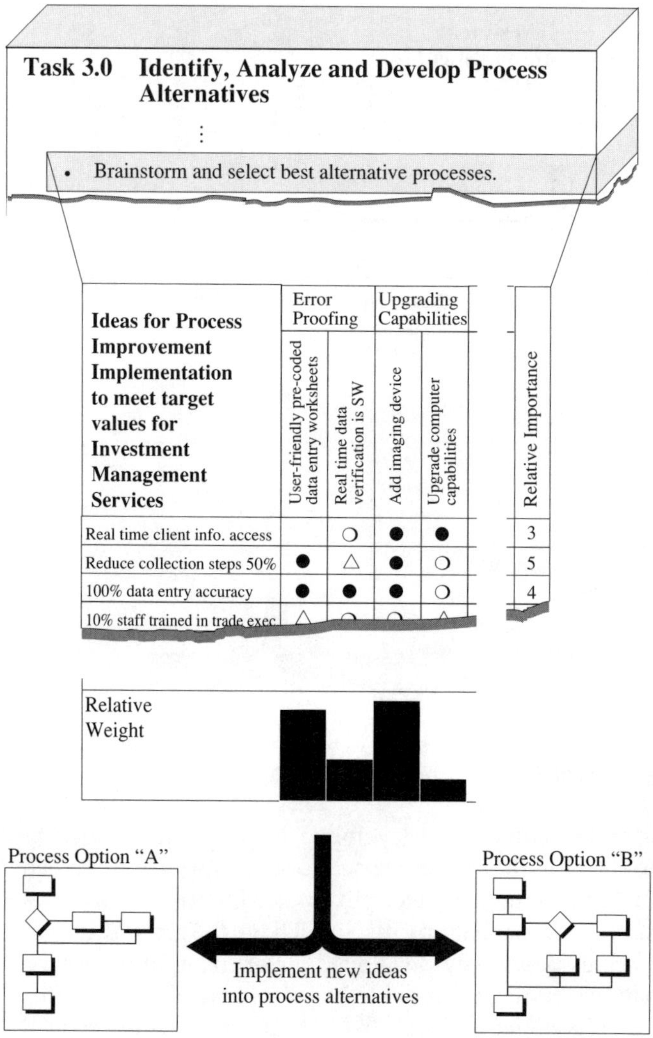

Figure 9-10 QFD provides structure to brainstorming and developing process alternatives.

9.2.4 *Selecting the Path: Task 4.0 — Improved Process Recommendation*

Having developed the recommended processes for analysis, the working teams should look at the following:

- Final improvements;
- Full definition of the recommended process;
- Refinement of cost, time, accuracy, and other performance measures;
- Resource allocation analysis;

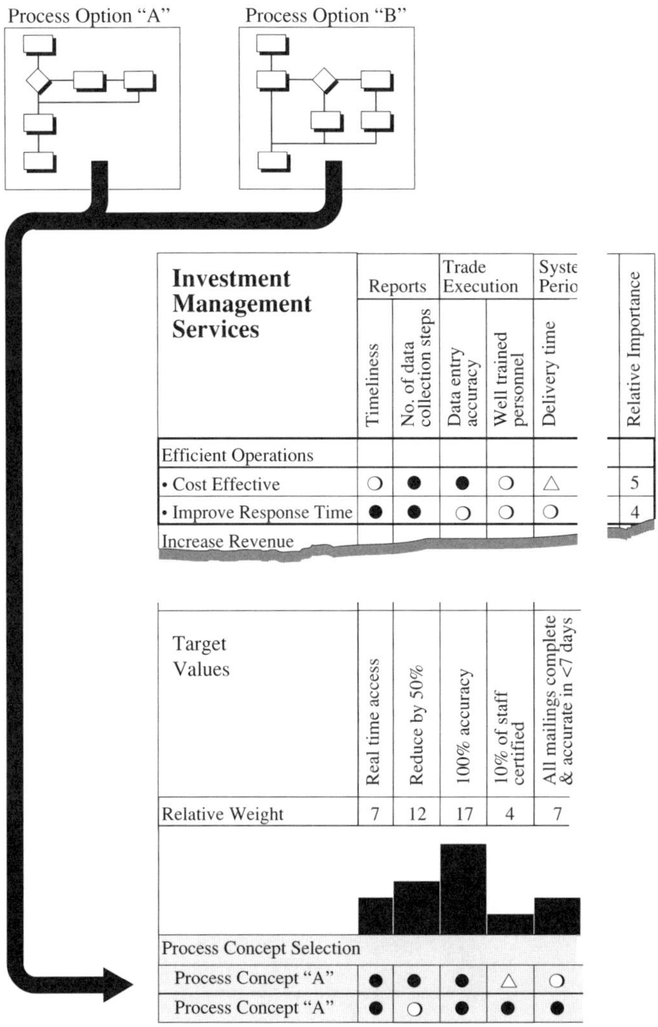

Figure 9-11 QFD provides a methodology to take the brainstormed and developed process alternatives and relate them back to the overall system through the concept evaluation.

- Risk assessments;
- Business operations interface analysis; and
- Development of an implementation strategy.

These steps set the working teams up for true implementation of the process improvements. As Chapter 6 ("Developing a Strategic Plan from the Core QFD Matrix") showed, the steps leading up to implementation are important, but the creation of a strategy for implementation is just as critical.

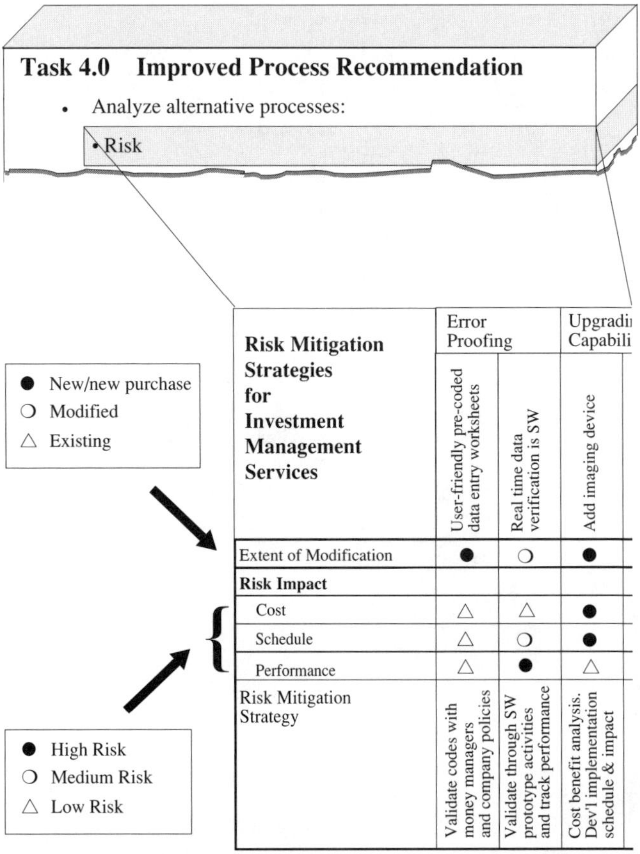

Figure 9-12 An example of a tailored QFD matrix to analyze risk and develop risk-mitigation strategies.

The QFD matrix comes into play in looking at the risks involved in implementing the new processes and in developing risk mitigation strategies. Often companies look at new methods of doing business without examining the potential risks that are inherent for radical or incremental change. Figure 9-12 shows how a tailored Core QFD Matrix can be used to look at the risks associated with process modifications, and the strategy that can be developed to mitigate that risk (shown at the bottom of the figure).

The final step for Task 4.0 is for management to approve the recommended course for improvement, which includes approving resources, approving the integration plan and supporting the working teams and the rest of the company in the process re-engineering.

Although the working teams could stop at this juncture, the management should require one more task from the teams: how to maintain continuous process improvement.

9.2.5 Don't stop now: Task 5.0 — Implementation and Continuous Improvement

Task 5.0 is the full implementation of the process improvement recommendations. The plan for implementation, if created as suggested in Chapter 6, should contain feedback mechanisms for monitoring progress of the effort. This is critical if the organization is to track the process improvement and be able to make real-time changes to the processes during the implementation phase.

Task 5.0 also involves the creation of a plan for implementing "continuous improvement." Emphasis should be placed on the word *continuous*. Processes are often put into place and, after a period of time that can range from several months to many years, those processes become etched in stone as the way to do business. As a result, it is difficult to change the process both from an organizational and from a human interaction perspective. People who have grown up with the process see it as not only a way of doing business, but also integral to their daily activities. Even the "new" processes can become outdated and the organization can slip back into inefficient methods and procedures. Change is constant and the effective organization should be prepared to continue with change as needed.

It is important then, in defining process improvement, that a methodology is followed that can identify weaknesses in the current way of conducting business, without threatening human reliance upon familiar job definitions and requirements. This is where continuous process improvement must be in place.

Continuous process improvement is established by validating the process periodically, defining and illuminating process problems, benchmarking, and providing training in advance on the improved process. These are the feedback mechanisms that allow a company to constantly adjust where necessary to the changes in customer needs and process techniques, including automation. As the company keeps its hand on the pulse of the customer, it should be easy to trace any changes in customer needs through this new internal process structure to determine if any procedures need to be re-evaluated, changed, or scrapped in favor of a new process to meet the customer needs.

9.3 QFD focuses process improvement

The above represents how QFD is involved in the five tasks and the steps involved in each of those tasks for process innovation. Process improvement or process re-engineering can be done without using QFD, but QFD analysis allows the organization to quickly and methodically analyze itself to obtain process improvement where it is needed. Whether QFD is used in the design of a new product, or in undertaking process improvement, process re-engineering, or business re-engineering, an organization wants to know the straightest line to results. Effective use of QFD prevents that line from zig-zagging and gets the organization to the mark.

9.4 Chapter summary

1. QFD works well in process improvement whether on a grand scale such as business re-engineering, or in pilot programs.
2. A process has three parts: (1) Inputs; (2) Activities; and (3) Results. Processes also can be arranged in a hierarchy: Macro-processes are broken down to Subprocesses, Subprocesses to Activities and Activities to Tasks.
3. The steps to process improvement are

   Task 1.0 — Business framework and establishment of critical processes for improvement;
   Task 2.0 — Empower working teams;
   Task 3.0 — Identify, analyze, and develop process alternatives;
   Task 4.0 — Improved process/Recommendation; and
   Task 5.0 — Implementation and continuous improvement.
4. QFD is involved heavily in process improvement to focus on the critical processes, trace requirements in the steps, and create an integrated plan.

chapter ten

Management and participant perspectives

The man who travels alone can start today, but he who travels with others must wait until they are ready.
Henry David Thoreau

10.0 Understanding the players and the agendas

Many management consultants and pundits talk about "managing change" when dealing with almost any effort to rethink how a corporation operates or how to employ techniques that involve a fundamental departure from the way the organization currently operates. "Change management" is a term that in itself has become somewhat mutable in meaning. The authors see QFD as a method in itself that involves "change management" because the fundamentals of any QFD approach require a new way for the organization to deal with customers, designs, services, and management. Therefore, having introduced the concept of QFD the authors believe it is important to talk about managing change not in terms of abstract concepts but rather in the context of implementing QFD. It is important to understand what environment QFD should operate in prior to discussing the facilitation guidance for the method itself. Unless an organization understands its own environment, and how it may be thinking about changing the roles and agendas of management, it is difficult to design an effective QFD implementation program.

Every organization has some basic management structure that is easily recognizable by the management organizational chart or by its basic construct for decision making. For example, nearly every organization has a single leader or a board that acts as the ultimate decision-making tool of the company; some sort of departmental or top-level management structure for disseminating decisions into the company (which may also include a middle management depending on the size of the company or organization); and then those individuals who are often classified as employees or workers who carry out the tasks of management and are chiefly responsible for producing the organization's service or product. Based upon these considerations most organizations appear to have similar characteristics that one would believe could be easily used in setting up a change management vehicle for incorporation of ideas such as Quality Function Deployment.

However, anyone who has worked in more than a single organization, or any management consultant who has visited a multitude of organizations, knows that similarity between management structures in corporations ends with what might appear on paper. Every organization has its own history, cultural dynamics, visible and hidden decision-making processes, and philosophies about the interaction of its members. Therefore, to impose any single philosophy of change upon an organization would not only defy the basic premise that change must be flexible and adaptable, but would succeed only in a limited number of cases.

Based upon the above premises, the authors believe that QFD should be viewed as somewhat of a chameleon. It is a mutable philosophy that can be tailored for the individual needs of any organization. The authors believe that the approaches taken in this book will be useful for a vast majority of organizations attempting to apply them. However, no organization should feel so constrained as to not alter the QFD tool with its own twist, spin, or interpretation that would work best within the management confines of the particular application. Although it would seem somewhat contradictory to now talk about roles and responsibilities when implementing the QFD analysis, the authors are merely putting forth references and information that they have come across in their own consulting experience for further consideration. It is hoped that the reader will recognize certain scenarios or concepts that may apply to his or her own organization and think about methods to prevent traveling down an unproductive path or head-off unnecessary expenditures, unnecessary time delays, and unnecessary organizational frustrations. Having stated this disclaimer of sorts, the following represents suggestions on the basic perspective that should be taken by the groups associated with the implementation of QFD.

10.1 *Executive management perspective*

Given sufficient time, the executive management of any organization can research the annals of business literature, both recent and probably ancient, and find some reference to support any management style that they so desire. For example, with specific regard to current trends, Michael Hammer recommends that executive management adopt the philosophy of "obliteration" when dealing with radical changeover (re-engineering) of a corporation.[45] Hammer and Champy say:

> Business re-engineering means putting aside much of the received wisdom of two hundred years of industrial management. It means forgetting how work was done in the age of the mass market and deciding how it can best be done now. In business re-engineering, old job

> titles and old organizational arrangements — departments, divisions, groups, and so on — cease to matter. They are artifacts of another age. What matters in re-engineering is how we want to organize work today, given the demands of today's markets and the power of today's technologies. How people and companies did things yesterday does not matter to the business reengineer.[45]

Hammer and Champy advocate a "ground-up" approach to management and the way that corporate America does business. Change is radical in the Hammer/Champy school of management and there are many followers to sing the praises of doing business in this "new" way that advocates throwing everything out and starting over. The authors have shown how QFD can fit into a business re-engineering effort (Chapter 9) and enhance the gains that are sought by that method.

Advocates of Total Quality Management, near the other extreme, recommend that executive management take a "teaming" approach with its organization and employees when implementing a change agenda. This approach is geared more towards changing the culture first and then the business operations. It is a more gradual approach that denies the "bet the company" approach of Hammer and Champy. QFD can also produce solid gains in any quality program, whether called TQM, TQC, or any other acronym for a system based on Deming and Juran.

Robert M. Tomasko, in *Rethinking the Corporation*, takes a look at reshaping management from an "architectural" viewpoint:

> In a number of ways, the field of architecture provides a good source of perspective an the task of reorganization. The architect provides a master plan for each project and then coordinates the work of many skilled trades and disciplines. While there is room in the design process for art and creativity, few architects have the freedom to start with a clean slate and impose their design on the environment . . . Certainly, I am not suggesting that the job of corporate reorganization be contracted out to a nearby architect . . . But there is something to be gained by letting the design mentality of the architect guide, at least metaphorically, the work of the executive or manager as he or she reshapes the corporation.[47]

Tomasko takes a more structured approach to management than Hammer, but is still advocating a major change in the way that business should be managed. QFD can also be part of the "architecture" of change for an organization, from a small program to the entire company.

None of these approaches should be thought of as "right" or "wrong." Value judgments are often superfluous when talking about how the executive management of a company goes about running day-to-day operations or planning for change.

The important consideration for the reader of this book is that QFD can be embraced by any of the overriding philosophies of management espoused above. The authors have seen implementation of QFD work to some degree with almost any management style as long as certain steps are taken. Management styles and business approaches are important in running a business effectively. QFD is part of any new management organization at some point, because it, too, addresses change. However, QFD gives immediate results and helps an organization align for success within the bounds of whatever management course is taken.

In order for a QFD to work at an organizational level, there must be support from the executive management. But what is support? Table 10-1 lists some elements that a successful executive management team can use to support the implementation of QFD. Many of these ideas seem to be self-explanatory or common sense. However, the authors have discovered that there is a great deal of confusion that exists in the work place when discussing terminology. The executive management individual or team should take a leadership role and define the purpose of the QFD implementation, without mandating participation. Volunteers work better than prisoners.

Management should also be as up-front as possible with the purpose of the QFD implementation and be willing to accept recommendations and outcomes that may result from the QFD process. There has been a lot of lip service given to the concept of "empowerment" in both organizations and in the popular press, but many executive management teams think by merely stating that the employees are now "empowered" without giving parameters, definitions, and goals means that empowerment will somehow sprout like new grass shoots in the spring. Empowerment is a two-way street. Management must understand what power they are asking employees and middle management to accept and utilize, and those employees and middle-management personnel must understand that with power goes *responsibility*.

With regard to the actual implementation of QFD, executive management should make a decision as to whether or not they are going to participate directly in the QFD, or let the project team carry the day and report their progress back to management. Sometimes direct participation of executive management stifles the creativity and free-form association necessary for a QFD to succeed — especially when executive management is accustomed to making decisions and can unduly influence the outcome of a QFD effort.

In one financial corporation, a QFD pilot program involved two teams working on improving two critical processes. The teams presented their findings to the president and executive staff showing how employees would be taking greater responsibility for ensuring execution of transactions, thereby decreasing or eliminating the need for management to perform checking or monitoring. The president's response was to not only support the initiative,

Table 10-1 Management's Role in Implementing a QFD Project

- Act as Team Consultant.
- Define Purpose of the QFD Project.
- Define expected results.
  - Deliver products
  - Describe breadth and depth of required information
  - Describe how results will be used
- Define Project Scope.
  - Boundary conditions
  - Ground rules and assumptions
  - Constraints
- Provide guidelines for schedule and team participation.
- Support team formation and selection of facilitator.
- Provide adequate resources.
- Attend Kick-off Session.
  - Affirm management commitment and on-going support
  - Commit to attend informal sessions as requested by the team
- Schedule and attend periodic review.
  - Validity check during development
  - Understand team status and provide additional resources as needed
- Support implementation of recommended solutions.

but to inform the teams that this would free management for other responsibilities — essentially, agreeing that abdicating the power to employees for self-monitoring made management more powerful, not less.

10.2 *Management perspective*

By "management" perspective the authors are referring to that layer of management that is responsible for many of the day-to-day operations of a company, project, or program. These managers are not responsible for the ultimate decisions made on a global basis for the organization, but they *are* key players in communicating the executive decisions to the rest of the company. This encompasses a wide variety of management responsibilities and tasks, but the authors have found that the role of management is continually changing. This also applies when attempting to implement QFD at any level in an organization. Today, management function is evolving from traditional responsibilities of planning, organizing, coordinating, and controlling to three major categories: interpersonal, informational, and decisional

Table 10-2 Observations from Middle-Managers during QFD Sessions

Areas of Concern

- As a manager, it was difficult not to dominate, interfere or lead discussion.
- Adequate time should be provided to prevent being driven by the schedule.
- Best results are obtained when all team members fully participate.
- As the team begins to own the program, managers may feel a loss of control.
- Time is required to overcome a sense of vulnerability by all team members.
- Good listening skills are key to success.

Positive Results

- Team members became empowered accepting more authority and responsibility for program results.
- Project team acquired a greater understanding of the critical issues and captured the knowledge to share with future team members.
- Technology transfer was driven more by communication than data packages.
- Iterative activities were refinements rather than repeats, they were connected iterations with focus and purpose.
- What started out as an act of will ended up as a scientific process, an organized integrated process that realized all resource groups.
- Created channels for communications between team members and customers.
- Most managers could predict 40 to 50% of the results without the help of the team. By allowing team participation, they get 100% of the project scope and an enthusiastic, committed team.

roles. The interpersonal role requires the manager to act as a figurehead, a leader, and a liaison. The informational role requires the manager to monitor and disseminate pertinent information and act as a spokesman for the multifunctional teams that may be involved in a QFD. Finally, the decisional role requires the manager to act as an entrepreneur, a mediator, and a negotiator with other departments and often executive management itself. Table 10-2 includes some of the perspectives that might be adopted by middle management when engaging in a QFD.

Many of the same elements that executive management should be considering are also part of the perspective middle management should take when implementing the QFD. For example, management should be focused on the purpose and outcome for the QFD. This concept is often overlooked and teams sometimes plunge into QFDs without proper introduction or focus as to what the outcome of the QFD will be. This can lead to unnecessary detailing of the QFD process; and, more commonly, bogging down the details of the QFD, which sometimes results in an abandonment of the

Table 10-3 Example Questions That Can Be Used to Facilitate the Development of a QFD Project Scope

Project Objectives:

- What is the purpose of the QFD Project?
- What are the expected results?
- What are the expected deliverables?
- How will the resulting information be used?
- Who will use the results and when will they be used?

Operating Considerations:

- What marketing information is available to team members?
- Should a competitive or market analysis be done in conjunction
- with the QFD activity?
- Who will coordinate customer interfaces and information?
- Are there any industrial trends or technology information that would support the team AFD analysis?
- Are there any internal activities that the team should coordinate?
- Are there any resource constraints such as budget, schedule or critical personnel?
- What mechanism should the team pursue if conflicts in necessary resources or issues arise?

Resources:

Personnel

- What departments need representation to fully analysie the QFD issues?
- What team members should participate, full or part time?
- Do these team members have the adequate marketing, research, operations, manufacturing, test and distribution knowledge to complete the analysis?
- Who should facilitate this team?
- Is training required prior to participation?

Information

- Is the following information available for team review?
  - Competitive Analysis
  - Additional customer information such as customer complaints, surveys, or focus group results
  - Product and service strengths and weaknesses
  - Applicable technologies
  - List of appropriate customer constraints
  - Cost data

Equipment

- What location and space is available for team coordination? Are computer resources needed and available?

Short Term Project Objectives:

- How will team success be measured?

project. Table 10-3 contains a list of questions that can be asked to help define the purpose of the QFD, the direction taken and the ultimate outcome expected.

10.3 Team perspective

Upon selection of the QFD team (Chapter 11 contains facilitation suggestions for team selection), management must endow the team with purpose, responsibility, and focus. Putting that collection of individuals together in a room does not make them a team. Ample time must be given to the team to work through the dynamics of the QFD, understand their roles and responsibilities, and the anticipated time commitment that will be necessary. In some organizations the team is committed to the project or the task 100% and is relieved of any additional duties aside form completing the QFD and working out the problems associated with the program. In other situations, especially when organizations are smaller or the team is working on process improvement and has been selected because of their expertise in a specific area, the team is only partially committed to the QFD or process improvement and has additional tasks to complete as part of their daily responsibilities. Whatever the case is in the particular application of the QFD analysis, this time commitment must be clearly understood by all members of the team.

The authors have discovered that teams more quickly gel into functional organizations if certain conventions are followed. Those conventions are briefly outlined in Table 10-4. As the reader can see, many of these conventions are straight forward, common sense suggestions that can be easily implemented without a cost or resource commitment that burdens the company. However, their benefit in holding the team together as a unit cannot be underestimated. For example, the creation of a "war room" or team location where the team can meet, discuss elements, and schedule tests is an effective tool to facilitate team unity and progress. The authors have found in their experience that when the team does not have a specific location to conduct their activity, they can often be distracted, feel isolated, or fail to communicate with all the other team members. Again, this is a small consideration but reaps large benefits in terms of team cohesiveness and progress.

There is no magic potion that makes a group of individuals a "team." No spell can be cast to create harmony and openness among team members. Building a team is hard work and the team must come to rely on its members. Teams do not materialize after a single meeting and teamwork is usually learned, not innate. The important consideration for the team is to be focused of purpose and willing to learn. Of all the responsibilities a team has, one that should be of paramount importance is for all team members to respect the opinions of each other team member.

One final rule comes from *Everything I Needed To Know I Learned In Kindergarten*, by Robert Fulgram — **No hitting**.

Table 10-4 Suggestions to Support Team Development and Focus

- Create a "War Room" — A common location where information can be posted and exchanged with a telephone, writing pads, overhead projector, copy machines and computer system for real-time documentation.
- Provide each team member with documentation defining QFD project scope, expected results and other pertinent product and process information.
- During the initial meeting define roles and responsibilities and have all team members confirm time commitments.
- Provide telephone lists of all team members and necessary customer contacts.
- Provide a mutually agreed upon schedule, listing team meetings and locations.
- If possible schedule meetings at the same place and times during the day to avoid confusion.
- Find common locations for information exchange or coordinate electronic mailing.
- Provide snacks when possible.
- Use a lot of humor during sessions to keep motivation and team enthusiasm high.
  - May want team members to take turns kicking-off each meeting with a funny comic or joke.
- Allow team members to take turns facilitating. It will make them all better participants.
- Prior to each session, spend time discussing any concerns and potential remedies.
- Use action lists to document progress in resolving conflicts.
- Give members adequate time to develop, analyse and refine information prior to meetings so that the meeting time is focused on sharing information and reaching consensus versus arguing over unsubstantiated hearsay.

10.4 Facilitator perspective

There must be an individual who facilitates the creation of the QFD matrix, and who is aware of the executive perspective, management perspective, and team perspective as outlined above. The authors have found that time after time a QFD has failed because there is no clear individual who is either acting as a catalyst to complete the QFD analysis or as a team leader to focus the efforts of the individuals. This is not to say that in all cases a professional facilitator is necessary in order to have a successful QFD analysis. Rather, the authors have seen individuals emerge from a team who often fill the role of facilitator, or catalyst, without dominating the group. Sometimes these individuals are not the natural "leaders" of the organization but rather possess negotiation and management skills previously undetected. Table 10-5 illustrates characteristics that are essential to the facilitator perspective for a successful QFD analysis.

Table 10-5 Expected Capabilities of a Team Facilitator

Manage

- Define, coordinate, schedule and track project.
- Run effective meetings.
- Integrate various pieces of information into an integrated approach or thought
  (Conceptual or Systems Thinker)
- Select and tailor the matricies to maximize productivity and documented knowledge.

Instruct

- Educate and inspire team members on the QFD process.
- Clearly communicate the project need, scope and necessary activities.
- Develop and train team in group dynamics and decision making.

Motivate

- Translate their passion and vision for change into action.
- Build trust in a team environment.
  Provide rewards for group progress.
- Provide through visibility to management and company personnel.
- Develop creative activities and new approaches to keep sessions interesting and motivating.
- Lead with charisma and style.

Negotiate and Arbitrate

- Monitor group participation and behavior.
- Balance team participation and prevent domineering members from biasing the outcome.
- Identify and settle disputes to ensure problems are being solved and people stay committed to the process.

Companies will sometimes use both external and internal facilitators when conducting QFD analyses. A methodology the authors have seen that is very effective is to utilize the services of an outside facilitator to initially take the team through the first level of QFD and to ensure that the group stays focused on its task and the expected outcome of the QFD. From there, individuals on the team can continue the efforts required to complete the task, of which the QFD may only be a part. In this way, the team ultimately takes ownership of the success of the program, but has been given a "jump start" by the introduction of an outside facilitator.

The facilitator of a QFD mirrors more closely the role of a judge or a mediator with occasional digressions into the realm of cheerleading. A facilitator cannot be an advocate of a particular design or service approach and must ensure that the team is properly utilizing the QFD tools to maximize benefit and minimize bias. Facilitators often have to table conflicts and are responsible for keeping the team moving toward its goal. The authors cannot emphasize this enough. Often facilitators, especially those internal to the organization, can become as bogged down as the team itself in terms of failure to resolve conflicts or keeping the team moving through the various elements of the QFD matrix.

The benefits of having a facilitator who has lead several teams in a QFD analysis is that the facilitator then knows both the whole process and the stages that a team goes through in a QFD. Those who have never gone through a QFD before often do not have the perspective of seeing the entire process work. This makes it difficult to instill confidence in the team to trust the decision making inherent in the QFD process. Additional facilitation guidance is given in Chapter 11.

10.5 QFD is not an excuse for abdication of management responsibility

The reader should keep in mind that although QFD can fit into many management styles, it does not relieve management of its primary role: to manage. Management controls the way that business is done in the organization, and no customer-focusing methodology can replace the responsibility of management to still control the operation. QFD produces results when used properly, and management should always know the extent to which they are willing to apply QFD or methods such as RSM™, and know their own limits for change. QFD is not immune to sabotage. In one company doing a QFD, a key employee was extremely upset with management and withheld key design information from the team to "teach management a lesson." The QFD produced some results, but the animosity generated by this employee made it difficult to continue the course dictated by the QFD prioritization. Unfortunately, there was a deeper problem that needed to be dealt with at this organization than meeting the customer needs.

Ultimately, when engaging in a QFD effort, management may want to keep in mind what one of our most honored forefathers said about change and benefit: "No more good must be attempted than the people can bear." — Thomas Jefferson.

10.6 Chapter summary

1. QFD can be thought of as a "change management" methology because the fundamentals of any QFD approach require a new way for the organization to deal with customer, designs, services, and management.
2. QFD can be tailored to the specific needs of the organization.
3. QFD can be implemented to some degree in many management styles as long as certain steps are taken:
   - Support from the executive management;
   - Honesty about the purpose of using QFD;
   - True empowerment must be practiced; and
   - Management must make the decision early whether to be directly or indirectly involved with the QFD implementation.

4. Management must act as leaders and liaisons. Managers must take more of an informational role, act as entrepreneurs, mediators and negotiators in the QFD implementation.
5. Teams do not "just happen" but must be put to work solving problems together in order to function and exchange information. Team behavior does not happen overnight.
6. There are many characteristics needed for facilitating a successful QFD — a good sense of humor and being a catalyst are two critical characteristics.
7. Ultimately, QFD is *not* a management style.

chapter eleven

Facilitation guidance

Tell me and I forget
Show me and I remember
Involve me and I understand.
Confucius

11.0 Putting QFD to work

As a friend of ours once said regarding the difference between theory and practice: "It is one thing to visualize having a Mercedes; it is quite another to actually have the keys!" The same analogy applies to reading about QFD and actually applying it successfully. When used correctly, QFD is a powerful tool that can utilize the strengths of a team and save enormous amounts of time, effort, and money. However, the authors have found that when failure of a project is blamed on a QFD analysis, or when QFD is abandoned by an organization or a team, the cause is not that QFD is a flawed analysis procedure. The cause is usually a failure of the organization to support the process effectively, or the failure of a facilitator to guide the team to conclusion.

This chapter discusses how to best facilitate a QFD analysis or program, based upon the application of the Integrated QFD Approach™ and RSM™, and gives guidance on pitfalls to avoid and techniques that the authors have found useful in designing and facilitating a successful QFD analysis. By no means are we implying that a single "right" application of QFD and its facilitation exists; rather, the authors believe that a planned approach can be used to help resolve some of the problems that recur in facilitation of both single QFD efforts and planned organizational implementation of QFD, and that this chapter contains information that is useful no matter what QFD approach is being used.

11.1 Roles and responsibilities: planning for success

Chapter 10 discussed the perspectives of the groups that are involved in a QFD effort, and it is important to understand the dynamics of the skill groups as well as the individuals involved in a QFD. As a facilitator, either in-house or as a consultant, there is often the opportunity to help design the

Table 11-1 Questions to Facilitator Should Ask to Determine Team Direction and Establish Project Expectations

Objectives and Goals

- What is the purpose of the QFD exercise?
- What is the expected results?
- How will the resulting data be used?

Customer

- Who are the primary and secondary external customers of the product or service?
- Who are the primary and secondary internal customers of the product or service?
- Who are the driving customers and why?
- Who is the end user of your product or service?

Project Scope

- What is the scope of the QFD analysis?
    - What is included?
    - What is excluded?
- What are the ground rules and assumptions?
- What are the constraints?
- What type of information and level of detail is needed to accomplish the expected results?
- What elements of the integrated QFD Approach need to be accomplished and by when?
- Should any of the charts be tailored to be integrated into other documents?

Information

- Is the following information available to team members:
    - Marketing information?
    - Competitive analysis?
    - New technology?
    - Cost data?
    - Alternative concepts?
    - Critical issues?
    - Potential failures?
    - Product strengths?
- Who will coordinate information?
- Who will gather information?
- Who should be contacted and in what manner to gather additional information or validate customer perspective?

Team Members

- Can the current team members adequately address all elements of the integrated QFD approach to be accomplished?
- Who else is needed and when?

Schedule

- When are customers available for participation?
- What activities will be accomplished prior to their participation?
- What are the best schedule alternatives for team members?
- What is best schedule for management reviews?

effort from the start, which includes selecting the teams and the project. To do this, a team facilitator needs to know the scope of the project. Table 11-1 lists some questions that the facilitator should ask, both of the management and the team, to determine the direction of the QFD and in order to begin to set expectations. Reasonable results from the QFD should be discussed in advance. For example, if the QFD is being done on a new product concept, the results or expectations might include:

- Discovery of market niche;
- Determination of product direction;
- Discovery of any potential technical gaps;
- Creation of a working team;
- Mapping of an integrated plan for next phase of development; and
- Recommendations for go/no go decision.

Although these results or expectations may seem straightforward, remember that the success of a QFD effort requires all of the participants to be looking at the same map and agreeing on the same basic destination. Writing the team goal or expected result down eliminates misunderstandings and prevents creation of unrealizable results. If the executive management is looking for a go/no go decision from the team, while the team leader or facilitator knows that the initial QFD can only identify data gaps and create a plan for gathering more information to later make a go/no go decision, then the QFD may not "succeed" because of the nonaligned expectations. The facilitator must know what to facilitate toward as well as who to facilitate.

Facilitators must also understand the time constraints. Moving a team through a QFD on a single product at a concept level requires less time than taking a group through a complicated 50 × 50 QFD matrix and developing an integrated plan for all subsystems. Make sure that all participants, and management, know the time constraints and set the goal within the time allowed. Along the same line of logic, realistic goals must be set for the team. The authors once heard a motivational speaker say that all goals should be set higher than what the team believes can be accomplished because workers typically underestimate their capability and are not used to working to their maximum efficiency. While sounding lofty and idealistic, the authors believe that following such wisdom all the time is tantamount to planning for failure. A QFD team should know what is the goal and what is a realistic schedule for completion. The authors have found that most people *do* work hard under the proper circumstances and are in a position to properly evaluate the time necessary to do their jobs.

11.2 Roles and responsibilities: selecting the team

Sometimes a facilitator may be asked to help select the team, especially in a pilot program (how to construct a Pilot Program is discussed in more detail in Chapter 12). Selecting a team can influence the success of a QFD

Table 11-2 The Three Types of Individuals in a QFD Session

The Prisoner

A person who does not want to be involved and does not expect added value through participation. There is no commitment and usually resentment when requesting individual to participate.

The Vacationer

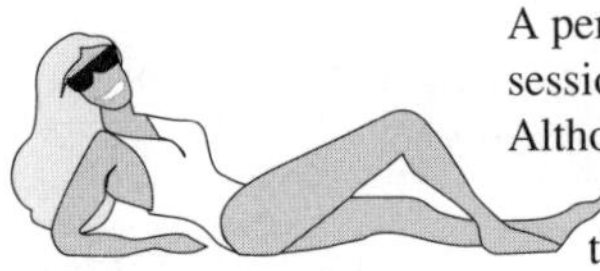

A person who would rather be in the QFD session than at their desk performing other tasks. Although they enjoy the participation they are not necessarily committed to implementing the results and therefore their input is suspect.

The Volunteer

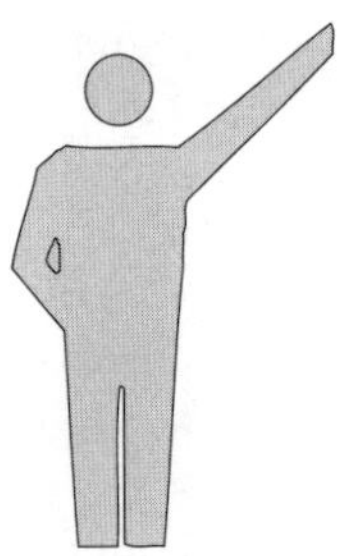

A person who understands the benefit of establishing project focus on customer needs and methodically questions each resulting step in the process. A willing participant.

analysis. Teams will have two components: technical skills and personal skills. A facilitator must balance these two elements in any team, and find ways to maximize the potential for success. Is this "stacking the deck?" *You bet!* In today's competitive environment any organization that does not take advantage of its members' skills puts itself at a disadvantage. The goal of facilitating a QFD is not to make it an exercise in egalitarian participation but to accomplish a specific goal in a specific amount of time that supports the organization. Not all people are good team players, and the facilitator should recognize that a dynamic team focused on success will be easier to facilitate than one composed of individuals not intent on working together.

In terms of personal skills, a team can be composed of three general types of individuals: Volunteers, Vacationers, and Prisoners. Table 11-2 illustrates the difference between these three groups. Obviously, a group of "Volunteers" would comprise the ideal team, but this is not always possible to achieve. Even people who "volunteer" for a team may be there to get away from daily tasks or be there at the "request" of a supervisor. To classify a team in these three categories may be somewhat simplistic, but the point is that the facilitator is better able to move the team if he or she knows the motivation for participation of individual team members. It does not hurt to even ask each member the motivation for their participation.

The second element of a team is technical ability or skill level. If the project has been properly scoped and the goals or results defined, then the team leader and facilitator can determine what skill mix is necessary to address the technical issues of the project. It is easier to evaluate technical skills than personal skills, but one method that has worked for the authors is to utilize the Human Resources department (if the client has one) and the team technical lead to determine a ranking for evaluating a selected team mix for the technical and personal attributes to resolve the tasks that will be generated from the QFD. Figure 11-1 shows how a modified Core QFD Matrix can be used to determine team attributes and individual skills in relationship to the needed skills for project evaluation and success.

11.2.1 Involving the customer early

Customer involvement at the inception of a QFD can also prove rewarding and cut development time by involving the customer in the design process. But the facilitator should be tuned into the purpose of the QFD and recognize the dynamics of the process before inviting the customer to participate. The process of building a Core QFD Matrix can expose the customer to internal conflicts, technical weaknesses (as well as strengths), and the team building process which can, at times, look chaotic. A middle-ground approach to involving the customer in the project directly can be to conduct the initial QFD with the in-house team first to expose the team to the process and iron-out some of the roughness of the process. Then the customer can be brought in to "review" the initial matrix and give feedback on whether or not the team has properly identified customer attributes of the product or process. This is a safer approach to customer involvement but still reaps the benefits of direct customer feedback. Table 11-3 lists pros and cons of customer involvement.

As mentioned in Chapter 4, there are a myriad of methods for discerning the voice of the customer and all of them should be considered when facilitating a QFD analysis. One key consideration is to choose those methods that are within the scope and resources of the particular project using QFD. Human Resources, Marketing, and Technical departments can all have input in selecting customer measurement methods.

Finally, the size of the team should be taken into consideration. The ideal number for a QFD team is less than 20 individuals. For most projects, programs, and systems, teams can be assembled consisting of less than 20 members and the necessary skill mix can still be provided. When teams become larger than 20, the facilitator is hard-pressed to maintain order during the process and it is easier for the team to become gridlocked or bogged down. If the program leader or management is concerned that

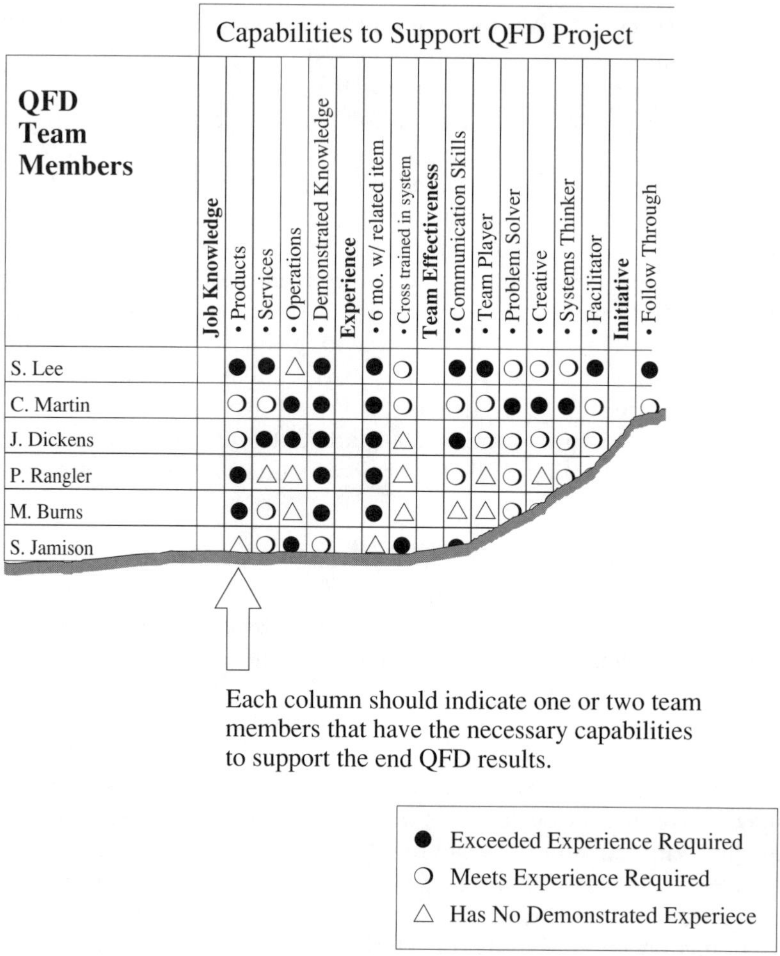

Figure 11-1 Modified Core QFD Matrix for team member skill evaluation.

insufficient resources may result with less that 20 members, provisions can be made to have off-line communication with other experts in the organization so that the team can remain small, but have access to information greater than the immediate knowledge base of the team. Bigger is not better when creating a QFD team.

11.3 *Roles and responsibilities: facilitator skills and responsibilities*

What makes a good facilitator? There are many authors who have written books about how to do team building and lead a project or effort. Some of the techniques are more useful to some individuals than others because there are probably as many styles of facilitation as there are individuals. Do

Table 11-3 Pros and Cons of Customer Involvement

Pros of Customer Involvement	**Cons of Customer Involvement**
• Establishes true image of customer focus. • Provides understanding of end operational use. • Establishes true image of customer focus. • Provides understanding of end operational use. • Allows for real-time feedback of strengths and weakness of existing and new system features. • Validates prioritization of needs. • Provides customer with understanding of impact when requesting a change to a design or service. • Avoids misconceptions and providing features that the customer does not want or would sacrifice for cost.	• If not well organized, product development can appear chaotic and unfocused. • If not well organized, product development can appear chaotic and unfocused. • May invite expectations that cannot be met. • May expose potential weakness of current capabilities. • Raises potential legal questions for patent ownership. • May expose customer ot internal "turf" battles. • Customer may feel vulnerable if they cannot clearly state needs and team has not allowed them adequate preparation time and information.

not hesitate to read as many books on facilitation and team building as you can find. What may work for one situation may not work for another and a facilitator must be flexible to accommodate the situation at hand. Adopt the style that works for you.

Successful facilitators will have some common characteristics that enable them to drive a team to a goal and get results. A successful facilitator will be able to build consensus from among the team members, identify and resolve conflict, monitor and balance power, pick up the pace when discussions bog down, and be a good listener and keep the team focused on its goals within the time frame allowed. Table 11-4 gives some of the attributes of a successful facilitator.

One of the important lessons the authors have learned in facilitating is to remember that as a facilitator *you are not a dictator*. The facilitator's role is unique among the team members. Even if the facilitator comes from the team itself, he or she must assume the role of catalyst — helping to combine the expertise of the individuals of the team into a synergistic force to solve a problem. A facilitator should guide the QFD analysis based upon facilitation experience but should not unduly influence the decision-making process. Team members must feel free to openly discuss concerns, design constrictions, process requirements, and customer requirements. A facilitator reminds the team of the destination represented by the defined goal, but does not *tell* the team how to get there or what form of transportation to use. The

Table 11-4 Some of the Attributes of a Successful Facilitator

- Good communication skills.
- Patience.
- Broad-based knowledge of subject.
- Friendly.
- Capable of speaking "language" of team.
- Confident.
- Good listener.
- Responsive to team's needs & ideas.
- Adopts quickly to change.
- Animated.
- Boundless energy.
- Good sense of humor.
- Fast learner.

good facilitator is a *guide*, and when guiding a team through treacherous or uncharted territory a facilitator must have a large supply of provisions (techniques) at his or her disposal to handle any situation that may confront the team. This is why a flexible facilitator, rather than one that adheres to a single approach, will be more successful in guiding a QFD team.

11.4 *Resolving conflict*

One of the most challenging tasks of the facilitator is to resolve the inevitable conflicts that will arise during the QFD analysis: conflicts that can occur often and over varying issues with a wide range of intensity. The authors have found that during the usual course of a project team using QFD, the team will progress from doubt about the methodology or its usefulness, to chaos as the normal learning or program techniques are broken down. The next step is usually debate (the authors try to keep this from escalating to violence!) as the methods for proceeding are discussed at a technical level, and finally the group usually reaches a stage of gratefulness as they see the progress and benefit (see Figure 11-2).

During this project metamorphosis, the facilitator will be required to table conflicts in various forms. It is not possible to itemize the limitless possible conflicts that can arise, but it is helpful to recognize what can be termed conflict styles, represented in Table 11-5. The one single key to understanding conflict is to recognize the source of the conflict and the perspectives of the conflicting parties. Is the conflict a result of inadequate information? Does one party feel threatened? Is there a power struggle?

Many conflicts can be resolved quickly by equalizing power among team members. Power is basically the ability to influence others. Table 11-6 lists some of the things that can give a person power, and many of these elements can be present when facilitating a QFD analysis. In facilitating a QFD, attention should be given to various ways to level the playing field when faced

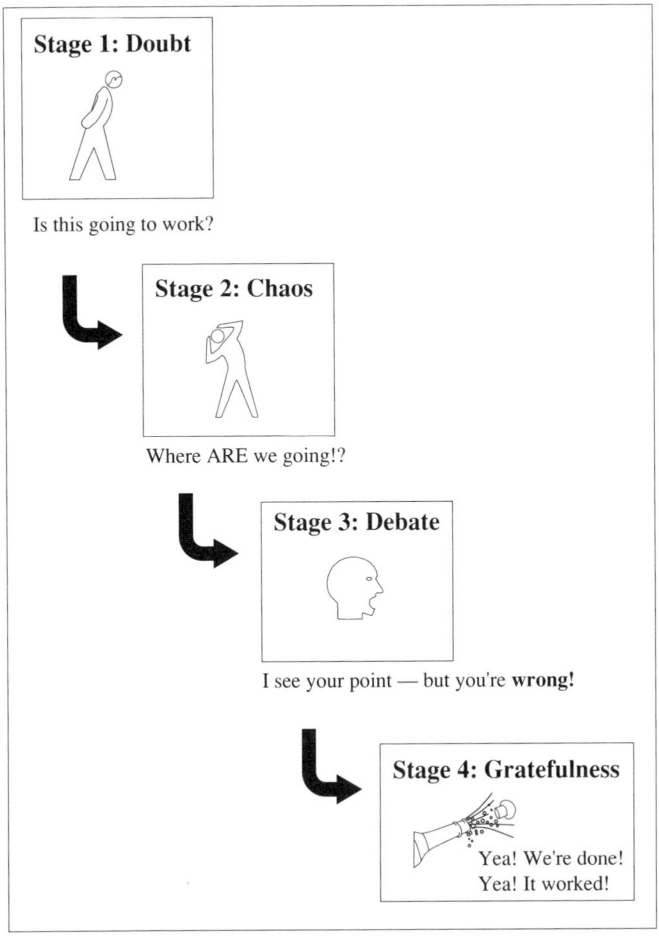

Figure 11-2 The stages of a QFD project from the perspective of the team.

with unequal power situations. One of the most effective ways of equalizing power, for example, is to make sure that an individual with superior knowledge on a subject shares that knowledge with the team. Another way of equalizing power is to give every person one vote in ranking importance.

Conflict is not always resolved in a sharing of power or as the result of a reasoned discussion and the facilitator should be prepared for such situations. While one of the authors was leading a QFD for a defense contractor, a dispute arose over a mission requirement. The military customer in this case had several representatives in the room who were disputing whether the customer need identified merited a "3" or a "5" in the comparison ranking. Despite the efforts of the facilitator to focus the heated argument on the technical elements separating the two customers, the disagreement escalated until one customer said: "If I do not get a 5 on this issue, I'll have your job!" The customer desiring the 5 rating outranked the other customer

Table 11-5 Some Conflict Styles of People

Aggressive / Confrontive
- Strong need to control situations and/or people.
- Directive and judgemental.

Assertive / Persuasive
- Stand up for oneself without being pushy.
- Proactive approach to conflict.
- Willingness to collaborate.
- Depend heavily on verbal skills.

Observant / Introspective
- Tendency to observe others and self analyze.
- Adopts counselling and listening.
- Cooperative, possibly conciliatory.

Avoiding / Reactive
- Passivity or withdrawal.
- Need to avoid conflict.
- Accepting and patient.
- Often suppresses strong feelings.

Table 11-6 People Can Have Power for Different Reasons

Power
- Definition: the ability to influence others.

What gives a person power?
- Money
- Expertise
- Ability to punish non-cooperative behavior
- Position
- Stability
- Willingness to take risks
- Self esteem
- Contacts
- Earned respect
- Owed favors
- Appearance
- Political rank
- Command of language
- Charisma

representatives in the room. The requirement was given a 5 and the conflict was, for that moment, resolved! The lesson here is twofold: (1) the customer can often have internal disputes on what is a key requirement, which may neither be apparent in the requested specifications nor communicated until the QFD is accomplished, and (2) despite the fact that the dispute was resolved in a somewhat undemocratic way, the customer resolved the internal dispute and the project could proceed with a clear definition of the importance of that customer requirement.

The facilitator should remember that disagreement is neither necessarily bad, nor is all conflict to be avoided. When facilitating a QFD analysis the various team members will bring their own opinion of what is important to the customer based upon knowledge, skill, and expertise. Marketing individuals should have a better hand on the pulse of the customer while the electrical engineers will know the best circuitry to use. The facilitator should pay attention to the individual skills of the team members and let the most knowledgeable individual on a particular subject or problem air concerns or solutions. Discussion is healthy when conducting a QFD. In fact, the very nature of a QFD analysis will generate increased communication and interfacing between individuals and groups who otherwise may not communicate on a project at all.

11.5 Working through the QFD matrix

Chapters 1 to 4 introduced the concept of QFD and how it fits into RSM™ and an integrated approach to quality systems, and discussed in detail the elements of the matrix. But the most difficult task is not understanding the matrix on paper, or even how it fits into other systems or combines with other tools, but how to actually walk through a matrix with a team. The authors earlier discussed some of the facilitation techniques that can be used in setting up a successful environment for a QFD. The next step is to recommend some procedures for doing a QFD.

11.5.1 Developing customer needs

All team members should participate in developing the customer needs. Many QFDs conducted by the authors have also included the customer and suppliers where appropriate, especially in product development when customer feedback is critical to features of the design.

But how to develop the customer needs? The authors have found brainstorming to be an effective tool to not only initiate the dynamics of team building but to create an unfettered approach to looking at customer requirements. One brainstorming method that has worked very well for the authors is to select two or three members from the team to act as scribes, armed with felt markers and Post-It™ note pads. The group then calls out customer attributes for the product, process, or service and the facilitator points to a scribe to record the idea on a note and stick it on the wall or on an easel pad. This process loosens the group up, is interactive and is effective for establishing the initial customer needs. The authors have used brainstorming sessions such as these involving as many as 200 individuals broken into 20 groups. The authors would recommend using this type of brainstorming technique in conjunction with the gathering of specific customer-based information to yield the initial customer needs.

Generally, after brainstorming the needs, the next step is to arrange the customer needs into affinity diagrams. This can be done initially by having the scribes each select another team member to arrange the notes, but with the following constraint: the arrangers should not talk while grouping the ideas. A time limit appropriate to the number of tasks should also be imposed (for example, 15 min for a board of 50 ideas). The restriction on communication during the initial affinity diagraming prevents the team from digressing into a discussion on the merits of an idea, and speeds the task of grouping the ideas. After the initial affinity diagrams are complete to the satisfaction of the arrangers, or time has run out, the team should then discuss the groupings and try to create a title or descriptive attribute for each group. If there is sufficient time and categories available, the next step would be to create a tree diagram of the group attributes. Grouping the customer requirements in this manner allows for a visual breakdown of the requirements into levels of detail, as was done for the video game example in Figure 4-7 in Chapter 4.

The team should review the requirements to determine if a perceived customer attribute is really a technical solution. This ability to discern between an actual customer need and a technical requirement is more an acquired ability to rethink the problem than an innate gift, especially among technically-trained individuals who are trained from the start to examine a problem and immediately develop technical solutions. Creating hierarchial trees for the customer needs in the process of grouping in affinity diagrams often leads to the elimination or identification of technical solutions. One way to check to determine if the customer need is truly a customer need and not a more limiting technical solution to a larger desire is to go through the checklist in Table 11-7. If the perceived customer need *is* a customer need, it can often be expressed in the form of an action, as opposed to a description. One of the easiest ways of demonstrating customer needs vs. technical solutions is to use a simpler example to first brainstorm, such as what people want in snacks (a popular topic when facilitating long QFD sessions!). In Figure 11-3, the sample characteristics of what people want in snacks is shown. For example, "crunchy" may be a customer need for a snack, and the technical requirements for meeting that quality of "crunchiness" can be peanuts, chips, celery, etc. The trait of being "crunchy" can be met in a variety of ways, from a healthy snack such as celery, to a junk food fix such as potato chips.

The point of this exercise is to start to retrain the team to think differently about what is a technical requirement and what is a customer need. If a technical requirement is listed on the customer needs side, then the team is

Table 11-7 Checklist for Determining Customer Needs (whats) from Technical Requirements (hows)

- ❑ Is the concept broad enough to cover many subcategories?
- ❑ Is it an action or does it have an action verb associated with it?
- ❑ Does it preclude other technical requirements?
- ❑ Does it describe a customer want, desire or need rather than a thing, object or device?
  If there is a single technical "solution" associated with it, then it probably is not a need, but a technical requirement.
- ❑ Does it encompass a quality?

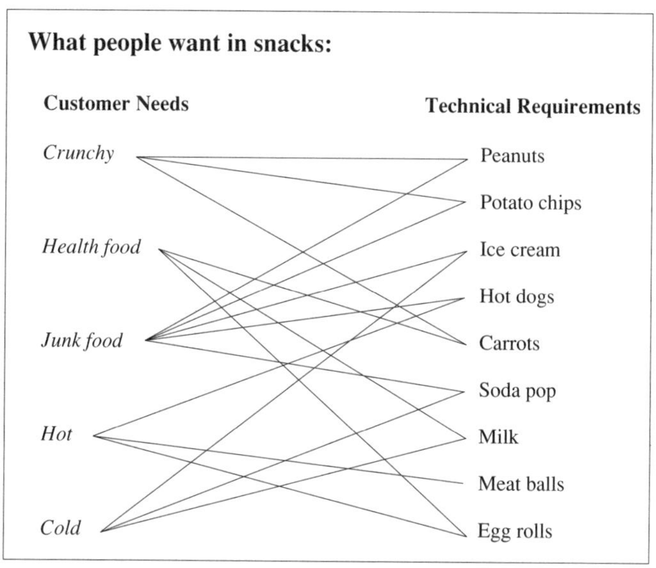

Figure 11-3 Some examples of Customer Needs vs. Technical Requirements.

unnecessarily limiting the options for solution for the QFD analysis. Changing a technically-trained person's learned approach to problem solving is not easy and often does not happen in a flash of cognition. During a 1-day QFD overview seminar taught by the authors, one team composed primarily of engineers and technicians from civil and environmentally-oriented firms decided that they would choose as their subject for a QFD analysis the creation of a recycling program for a municipality. As the team brainstormed the customer needs, they immediately assumed that the municipality would run the program from marketing, to collection, to disposal, and that it would cover all plastics, metals, and papers. Consequently, they developed customer "needs" such as "collection trucks," "recycling facility," and "separation facility." After arguing over the type of trucks and the difficulty and cost in building a recycling facility, the authors pointed out that these were technical solutions and were limiting responses to their customer's true needs. What if the municipality did not want to maintain collection trucks? What if the perceived need for recycling was less than total plastic, metal, and paper recycling? What if the municipality did not want to expend the money for a recycling and separation facility? After pondering these questions, the team came up with broader customer needs such as "collecting goods to be recycled," "cost-effective recycling," and "effective program information dissemination." By doing this the team opened options for the municipality by describing a broader need that could be met by several technical solutions. Contract collection and transportation to a central recycling center could be considered in addition to building a system from scratch because it would be a technical solution to "cost-effective recycling" and "transportation system" for "collecting goods to be recycled." The engineers, rather than considering "what" the customer wanted, immediately went to "how" to resolve the perceived problem.

Because people often interpret information and ideas in different ways the authors recommend that a definition list be maintained during the creation of a matrix. The customer needs or "whats" should be defined at the time of creation and discussion so that a consistent meaning is carried through the matrix development. It can be helpful to publish the list of definitions and give it to the team members for reference.

An example of how important it can be to make sure the team members are reading from the same page when it comes to definitions of customer needs involved a customer whose own personnel interpreted a key customer need very differently. The project involved a complicated ground support and spacecraft system being developed in the late 1980s. One of the authors was conducting a QFD analysis with the customer present. All of the team members gave the customer need of "Mission Success" a five in importance, but when the author asked what was mission success, a colonel offered a definition with which he thought everyone was in agreement. A captain strongly disagreed, stating that the mission goal as defined by the colonel was not capable of being accomplished with existing technology. Heated debate followed and eventually the resolution was to take on additional action items to discover what the true mission success parameters were, and whether the colonel's definition was obtainable. This is a critical lesson for a facilitator: if the definitions are vague, chances are your product will not be focused and the benefit of a QFD analysis is greatly reduced.

11.5.2 Developing the technical requirements

Brainstorming the technical requirements is done in a similar fashion to developing the customer needs, but with a different question being asked

Table 11-8 Checklist for Determining Technical Requirements (hows) from Customer Needs

- ❑ Is it only one way to satisfy a customer need?
- ❑ Can it be controlled?
- ❑ Can a target value be developed for it?
- ❑ Does it satisfy more than one customer need?
- ❑ Is it measurable?
- ❑ Can it be broken into subcategories that can be controlled?
- ❑ Is it more descriptive than active?
- ❑ Is it part of a "specifications" document?
- ❑ Is it a thing, object or device?

when selecting what may be a technical requirement. To begin with, the facilitator and the team will probably realize that in brainstorming the customer needs they also developed a list of technical requirements, although not a complete list. Technical solutions are also divided into a hierarchical order to foster grouping. The error that the authors see the most at this stage of developing the matrix is getting too detailed too fast. In facilitating the initial QFD matrix defining the customer "whats" and the technical "hows" the authors recommend a matrix no larger than 25 customer requirements by 50 technical solutions. The tendency for most teams that are uninitiated to the QFD approach is to literally try and create the detailed design in the first matrix. This is why brainstorming the technical requirements and using affinity and tree diagraming to create a hierchical list is essential to constructing as small a first matrix as is possible and still give enough detail for meaningful distinction for prioritization when completing the matrix.

Table 11-8 describes a checklist when examining the technical requirements for fulfilling the customer needs. In a larger matrix it is often helpful to brainstorm the "hows" for each of the customer "whats" prior to using affinity and tree diagraming. This approach allows free association of technical requirements that will meet each customer need, and then allows the team to examine the entire list of "hows" for affinity diagraming, helping to establish a hierarchy and eliminate duplicate technical requirements.

The development of the technical requirements can be a tedious process on a complicated high-technology product, or require less exactitude when examining a service. Facilitators may often find it difficult when to determine "closure" on the process. How detailed is detailed enough? How are you certain the technical solutions have all been discussed and appear on the matrix? There are no hard-and-fast answers to these questions and suffice it to say that the answers also vary depending upon the level of detail of the project, the size of the matrix and the time available to the team to create the initial QFD matrix. Usually the answer is self-evident because the team will be using its collective knowledge to examine the solutions. Gaps unidentified in the initial matrix will also be identified as the level of detail is increased. The authors have also found that teams usually do not lack the expertise or knowledge when it comes to defining technical requirements because that is the way they have been classically instructed, especially the engineers and computer science experts. Proper definition of the customer requirements is more key to the solution and will drive the creativity of the team in meeting the technical challenges.

11.5.3 Developing the prioritizing of the "whats" and competitive benchmarking

Chapter 4 discussed the mechanics of prioritizing the "whats" and creating a scale for competitive benchmarking of how well an organization currently meets those needs and how much it intends to improve in those areas in a given time period. This section discusses some of the ways to accomplish that benchmarking and how to avoid pitfalls when facilitating prioritization and benchmarking.

The first point to remember when ranking the customer needs, Column 2A of Figure 11-4, is that the ranking is relative to all of the customer needs. Team members will often voice concern that by giving any element a low ranking is saying that the need is unimportant to the customer. This is not the point of ranking. All customer needs should be addressed, but the use of ranking in the matrix is to create prioritization of the needs in relationship to their impact on the customer. The customer may want 25 features in a new product or service, but the relative importance of those 25 varies in terms of cost, or availability, or functionality. The ranking of the customer needs gives a measure of which customer needs are *the* drivers to the project and which play a lesser role. Despite this explanation, team members are sometimes still insistent on giving so many items a "5" that the ranking is meaningless. When this happens, the facilitator should take the following approach: limit the number of 5s, 4s, and 3s the group can use to rank the elements. This forces the team to more carefully scrutinize their decision-making process and carefully evaluate marketing and customer information gathered. If the ranking process still fails to meaningfully differentiate any of the customer needs, then the matrix may not be at sufficient detail. The solution is to go to the next level of detail for the tree diagrams for the customer needs. Another problem may be that the customer information may not be accurate enough to allow reasoned opinions as to relative importance. If this is the case, the QFD should be put on hold and more customer input should be obtained. The facilitator may find that on occasion the product or service is such that the identified "whats" are all critical to the success of the program, or are nearly identical in importance to the customer. This realization in itself yields important information to the team. They now know that failure to satisfy

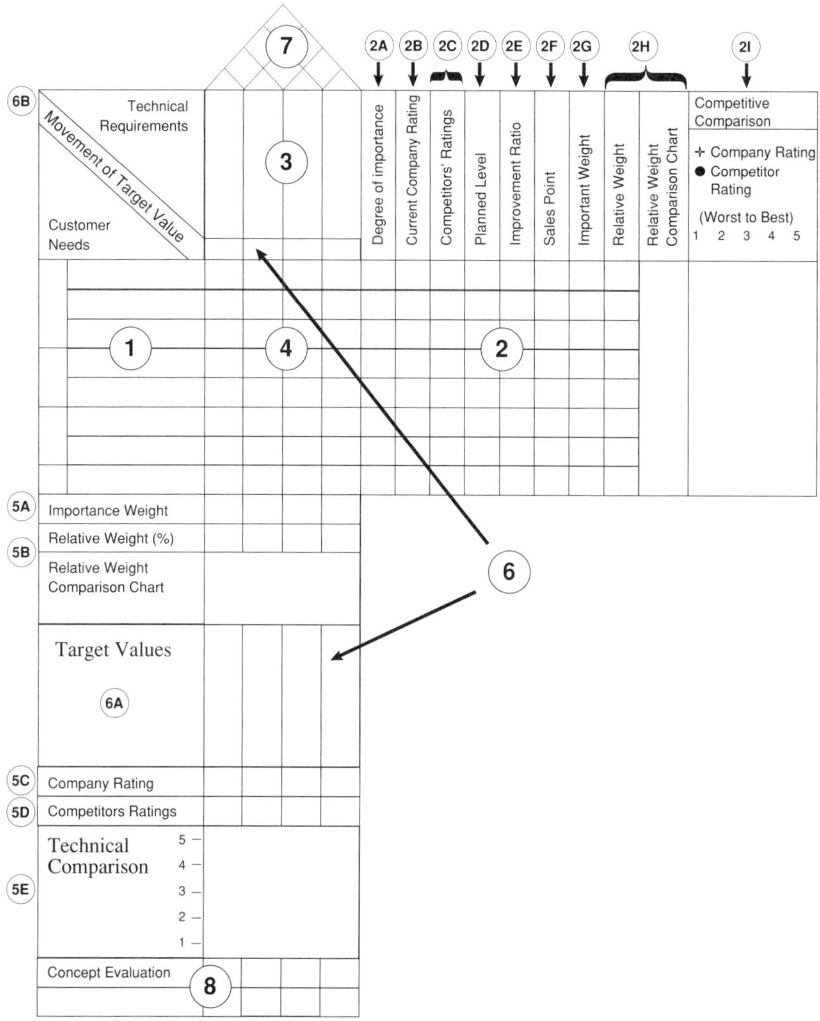

Figure 11-4 Detailed Core QFD Matrix guide.

even a single customer need fully could create failure. Based upon this information the team may want to re-evaluate its intended plan, or focus on the technical details to ensure that all customer requirements are being met to the level of importance necessary for success of the service or product.

When facilitating the actual numerical completion of the matrix, there are different schools of thought about how to arrive at a single number for each customer need. Is there a vote with a majority rule? Do you take the arithmetic mean? How about an average of all the values from the team? There is no right or wrong way to obtain values, but the authors have developed a method that seems to work better both in obtaining realistic comparative values and moving the team quickly through the task without bogging down in unnecessary arguments over values. That method is to distribute voting cards numbered one to five to each team member. The facilitator then goes through the list of customer needs calling for a vote on each without discussion. The facilitator then selects the most common number (mean) from the team vote, but takes notice if there is a wide discrepancy between members of the team, or if all team members are within a point, except one or two. A note should be made on these key votes and they should be revisited after the initial vote has been taken for all of the requirements. The facilitator should then open up the team to discussion on those items where significant discrepancy was found among the team members and try and determine the basis for the votes. In this way, the team obtains special knowledge about the customer or the requirement can be obtained, and they can decide if there is a need for further research on that requirement or if the discrepancy is really not important. In actuality, the authors have found that most of the time there are only a few items that produce a wide variance in ranking, and these items can be discussed and re-evaluated to the satisfaction of the team. The facilitator should be concerned about extremes of one vs. five, but not two vs. three.

With regard to the competitive benchmarking elements of the matrix, these elements obviously must be carefully considered to avoid inaccurate ranking of both a competitor's ability and the organization's current ranking and future level desired. During team evaluation of these elements special emphasis may be placed on information from individuals who have expert knowledge in a particular area or who have been requested to gather competitive data. However, the authors have found that it is still helpful to have opinions from other team members who may have valuable insight and input that should be considered by the team. Rarely have the authors suggested that certain members of the team not vote because they lack "knowledge" on a technical issue by being from marketing, or should not comment on the customer needs because they are from production control. Data or opinions can be discounted based on greater knowledge, but they should not be withheld because of a perceived lack of expertise.

Sometimes an organization may find it difficult to benchmark new products or services against perceived competitors. In this case the team can either not do the competitive benchmarking and rely on the initial comparative rankings or use similar products to benchmark against.

When selecting a customer need as a sales point, the facilitator should encourage the team to keep major and minor sales points to one or two for every ten customer needs. Teams can become confused on what is a sales point and designate items that are high-ranked as sales points, defeating the purpose of this element. As explained in Chapter 4, a sales point is something the customer may not really desire, but the existence of it brings a "Wow!" from the customer. Using this approach there are really very few sales points in a matrix and sometimes it is necessary to actually limit the team to a certain number to prevent overranking key items.

11.5.4 Developing the correlation matrix

By the time the team gets to the correlation matrix of the Core QFD Matrix (Section 4 of Figure 11-4) a team dynamic should be developing and some of the obstacles to free exchange of information should have been removed or decreased. Completing the correlation matrix will usually reveal any final hidden agendas among team members or the departments they represent. The facilitator should be prepared for some vigorous discussions in completing the correlation matrix.

The mechanics of completing the correlation matrix are straightforward. Each team member should be given voting cards with one, three, and nine. The next step is to conduct voting with the group using the same methodology as ranking the customer needs, but a tool must be used for correlating the "hows" to the "whats" as initially discussed in Chapter 4. The tool is the sentence, "If I control (insert a technical requirement or "how"), I am meeting my customer's need to (insert customer need or "what")." This sentence provides a method for comparing the technical requirement ("how") to the customer need ("what") that standardizes the outcome of the matrix and ensures that all members of the team are viewing the relationship in the same light using the same definitions. A high correlation between a "how" and a "what" (a value of nine) would mean that if the team can "control" the technical requirement, then it will enjoy success in meeting that particular customer need. A correlation that is not as strong would merit a three; a correlation that is only slight would be a one; and no correlation would be a blank.

The facilitator should drive the team through the correlation matrix rankings using the same procedure as in the customer rankings. No discussion should be allowed for the first pass through the matrix, and the facilitator should take note of wide variations in individual correlations. The team should then be taken back through the matrix to discuss those correlations in detail: (1) to determine the nature of the individual team member rankings, and (2) to see if a consensus can be reached. The importance of the first pass through the matrix prior to discussion cannot be underestimated. If the facilitator begins to allow involved discussion and debate over rankings from the start, the process bogs down and becomes increasingly tedious. This is not to say that using the first-pass-through approach is not without its heated discussions when the team returns to those items of wide interpretation! It is also possible that the team may disagree on many correlations, at which point the facilitator must be certain: (1) that definitions are clear, (2) that the "If I control . . ." rule is being applied correctly by all team members, and (3) that all relevant technical requirements have been properly identified.

After completion of the correlation matrix, the relative weighting in Section 5 (see Figure 11-4) is a calculated value, and should not require any facilitation, or the team has really become contentious!

11.5.5 Developing the target values

The target values (Section (6) are linked to the technical requirements themselves. Chapter 2 discussed the genesis of the target values as related to the technical requirements. Target values must be measurable in order to have value and meaning to the teams' efforts. In developing the target values, care should be given to determine realistic values, or values that have a basis in the requirements for the project. Chapter 4 gave examples of target values from the video game example, and how those values had a direct tie to a technical goal, which then supported the customer requirements. In facilitating target value development, the facilitator must encourage the team to first define those goals in measurable terms: quantitative amounts, target numbers, percentages, etc. Qualitative target values such as "good" or "better" or other modifiers are usually not measurable and, therefore, do not give the team a direction or measurable standard to compare against. Often manufacturing or industrial engineering can give excellent guidance in developing the target values. In the situation where a specific product is being developed, or a product is being manufactured in conjunction with a Request For Proposal (RFP) or Request For Quotation (RFQ), there are specifications spelled out in documentation or engineering requirements that can be used appropriately to develop target values.

The team should commit a block of time in developing target values that is commensurate with the number of "hows" in each matrix developed. Target values should not be glossed over. They can become the goals to be achieved in the development of an integrated plan, and often become the benchmarks of the product itself. Great care should be taken to keep target values in line with the technical requirement.

If no target values can be developed for a technical requirement, the team must realize that the technical requirement cannot be measured and, therefore, the team will not be able to track compliance of the technical requirement to the customer requirements. If the team truly cannot develop target values for a "how," then the team must consider the following:

- Is the technical requirement at too high a level, and, therefore, should another level of detail be considered?
- Is the technical requirement appropriate, but capability or monitoring ability is lacking?
- Should the technical requirement be abandoned in favor of another criteria that can be measured and still satisfies the customer needs?
- Is there a need for more field work, bench testing, data gathering, or concept development before a target value can be developed?

If there are only one or two elements for which target values cannot be developed, the team should ask three questions: (1) How critical are these elements? (2) Can substitutions be made for the technical requirements? (3) Can the QFD continue with an action item in place to develop measurable target values? In any event, the exercise of developing target values is valuable to the team for the goals that can be developed, as well as for discovering if there are technical or data gaps that must be filled.

As part of the task of determining target values, movement of the target values should be developed simultaneously to determine the trend for achieving continuous process improvement (see 6B in Figure 11-4). This task is usually straightforward and easy to determine with the group after the more difficult task of determining target values has been accomplished.

11.5.6 Identifying conflicts and correlations between technical requirements

One of the last stages in developing the QFD matrix is to determine if there are any correlations or conflicts between technical requirements. This task (Section 7) is accomplished by utilizing the four symbols in Figure 11-4 to illustrate positive-to-negative correlations (a blank in the corresponding matrix block between two technical requirements means there is neither a positive nor negative correlation). When facilitating a group through this section it is important not to gloss over any potential impacts of increasing one technical requirement that could have a negative or positive impact on another technical requirement. The correlations between technical requirements tell the team where potential program or project stoppers may be hiding, and which interactions to flag as having impacts on overall requirements. The "roof" (Section 7 of Figure 11-4) also can be critical in establishing relationships and dependencies in developing the Integrated Plan, as discussed in Chapter 6. Often Section 7 can become the means to identify key communication channels that should be formally defined and become part of the plan.

During the facilitation of the technical requirements, correlations of the exact nature of the impact between two requirements may not be identifiable as negative or positive without more detail or data. In this case the team should create a communication channel between the two technical requirement areas — especially if they are different functionally — to closely track their development and anticipate a possible negative interrelation.

Items that positively correlate may also be important in order to eliminate possible redundancies or duplications of effort. If the control of one technical requirement will ensure compliance with another, then the necessity of devoting resources to closely track both is eliminated or at least greatly reduced.

11.5.7 A facilitation exercise

There are many exercises, games, and role-playing methods that can be used to bring a team together and generate open discussion. The authors have found that for most organizations, the best team-building techniques get the team working on solving real problems as fast as possible. The

Integrated QFD Approach™ and creation of the Core QFD Matrix, gets a team moving quickly into the process by focusing on the task at hand. However, the authors have also found that teams work better when they get "mental breaks" that may involve game playing or related activities, depending on the length of the facilitation and the goals that are set for the team from the outset.

The authors recommend that any facilitator have at his or her disposal a good sense of humor and some comic relief for the team. These attributes will serve the facilitator well, because QFD is not for the timid nor the casual. It is an intense procedure that heavily involves people and requires a great deal of energy from the participants and the facilitator.

One of the facilitation exercises the authors employ — usually at the end of the first day after the team has gone through the basics of the Core QFD Matrix — is called the Spy Game. The Spy Game serves two purposes: (1) a mental break at the end of the first day, and (2) an example of how behavior can influence decisions. The Spy Game is a low-risk role-playing and puzzle-solving exercise that involves the entire team. The authors have used it with success for groups of 150 people divided into teams of 8 to 10. It breaks down communication barriers, gets people involved in a team environment, and provides a pleasant diversion at the end of a tough day.

The Spy Game should be introduced by first instructing the team that there are many ways that a person can react to conflict. Some people become introspective, some react strongly, and some even violently. The important point is that we all react differently to situations that create conflict in out lives, and sometimes that reaction can create negative effects in our dealings with other people, especially in the work environment. For example, our behavior towards conflict can be categorized loosely into animal types, just for the sake of illustration. Table 11-9 lists some animal categories to describe certain behavior we all may have seen in ourselves or others when reacting to stressful situations. For example, who has not been the "Owl" at one time or another, thinking you know everything and can solve the problem! Table 11-9 offers some humorous diversion by getting people to look at themselves and recognize behaviors they may have exhibited, but in an environment that does not create a threat.

The next step is to divide the group into at least two teams (if there are more than ten team members total). The facilitator should have already created index cards that have printed on them either "You are not a Spy" or "You are a Spy: Act like a _____" with one of the animals in Table 11-9 in the blank. The facilitator should have a ratio of approximately one "Spy" for every two non-spy cards. These cards should then be handed out randomly to the teams, with instructions that no one is to show anyone else his or her card.

After this is done, the teams should be given the problem: there are three alien ambassadors in a room and the team needs to match the alien with their name, food preference, and home planet. The clues are given in Table 11-10. The problem gets interesting when the rest of the General Guidance is also

Table 11-9 When It Comes to Behavior, We All Act Like Animals!

Type	***Behavior***
Jackal	Less intimidating than a Wolverine but more cunning with each strike; will bite from behind; makes you look foolish in a crowd to prove control.
Ostrich	Avoids conflicts and controversy; shuts up because either he is afraid of confrontation or wants to get you.
Peacock	Seeks admiration and respect by acting like an expert when he is not.
Monkey	Throws temper tantrums in an almost automatic response to feeling thwarted.
Cat	Self-righteously blames, accuses and finds fault with everything.
Cow	Super-agreeable; agrees with any commitment but rarely delivers.
Sloth	Not want to hurt anyone; delays decisions until need disappears.
Vulture	Dampens any suggestions regardless of merit; quick to point out something won't work; convinced he has little power over his life and is bitter; no one can be trusted.
Wolverine	Abusive; abrupt; intimidating.
Owl	Expert on everything; know it all.
Dinosaur	Thinks everything was better before; uncomfortable with change.

given to the teams (Table 11-10): Those who are spies must prevent the puzzle from being solved and not get caught! Give the teams no more than 15 min to solve the puzzle, and also provide them with the "Puzzle Matrix" in Figure 11-5. Having done this, sit back and let them at it!

Most teams will bog down in either trying to find out who the spies are, or struggle with the Puzzle Matrix and not find a solution. The authors have found that the puzzle matrix is much harder to use to solve the puzzle than the common sense solution provided in Figure 11-6. The solution is easily obtained by creating a 3×3 matrix with the column titles being: Food Type,

Table 11-10 An "Out of This World" Problem: Solve the "Spy Game"

The Problem

You have been asked by the Intergalactic QFD Council to take three Ambassadors from other worlds to dinner. The three Ambassadors are in a conference room together, sitting next to each other at a table.

You were told that this is a highly sensitive situation because none of the Ambassadors speaks English. Prior to entering the conference room, an aide tells you that the universal translator failed shortly before you arrived.

You know that one of the Ambassadors is called Grop and one likes to eat Flying Zorpals. The aide tells you the following:

1. The first Ambassador at the table eats Green-eyed Dwips; the second at the table is Kling; and the last Ambassador is from Dornon.
2. The Ambassador from Abaxia likes Hoofed Zingats.
3. Snorp isn't from Zenon 5.

General Guidance

Goals: Solve the problem, get rid of your spies, and correctly identify their undercover behavior.

Rules:

1. If you are a spy, you must act like your undercover behavior to prevent the problem from being solved. You succeed if you prevent them from solving the problem without being caught.
2. The only way to get rid of a spy is to have a majority vote to jail the spy.
3. You may use the puzzle logic matrix to solve the problem.
4. You have 15 minutes.

Name, and Planet. The next step is to have someone read the three clues one at a time. After Clue 1, there is an entry in each of the rows: Green-eyed Dwips in Row 1, Column 1; Kling in Row 2, Column 2; and Dornon in Row 3, Column 3. After Clue 2 is read, everyone can see that a place for a Food Type and Planet must be open on the same line, which means that Kling eats Hoofed Zingats and is from Abaxia. This now means that Zenon 5 can be put in Row 1, Column 3; and the alien from Dornon must eat Winged Zorpals. Clue 3 now tells you that Snorp is not from Zenon 5, therefore it must be from Dornon and Grop is from Zenon 5 and eats Green-eyed Dwips!

The facilitator should offer any of the spies who were voted out an opportunity to defend themselves, and usually at least one of the teams has voted out non-spies who were just trying to help. The lesson learned can be related as follows:

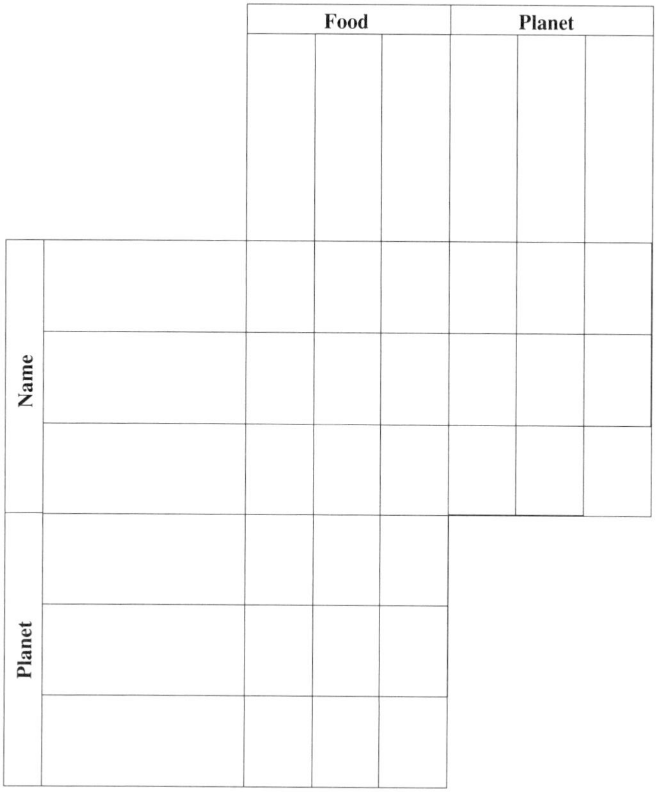

Figure 11-5 The "Puzzle Matrix" for solving the quandary of the Alien Ambassadors.

- We may act a certain way when in conflict because a position or idea has value to us, and we all should try to understand each person's perspective;
- People who are trying to help are sometimes perceived as "spies" because we may not trust them or their behavior; and
- Use the *right* tool: QFD may not be the right tool for every situation, and may not be needed in every circumstance.

More than anything the "Spy Game" should be used as entertainment and to make a point, but it should not be drawn out nor should it be taken too seriously! The message should be that in the coming days of the QFD analysis, all of the team members should try to understand the position of the other team members on technical, departmental, and customer issues, and no one should be "voted out" because of their position on an issue, or direction they think the team should take.

The above exercise is only one of many that can be used to break the intensity of the QFD analysis. Facilitators should keep in mind that each team may be different in its makeup and selected goals, and if "team building" and "organizational change" are important considerations as part of the scope of

Clue 1:

The first Ambassador at the table eats Green-eyed Dwips; the second at the table is Kling; and the last is from Dornon.

Food Type	Name	Planet
Green-eyed Dwips		
	Kling	
		Dornon

Clue 2:

The Ambassador from Abraxia liked Hoofed Zingats.

Hoofed Zingats		*Abaxia*

Therefore, Kling must be from Abaxia:

Food Type	Name	Planet
Green-eyed Dwips		
Hoofed Zingats	*Kling*	*Abaxia*
		Dornon

Clue 3:

Snorp isn't from Zenon 5.
Therefore, Snorp must be from Dornon and likes Flying Zorpals; and Grop is from Zenon 5 and likes Green-eyed Dwips.

Food Type	Name	Planet
Green-eyed Dwips	*Grop*	*Zenon 5*
Hoofed Zingats	*Kling*	*Abaxia*
Flying Zorpals	*Snorp*	*Dornon*

Figure 11-6 Solution to the Alien Ambassador puzzle.

the project, the facilitator may want to employ more human relations training, or even involve a team-building expert and organizational-change expert as part of the QFD agenda. Likewise, if the scope is to just get the matrix done and "get action," the facilitator needs to spend more time on the implementation and less on culture and team building.

11.6 Additional facilitation procedures

The above facilitation guidance instructions are not meant to take the place of professional facilitators with industry-specific experience but to help prevent some of the pitfalls inherent in facilitating a QFD analysis. The authors have not gone into detail with regard to Section 8 of the matrix, Concept Evaluation, because of the large amount of additional instruction required to validate design or service concepts, and because it begins to go outside the basic instruction needed for a QFD analysis.

11.7 Chapter summary

1. Facilitation of QFD is the key to its successful application.
2. Although there are many specific suggestions and instructions in this chapter on facilitating a QFD, the facilitator should define roles and responsibilities early in the process and manage expectations.
3. The customer should be involved early in the scoping process, but care must be taken to avoid the appearance of mayhem or disorganization that can sometimes exist at the beginning of the process.
4. The facilitator should use conflict-management techniques to table conflicts and balance power among team members.
5. Brainstorming is an important technique in QFD for exploring concepts and ideas.

chapter twelve

Developing a QFD pilot program

If the Earth had waited for a precedent, it never would have turned on its axis!
Maria Mitchell, Astronomer (1818–1889)

12.0 The need for a pilot program

"Who's first?"

All of us have heard those two words a seemingly infinite number of times beginning when we were lining up to go to recess in kindergarten, until recent memory while waiting in line at the local fast-food restaurant. Two words that probably have been met with anxiety more often than excitement and fear more often than anticipation. Americans are often obsessed with being first, and the rest of the world sees this obsession more clearly than we do. When asked about the biggest difference between Canadians and their southern neighbors, a Canadian once told one of the authors that Americans always have to be "Number 1." He commented how every American sporting event broadcasted shows maniacal fans sporting their index finger for their team, no matter what the record of the team or its position in any polls. "Sometimes, I think Americans would rather be number one than good!" was his summary comment.

As a nation so driven by the desire to be "first," it is somewhat of an ironic fact that in the area of testing new concepts and employing new methodologies, businesses in America are extremely reluctant to be the "first." This may be because of perceived risks that outweigh benefits, or the hesitancy to use untried methodologies, or maybe the not-so-unconscious resistance to anything that smacks of change. Whatever the reasons, it is often difficult for an organization to embrace a "new" concept such as QFD when it has not been used previously in the organization, or sometimes even in the industry as a whole. Businesses are adverse to risk for good reason, and the poor first-time application of any method can produce a variety of unwanted repercussions:

- Reluctance to attempt other innovative ideas;
- Adverse impact to the careers of those who advocated the change or methodology;

- Negative morale based upon a perceived failure;
- Processes being worse than before;
- Frustration from work accomplished without a clear benefit;
- Poor publicity or an adverse impact on customer relations; and
- Impact on the bottom line — lost revenues.

A potential solution to overcoming this fear or risk of being first is to initiate a pilot program for implementation where the risks can be defined and controlled, and the scope is large enough to quickly demonstrate capability, but narrow enough to prevent fear of any catastrophe. Change is inherent in implementing QFD, and part of making QFD successful is to manage that change in a relatively safe environment. A well-designed pilot program is the vehicle to prove the feasibility of using QFD in any environment.

This chapter is about how to set up a pilot program for QFD implementation.

12.1 The advantages of using a pilot program to introduce QFD

Chapters 10 and 11 discussed perspectives for a successful application of QFD and the role of the facilitator. This chapter goes into greater detail in how to set up a successful pilot program building upon the information presented earlier.

The first question is why do a pilot program? Haven't the authors said that a company must embrace QFD and employ it in a cohesive management and customer-oriented approach? The answer to this query is emphatically "yes." But, as mentioned in Chapter 10, not all companies will plunge into QFD with the exact same purposes in mind, nor the same resources to commit. The management environment in a small to mid-sized business built on entrepreneurial spirit may adapt QFD more quickly than an established institution (large or small) that is used to more incremental changes. Some companies may have already taken a stab at QFD and chosen not to adopt it at that time, and are reinstituting the process in an environment that is less than optimistic about its success.

The recommendations for a pilot program contained here are also not entirely limited to a small-scale implementation. A pilot program can be a huge undertaking if the management decides to include an entire product line, a key core process such as accounting, or a new product or service. As the reader will see in the following pages, the key to the success of a pilot program, like the key to the success of RSM™ and QFD themselves, is a proper scope, a necessary commitment of resources, and a viable road map.

A properly defined and executed pilot program avoids the pitfalls and negative aspects of implementing QFD. Some of the benefits of using a pilot program to implement QFD are presented in Table 12-1. The first benefit

Table 12-1 The Benefits of Using a Pilot Program for the Implementation of QFD

Pilot Program Benefits

Benefits	Description
• ***Focus***	Management and participants are focused on the parameters of the pilot.
• ***Positive Environment***	Team members are exited about trying new concept and will not be "punished" because project has no downside.
• ***Control***	Size of project is specifically limited in order to be manageable.
• ***Immediate Benefit***	Management, team members and other projects and programs will see a defined benefit within a specific time period.
• ***Training***	Proper implentation adds to the core knowledge of the team for future applications and allows time to customize the approach to maximize productivity of team members prior to implementing on a large scale.
• ***Cost-Efficient***	Resources committed can be easily defined and measured; Doesn't "bet the company" on outcome.
• ***Improves P.R. and Customer Relations***	Customers can be part of the pilot program.

mentioned in the table may be one of the most important: Focus. Design of the pilot program aimed at specific goals and bounded by clear parameters is the first step in setting up the pilot for successful completion. Management and participants know the scope and what is expected, and unreasonably high expectations and unpleasant surprises are avoided. A pilot program that is presented as an opportunity to excel or stand out for participants removes the stigma that "failure," however defined, will not negatively impact the team members. Pilot programs should be designed so that the participating members know that they are not being singled out for possible layoff, or being taken permanently from their regular assignments. Team members need a positive environment and a chance to succeed, not increased attention and pressure to succeed. Customer involvement can be a strong drawing card and promote better relations, but care must be used when involving the customer, as discussed in Chapter 11 (see Table 11-3). The list in Table 12-1 is by no means an exhaustive one when discussing the benefits of a pilot program, but those items in the table underscore the reason a well-designed pilot appeals to management when instituting QFD in the organization.

12.2 *Where to start*

What is a well-designed pilot program? Any pilot program, whether instituting QFD or testing a new polymer, involves utilizing the major elements of the process, program or methodology being considered for a larger application. A well-designed QFD pilot program incorporates all of the elements of using QFD in order to act as a model for wider or additional applications within the organization. For any organization using a QFD pilot program, one of the key goals of implementation should be demonstration of the methodology in a real problem-solving situation, and to act as a catalyst for future implementation.

There is no real magic to setting up a pilot program, but planning and setting goals are critical elements. The next sections discuss how to plan the program and to determine the goals for the pilot program for instituting QFD.

12.2.1 *Management commitment*

By now the reader is probably tired of hearing the term "management commitment" as a panacea for the success of almost anything. But the fact remains that without proper support of resources and leadership QFD will not succeed. Almost every article about organizations undergoing significant change or trying new techniques underlines the necessity of management support for success. For example, in a recent article on re-engineering from the *Harvard Business Review* the authors cited examples from several companies engaged in re-engineering and the tie of executive management commitment to the success of the re-engineering effort:

> Even with sufficient breadth and depth, a re-engineering project will fail without the full commitment of senior executives. In nine cases in our study, we evaluated the disparity between actual and projected impact of fully or nearly fully implemented projects . . . Five of these efforts have achieved their projected impact. And, interestingly enough, in four of these five successes, new chief executives were brought in before or during the projects. These senior executives understood how to lead an organization through a period of radical change . . . [48]

The authors noted that top executives involved in the study companies devoted between 20 and 60% of their time on the re-engineering projects, selling the benefits to the employees and supporting the re-engineering teams with resources and the best people.

While the purpose of running a pilot program is very different than a wholesale company commitment to a business re-engineering effort, the lesson of management commitment is not one to take lightly. When management openly supports a QFD effort and communicates to the organization that support, the chances of success of the QFD effort increase substantially. The team will face less resistance and more cooperation from company departments and people not directly involved in the QFD if the company's "line troops" know the importance the executive management is placing on the success of the program.

Does this mean that an organization's executives will have to spend 20 to 60% of their time directly involved in a QFD pilot program to ensure success? Not at all. Pilot programs are not meant to be as broad and encompassing as total re-engineering efforts, and a company may not be able to commit the kind of executive resources necessary on a pilot program. However, what is important is that the leadership of the company stay involved in the QFD effort and understand its progress, difficulties, and needs for success. This may mean periodic meetings where the executive management is briefed on the status and progress of the program, and, at a minimum, it also means a public show of support for the program and a commitment of the necessary resources (both time and money) for the pilot (determined at the time of go ahead with the pilot program).

12.2.2 Prerequisites to a pilot program

The most obvious prerequisite to any QFD pilot program is a thorough understanding of the application of QFD by management and a designated facilitator for the program, either recruited from within the organization or the hiring of an experienced outside consultant. Depending on the size and scope of the pilot project and the existing level of understanding of QFD and techniques such as RSM™, or other quality and business re-engineering techniques, more than one individual adept at facilitating QFD may be required.

Because QFD utilizes other statistical and analytical tools in its best application, the organization that is planning on implementing a pilot program in QFD should also be able to call upon additional expertise in a variety of areas from within the ranks. Table 12-2 lists prerequisites that the authors have found to be necessary to varying degrees to increase the organization's rate of success in pilot QFD programs. It is often the scope of the QFD that determines the skills necessary to accomplish the pilot program; however, those items listed in Table 12-2 are used in almost every application of a QFD pilot program. The list addresses two key categories: management prerequisites and application/analysis prerequisites.

Table 12-2 Recommended Prerequisite Skills Set for a QFD Pilot Program

Recommended Prerequisite Skill Set for a QFD Pilot Program

Management Prerequisties:

- Management by Planning (Gantt & PERT Charts)
- Mangement Tools
  - Affinity Diagramming
  - Interrelationship Graphs
  - Tree Diagrams
  - Matrix Charts
  - Process Decision Program Charts
  - Matrix Data Analysis
  - Arrow Diagrams
- Employee Involvement

Application / Analysis Prerequisites:

- Statistical Process Control
- Design of Experiments
- Basic Analysis Tools
  - Line Charts
  - Flow Charts
  - Histograms
  - Pareto Diagrams
  - Scatter Diagrams
  - Control Charts
  - Cause and Effect Diagrams

Management prerequisites are concepts such as management by planning (as opposed to reactive or crisis management), employee involvement, and management "tools" such as affinity diagramming and tree diagrams, for example. Management should also be aware of the process of planning using Gantt and PERT charts, and even the computerized project planning tools available at a relatively modest price and capable of operation on several of the most popular personal computer operating systems. These types of management skills are necessary because of the very nature of QFD, as has been explained in great detail throughout this book. To begin with, QFD is obviously a planning tool and one of its results is a plan of action for execution. Companies that are used to reactive management techniques will usually fail at a pilot implementation of QFD because of a failure to properly support and commit resources to the QFD effort. Organizations that understand planning management will have a much greater success rate with a QFD implementation program because they understand the process of planning upon which QFD is based. This does not mean that QFD is a slow or ponderous tool, but merely that reactive management organizations fail to take the proactive steps in a QFD regimen that are required for its success. Management must be involved in the QFD effort and *plan* for its success as opposed to isolating the QFD effort and only becoming involved when a crisis appears. QFD requires a rethinking of problem-solving strategy and the leadership of an organization must be in tune with this approach. If a company feels that more planning expertise is necessary before taking on a QFD pilot project, then the company should engage in a program of proactive management first in order to lay down the foundation for applying more advanced applications such as QFD.

Management must also have a ready understanding of basic tools such as affinity diagramming, interrelationship graphs, matrix data analysis, tree diagramming, matrix charts, process decision program charts, and arrow diagrams. Most organizations that have tried any type of process improvement, TQM, or other quality improvement programs are experienced in using these tools. Not all management personnel need to be experts or well-versed in these tools but the individuals that will be participating in the program would at least have an understanding of the purpose and use of these tools in order to successfully apply them where needed during the course of the QFD pilot program.

As mentioned above, widespread employee awareness is also desirable in instituting a QFD pilot program to eliminate suspicions, rumor, and resentment. This does not mean that the scope of the pilot program must be so wide as to include all of the employees. Awareness can be involvement to the extent that those employees not directly participating in the QFD pilot know that the program may involve them in the future and that they are to support the program with information and resources, when necessary.

The other type of tools required are application/analysis tools such as Statistical Process Control, Design of Experiments, Pugh Methods, etc. Chapters 7 and 8 showed how QFD utilizes additional techniques that may already be part of a company's battery of resources. These tools help apply the information that is developed in the QFD or are used to analyze data in support of the decisions that are made using the QFD matrices. Product and manufacturing organizations will probably be very familiar with these tools and no additional training or learning will be required. Depending on the sophistication and service provided by service-oriented companies, additional exposure to these tools may be required. Organizations that are planning to use a pilot QFD in a product development area or design function may also find it extremely beneficial to ensure a fairly high level of understanding of Value Engineering/Value Analysis, Failure Modes and Effects Analysis and Fault-Tree Analysis. These skills are also helpful in service organizations. The authors commonly use Fault-Tree Analysis in a service organization in applying QFD to process improvement.

A final requirement may appear to be obvious, but sometimes is overlooked especially in industries using a larger number of unskilled labor. A basic competency in writing, reading, and basic math skills is necessary to participate in the QFD matrix completion. Employers who are more apt to leave out workers who are not skilled in reading or math rather than training these workers in these basic skills are missing out on a possibly important information source concerning that particular worker's responsibilities. Sometimes the implementation of a QFD pilot program uncovers areas needing skill development and this can be a valuable benefit in itself.

Having a good inventory of these skills in advance of the beginning of the QFD pilot program ensures a broader understanding of QFD and increases the chances for success. Advance preparation in these areas also reduces the possibility of an interruption of the pilot program in order to train participants in the use of these tools.

12.2.3 Organizing the project

Planning for the success of the QFD pilot program begins with organizing the effort prior to actually conducting the QFD analysis, and also involves more than just creating a QFD matrix. Visualizing in advance the steps of the pilot program, the necessary preparation, and the goals and benefits will keep the program on track. Figure 12-1 illustrates a checklist of items comprising organization of the QFD pilot project. The following is a brief description of the importance of each of the steps in the checklist.

- **Purpose and Customer** — The first step in organizing the project is for the leadership to decide the purpose of the QFD and the customer(s) targeted. By focusing on the purpose of the QFD the management will begin to determine the scope and resources necessary to complete the pilot program. This step is often undertaken with the guidance of the QFD facilitator. Defining a clear vision of what the QFD pilot project will accomplish is much like creating a mission statement for the company, as described in Chapters 2 and 11. The leadership of the organization is cautioned not to pick a critical program as the pilot project, or a new product that the company is betting its future on without a full understanding of the commitment behind using QFD. Selection of these types of projects are more suitable to the implementation of QFD after the success of a smaller pilot program. Projects critical to the success of the corporation are beyond the realm of a pilot designation and are more in tune with a full implementation effort with substantial support. The authors believe that successful first-time implementation of QFD can be accomplished on such high-visibility and critical programs, but these types of programs are not indicative of the type usually selected for a pilot or initial effort because management input and technique varies and success is critical.

Checklist for Organizing a QFD Pilot Project

- ☑ Purpose & Customer.
- ☑ Define the benefits.
- ☑ Define expected results.
- ☑ Define boundary conditions, ground rules and constraints.
- ☑ Define and collect appropriate marketing data.
- ☑ Select project team.
- ☑ Educate project team members of QFD.
- ☑ Describe product, process or service.
- ☑ Scope of matrix analysis.
- ☑ Establish schedule, meeting agendas, meeting results.
- ☑ Periodic management review of results.
- ☑ Perform QFD analysis.
- ☑ Implement results through Integrated Plan.

Figure 12-1 Checklist of items comprising successful organization of a Pilot QFD Program.

- **Define Benefits** — The leadership, with the help of the QFD facilitator, should be able to fairly accurately describe the anticipated benefits of the QFD pilot program. The benefits should be broadly defined at this point such as model for wider implementation, define cost-savings methodology, redesign of existing product, define new service, etc. The nature and purpose of the QFD should give insights into what the expected benefits should be. At this point in the QFD pilot project the actual detailed project may not yet be selected, but the management should know the kinds of benefits expected. Deciding this in advance helps to focus selection of the actual project. Defining the benefits expected will also define the benchmarks by which to measure the success of the QFD pilot program.
- **Define Expected Results** — The results to be defined are linked directly to the benefits and can be described as the results of achieving a benefit. For example, the result of defining a cost-savings methodology would be implementation and the actual outcome would be reduced costs. Defining results will greatly increase the focus of the pilot program by providing real goals.

- **Define Boundary Conditions, Ground Rules, and Constraints** — When QFD pilot programs are less than totally successful — success defined by expected benefits — usually the project has run aground at some point on hidden reefs of budget constraints, personnel commitment, or other shoals not mapped out for the team in advance. Every project must have boundary conditions such as number of team members, extent of active support from other departments or functions, physical resources, and financial resources. A team that is undertaking the QFD effort needs to know these boundaries in advance in order to prevent inflated expectations or understand how far the project can be taken. Ground rules are also necessary: What is the reporting procedure to management? What mechanism is in place for resolving time commitment disputes if the team members are borrowed part-time from other areas? Can the Customer be brought in to the QFD analysis, and to what extent? Finally, any potential constraints must be identified such as time limitation for completion, processes or departments that are off-limits to either involvement or change during the pilot, and any other conditions that limit the potential actions of the pilot program team.
- **Define and Collect Appropriate Marketing Data** — Because the pilot project team will be making decisions about customer needs, sufficient marketing information should be available to the team and collected in advance. This is not to say that an extensive marketing effort must be undertaken in advance of the QFD pilot project, but as the authors discussed in Chapter 4, the needs of the customer must be understood. Additional marketing data may be required later, but the pilot team should be armed with as much relevant data as possible in advance.
- **Selecting Project Team** — Selection of the project team can be accomplished in the same manner as described in Chapter 11. The parameters of the pilot project as defined by its purpose and benefits makes the selection of the team usually an easy process. The authors have worked with human resources in companies to select members for a team based on skills needed, desire to be a part of the pilot project and expected outcomes for implementation of the recommendations or design outcomes.
- **Educate Project Team Members on QFD** — After the team has been selected, they should be apprised of the goals, benefits, constraints, and ground rules of the pilot project. The next step is to educate them in the QFD process itself in a 2- or 3-day workshop environment. This prepares the team for what is ahead and fosters the dynamics of team building. This also gives the facilitator a chance to observe the team and to establish a rapport.
- **Describe Product, Process, or Service** — Once the basics of QFD have been given, the next step is to start the process of defining what it is that the team will be doing as part of the pilot program. This step

relates directly back to the purpose and expected benefits and results of the program and offers the opportunity for the team to begin taking control of the outcome. The team should begin to use some tools like brainstorming, fault-tree analysis, tree diagramming, etc., to develop a clear, detailed picture of the project itself. Some of the considerations for selecting the project are

- Manageable by the team;
- Meaningful to the team members;
- Capable of completion in 3 months (recommended time frame);
- Visible to the organization;
- High probability for success; and
- Possible upgrades in service and product (impact).

- **Scope of Matrix Analysis** — The team, based on the constraints and purpose, should decide in advance the level of detail to which QFD will be used. How much of the Integrated QFD Approach™ is the team going to employ? To what level of detail are the processes going to be broken down? How many matrices will be needed? These are critical questions because the answers relate directly to the scope of the effort, the amount of time required, and the impact on the organization in terms of resources. For example, if the goal of the QFD pilot program is to determine the feasibility of a new product, several layers of analysis may be needed to determine if the concept design can be accomplished.
- **Establish Schedule, Meeting Agendas, and Meeting Results** — QFD is a method that relies heavily on planning, and a schedule should be established during the first meeting of the team. Computer planning aids are very useful in maintaining visibility into the project, and the pilot project should be geared toward producing an integrated plan no later than after completion of the first-level matrix. In addition, meeting schedules should be established for coordination of the effort between team members, as well as between team members and other departments, individuals or data areas of the organization. One important consideration that is often overlooked when setting up meetings is the purpose of the meeting. Meetings should be held to transfer information, resolve problems, or achieve other defined purposes. A scheduled meeting should not be held without a formal agenda. Figure 12-2 is a checklist for holding effective meetings.
- **Periodic Management Review of Results** — As part of the scoping effort for meetings, the team and management may also lay out a schedule for review and input by executive management and/or other management layers. Care should be taken to make these focused and directed meetings where the primary goal is for the team to educate the management on progress, roadblocks, etc. and not for management to undo the process by dictating actions or changing the direction of the effort.

Concept For A Successful Meeting

Pre-Meeting

- Plan who, what, when, where, & **Why, Why, Why**
- Secure facility and equipment
- Publish detailed agenda
- Verify attendance

Meeting Start

- Start on time
- Introductions
- Review / revise / stick to agenda
- Manage time / set limits
- Review action items from previous meeting
- Solicit full participation
- Record decisions

During Meeting

- Encourage participation
- Be non-judgemental
- Listen
- Ask direct questions
- Be succinct
- Balance power

Meeting End

- Identify tasks to be accomplished before next meeting
- Assign responsibilities
- Establish next meeting
- End on time

Figure 12-2 Checklist for holding a successful meeting.

- **Perform QFD Analysis** — Only when the above steps have been undertaken to the satisfaction of the team should the actual QFD analysis be done. The QFD analysis, and the implementation of other actions relating to QFD as described in Chapter 7, is the heart of the QFD effort, and will proceed successfully if the proper foundation has been laid by the team and management.
- **Implement Results Through the Integrated Plan** — Results without follow-through are worse than no results at all. A pilot program is not complete unless it is implemented, and this should be one of the key benefits or results of the pilot program. The result of the QFD matrix is not the QFD matrix itself; it is only a means to an end. The last effort

of the team should be to create an integrated plan based upon the results of the QFD analysis for implementation. This effort brings the QFD pilot program to fruition and will provide the springboard to wider implementation, creation of the product or service, or streamlining of the process.

12.3 *Pitfalls to avoid in a QFD pilot program*

Pilot programs can be subject to many pitfalls that can prevent their success, but by following the above procedure many of the potential problems are avoided. However, the leadership of the organization should be aware of some specific pitfalls that can more severely damage or stop a QFD pilot program.[49] The following are to be watched for and avoided, usually with the help of a qualified QFD facilitator:

- **Mandating QFD** — QFD should be encouraged, not ordered.
- **Perfect Data Syndrome** — There is no such thing as having the "perfect data" on any project, including one selected for a pilot program. QFD is a means of prioritizing based on less-than-perfect data, but using available *relevant* data. A pilot program that tries to gather all of the perfect data will stall and become bogged down in an indecisional morass. Use the best data for the task, even if it is not "perfect."
- **Misunderstanding of "Whats" vs. "Hows."** — See Chapters 4 and 11 for a complete analysis.
- **The Carbon Copy Mistake:** Merely copying what another department has done in attempting a pilot program, or what has been done at another company, without proper analysis and fine tuning for individual application is a sure recipe for failure. Although the process of QFD is repeatable, the application of it should be given proper analysis for the purpose it is being used and the outcomes sought. This is the reason for the significant planning and organizing prior to engaging in a QFD pilot program. It is beneficial to look at what others may have done, but do not copy unless you know what it is you are copying and the logic behind its application.
- **Lack of Planning for Customer or Market** — Purpose! Purpose! Purpose! This pitfall is easily avoided by understanding the purpose of the QFD and the intended customer.
- **Lack of Agreement on Definitions** — Make sure all team members understand the definitions of the terms being used in the project to avoid misunderstandings and miscommunications.
- **Large, Unfocused Projects** — This pitfall is avoided by following the key steps above. No pilot project should be selected that is incapable of being defined with specific benefits and results.
- **Maintaining an Internal Focus** — The company and the team should be aware of the time deadlines and the resources that need to be accomplished; and delay of the pilot program should be avoided to prevent disinterest and abandonment.

- **QFD Charts Too Large** — The pilot program should be focused and a familiarity with the analysis of the Core QFD Matrix should be maintained (see Chapters 4 and 11). Charts that are too large prevent focus and bog down the team.
- **Charts Mix Many Types of Items** — This pitfall can be eliminated by focusing on the scope and purpose of the QFD pilot program. Usually, this problem occurs if the pilot program is attempting to encompass too much breadth or depth without the appropriate resources to support such an analysis.

Following the guidelines for implementation set forth in Section 12.2 increases the chances of avoiding these pitfalls, but the team, management and the facilitator should always keep these errors in mind throughout the QFD pilot program effort. Much of the success of a pilot program is obviously a result of preparation, not magic or chance.

12.4 After the pilot program is completed

After the successful completion of the pilot program the team should be used to formulate the next phases of implementation. Should QFD be implemented on a widespread basis, and what is the best implementation scenario? Is the company ready for a larger effort? These are questions to be considered by the team and the leadership of the company. A successful QFD pilot program often mobilizes such interest in the rest of the company that management can reap incredible goodwill and energy to undertake other projects.

But one caveat about a successful QFD pilot program: *Success and visibility empowers employees, and management risks negative repercussions if it does not continue the effort.* The elements that ensure the success of a QFD program — visibility, communication, planning, and leadership — are powerful forces when unleashed in any organization and management should have a plan for directing these forces after they have been unleashed. As the old saying goes, "Careful what you wish for, you might get it!"

12.5 Chapter summary

1. A pilot program can ease the organization into QFD and provide a successful starting point from which to expand its use.
2. Tools such as affinity diagramming and tree diagramming are necessary prior to the start of a QFD pilot program.
3. Widespread employee awareness in desirable to prevent negative rumors.
4. Selection of the pilot program subject matter is critical and should be in line with the resources management is willing to apply to the program.

chapter thirteen

Case studies and examples

Knowledge must come through action;
you can have no test which is not fanciful, save by trial.
Sophocles (495–406 B.C.), Trachiniae.

13.0 Application of QFD produces results

Many management books focus more heavily on case studies that explore how a company has used a technique or employed concepts in its own organization. Engineering textbooks have more of a tendency to focus on theory with some practice applications included. This book has focused on both concepts and implementation of those concepts involving QFD analysis. Although the authors believe that this book provides a great deal of information, theory, and application recommendations, it is also important for the reader to know exactly how QFD has been applied in other organizations. This chapter takes a closer look at seven applications of QFD and related techniques in solving specific problems in: (1) a small business; (2) an educational setting; (3) the manufacturing process; (4) manufacturing safety compliance; (5) waste reduction; (6) environmental cleanup; and (7) new product development.

The examples given here are not meant to provide a basis for completely understanding QFD analysis and application, but to supplement understanding of QFD by providing reference points of application. These examples demonstrate that not only is QFD used, but used to great advantage in diverse businesses and organizations.

13.1 Die Cut Technologies: QFD and the small business

One complaint that the authors have heard expressed about QFD is that it is too complex for application in smaller companies and organizations. This is simply untrue. The authors have worked with the U.S. Small Business Administration in offering short courses on using QFD in manufacturing for small businesses. In fact, because of the lack of preconceptions about TQM and other management and manufacturing improvement methods, the authors have found small businesses not only capable of utilizing QFD, but embracing it much faster than some "technically advanced" large companies. QFD is not just for large companies — it is an egalitarian tool.

One of the small companies that has embraced QFD is a company called Die Cut Technologies (Die Cut). Die Cut used QFD to solve a critical problem, and that solution also provided for some astonishing growth.[50]

13.1.1 Background of the company

Die Cut is a 30-year-old small manufacturing firm with about 30 employees. During its development it had built a good reputation for being a low-cost provider of finished gaskets. However, when Scott Flores took over as president in 1987, he determined that the company needed to find a new identity in the marketplace if the company was to survive and grow in the 1990s.

13.1.2 Purpose of QFD analysis

Flores' basic strategy was to transform Die Cut Technologies from a manufacturing firm that competed in the low-cost gasket market, to a firm that could compete in the higher quality market of specialty products and service. Flores' market analysis determined that large prime contractors in the defense industry were looking for businesses that could meet or exceed high-technology requirements on time, every time. The high-tech gasket market was the place to be — but not without some major changes.

Although Die Cut began to move toward the high-tech market, it soon found that it was scrapping approximately $8000 in gaskets per month, a clearly unacceptable rate for a $2-million-a-year enterprise. Most of these gaskets being scrapped were returned from the customer for not meeting specifications, which the president referred to as a "stays sold" problem. In order to solve this problem Flores had been approached by machine manufacturers to buy an expensive gasket-cutting machine to replace his older machines and to give him "better quality." This quick, but expensive, solution was appealing to Die Cut, but Flores resisted the temptation to automate without understanding the root cause of the quality dilemma. Using QFD the president was able to prioritize his needs, solve the problem, and come up with some surprising information on his organization.

13.1.3 Results of the QFD

Flores suspected that a major cause for gaskets being returned was not outdated equipment, but employees lacking the skill level to do Statistical Process Control (SPC). Armed with this suspicion, Die Cut did a basic QFD matrix analysis of customer needs vs. focused requirements, illustrated in Figure 13-1. The matrix (basically a systems level) shows the simple methodology used to discover the critical needs and technical solutions, and also the technical goals that needed to be met in order to solve the problem. Two key requirements emerged: (1) the quality methods skill level for employees needed to be improved, and (2) improvement was necessary in the consistency for checking parts.

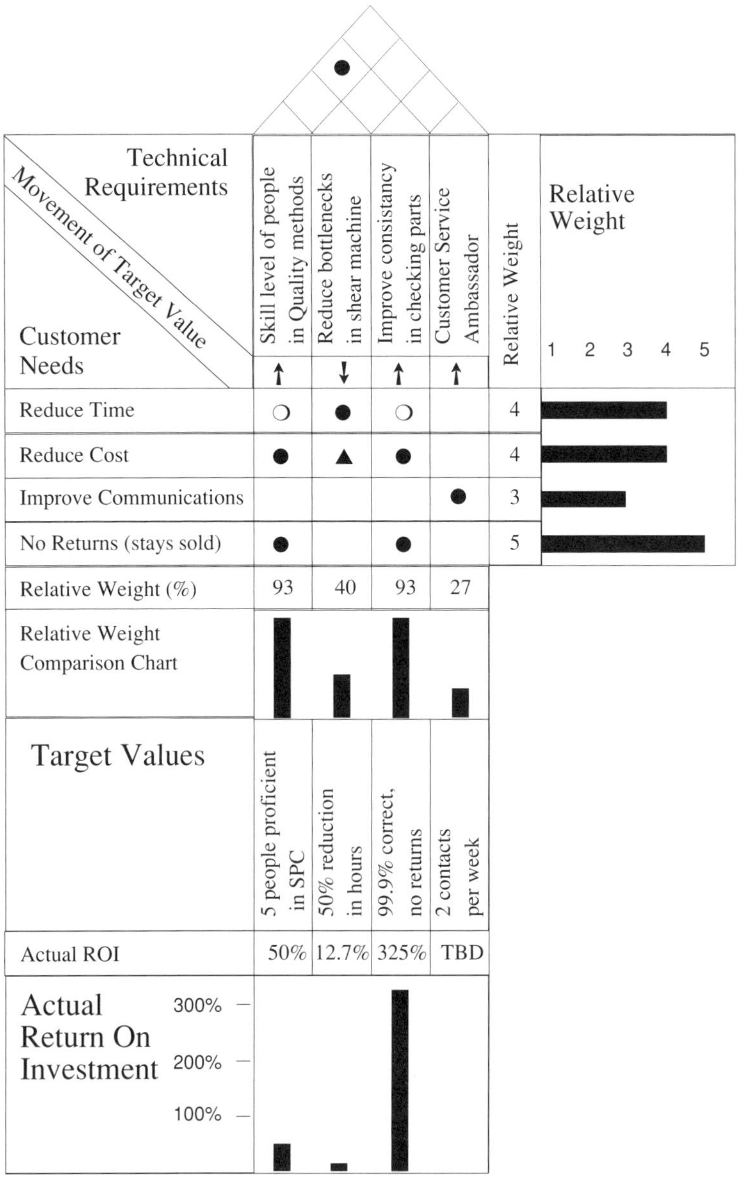

Figure 13-1 QFD as applied to small business manufacturing: Die Cut Technologies.

After spending approximately $12,000 to train employees in SPC, Die Cut not only reduced scrap to near zero by gaining the necessary consistency in checking parts — saving almost $100,000 per year — but also placed itself in position to become qualified as a vendor for several aerospace companies. This qualification for aerospace work netted an increase of 30% in sales. The return on investment for each QFD solution of is graphically represented in Figure 13-1 as well.

13.1.4 Remarks

Sometimes the obvious is not so obvious. Many company leaders jump to the conclusion that a new machine or increased automation will fix poor quality. Sometimes this conclusion is reached because production needs to be increased and employees often feel the stress of the pressure for faster production, expressing concern that reduced quality is the result. However, even if Die Cut had purchased a new machine, the result would have only been the capability to make bad parts faster. The underlying problem of insufficient training in SPC and metrics to monitor quality would have remained. QFD identified a persistent problem in the competitive 1990s — training — and showed how training could help support the needs of the customer, giving Die Cut the "biggest bang for the buck." This QFD analysis shows that the evaluation of a system does not have to be complex. The analysis focused the small business in getting to the root cause with hard facts.

13.2 University of Colorado, College of Engineering and Applied Science: development of an Integrated Teaching Laboratory (ITL)

QFD is not limited to application only in the commercial or industrial world. QFD analysis is very nearly a "pure" analysis method because of its flexibility. The authors have found that the applications of QFD may only be limited by the imagination of those who try it. This example involves the use of QFD in higher education in order to design an innovative engineering curriculum.[51]

13.2.1 Background

The College of Engineering and Applied Science at the University of Colorado at Boulder is in the planning stage for a new concept in undergraduate teaching involving the creation of a laboratory facility designed to encourage experiential, hands-on learning across all engineering disciplines in a problem-solving setting. The name of this creative plan is the Integrated Teaching Laboratory (ITL). Completion of the building housing the laboratory is set for early 1997, and the college is in the process of curriculum development.

13.2.2 Purpose of the QFD

The goal of the faculty curriculum team was to achieve a diverse experimental learning environment for their students that would take into account various learning styles and new theories on how people acquire and build knowledge through team learning. In addition to finding ways to increase the student's understanding and retention of engineering knowledge, the team also wanted to include methods to continually refresh that knowledge and create an enthusiastic environment for interdisciplinary learning across engineering disciplines. Most importantly, a primary result of the ITL was to instill in students the desire for life-long learning.

To translate these goals into realizable targets and requirements, the undergraduate engineering faculty performed a QFD to explore the needs of their customers and develop a new and motivating teaching environment.

13.2.3 Results of the QFD

The faculty defined their primary customers as students, with faculty, teaching assistants, and the students' future employers viewed as secondary customers. In brainstorming the customer needs, the faculty agreed that all of these customers demanded a learning environment that offered the necessary engineering subject matter, and skills such as communication and computer literacy. But if the ITL was to do all the things proposed by the students and faculty, motivation to learn was a critical need.

To help define this last need, the faculty asked the following question: "What motivates the students to learn?" If this question was asked of the business world who employ engineers, the answer would have been unequivocal — getting a job. However, recent surveys had illustrated that students learn the most when they believe their lives are moving towards making a contribution socially and that the subject matter relates to real world situations. Figure 13-2 shows the results of the QFD analysis from the ITL strategic planning session. The three main customer need categories were (1) Implicit Curriculum, (2) Subject Matter Knowledge, and (3) Motivation.

A somewhat surprising result (at least to the faculty) generated by this initial QFD session was that the high priority placed on mode of teaching instruction was *as important* to the customers as the content of information being taught. Some engineers, when thinking back to their college days, may not find this priority so surprising.

Once the strategic planning QFD was completed, the next step was to answer the question: What modes of instruction would best meet these needs, and what was the best way to integrate these issues? The solution would be to create an integrated experiental learning-based laboratory.

Another QFD analysis was performed to define the scope of an innovative facility designed to encourage learning over a broad range of engineering disciplines in a setting that strengthened creative problem-solving skills of students and groups of students. The ITL developers not only considered students as key customers, but also the faculty, donors, society, and employers. The top-level systems ITL QFD results are presented in Figure 13-3. The two most critical aspects of the ITL as defined by the QFD results as follows:

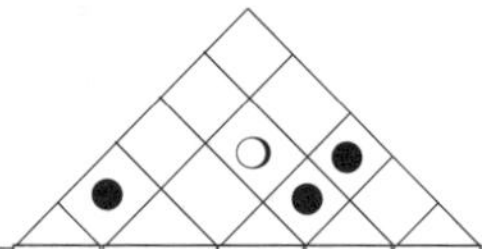

	Customer Needs \ Technical Requirements	Exposure to Advanced Technology	Mode of Instruction (Integrated Teaching Lab)	Examples from Real World	Course Content	Feedback and Evaluation	Relative Weight	Relative Weight
Implicit Curriculum	Communications Skills		●	▲	▲	●	3	
	Social Issues		○	○	●	▲	2	
	Computer Literacy	●	○	▲	●		3	
	Ethics		▲	●	○	▲	2	
	Economics	○	○	○	○		2	
Subject Matter Knowledge	Core Knowledge of Statics	○	●	○	●	●	5	
	Knowledge Application	●	●	●	●	●	5	
	Introduce Problem Solving Design	○	●	○	●	●	4	
	Critical Thinking	○	●	○	●	●	4	
Motivation	Relates to Real World	○	○	●	●	○	3	
	What is Engineering About?	○	○	●	●	▲	2	
	Social/Emotional Needs	○	●	○	▲	●	5	
	Better Employment Opportunities	●	▲	○	▲		1	
Importance Weight		156	273	183	273	249		
Relative Weight Comparison Chart								

Figure 13-2 QFD as applied to engineering education: the QFD results from the University of Colorado at Boulder, College of Engineering and Applied Science Integrated Teaching Laboratory strategic planning session.

		Learning Program	Student Involvement	Faculty / TA Training	Faculty / TA Rewards	Faculty Support	Support Services	External Interfaces / Affairs	Physical Environment	SW Selection, Use, Maint.	HW Selection, Use, Maint.	Systems Integration	Relative Weight	Unique Elements
Faculty	Education Effectiveness	●	●	●	●	●	○	○	●	●	●	●	5	○
	Enhance Professional Opportunities	▲		○	○	○	▲	○	▲	○	●	○	1	
	Faculty Recognition		▲	○	●	○		○	○	○	○	○	3	
Institutional	Attention to Undergraduate Education	●	●	●	●	●	●	▲	●	○	●	●	4	●
	Enhance Reputation	○	○	○	○	○	○	○	○	○	●	●	2	
	Increase Resources	▲	○	○	▲	●	○	○	○	○	●	●	3	
	Meet Constraints	●	▲	○	●	●	○	▲	●	○	●	●	4	
Student	Better Engineering Education	●	●	●	●	●	●	●	●	●	●	●	5	○
	Facilitative Learning Environment	●	●	●	○	●	●	●	●	●	●	●	5	●
	Prepare for Life Long Learning	●	●	○	▲	▲	○	○	○	○	○	○	4	
	Importance Weight	376	354	302	250	311	235	235	321	251	357	393		

Figure 13-3 Illustration of system level QFD charts analyzing the development of an Integrated Teaching Laboratory for the University of Colorado at Boulder, College of Engineering and Applied Science.

1. **Systems Integration** — Defined as the integration of curriculum development, lab experiments, facility experiments, and software and hardware networking.
2. **Learning Paradigm** — Defined as the process itself of creative learning including problem-solving, learning environment and adult-learning techniques.

Each of these categories was further decomposed for more detail in order to set a plan of action into place as illustrated in Figure 13-4. The results showed the urgency in developing an innovative curriculum concurrently with the construction of the facility and definition of the hardware, and software requirements.

As a result of the QFD analysis, the faculty agreed that "pilot" programs for instituting the new teaching curriculum should start immediately. The faculty developed new curriculum enhancements including a first year engineering projects course, new course demonstrations to enliven lectures using interactive work-station-based computer simulations augmenting physical experiments, and laboratory modules designed as experimental, open-ended adjuncts to theory courses. Finally, courses were developed that encouraged multidisciplinary team-oriented activity that emulates the way engineers function in the field. The faculty also developed an integrated plan for

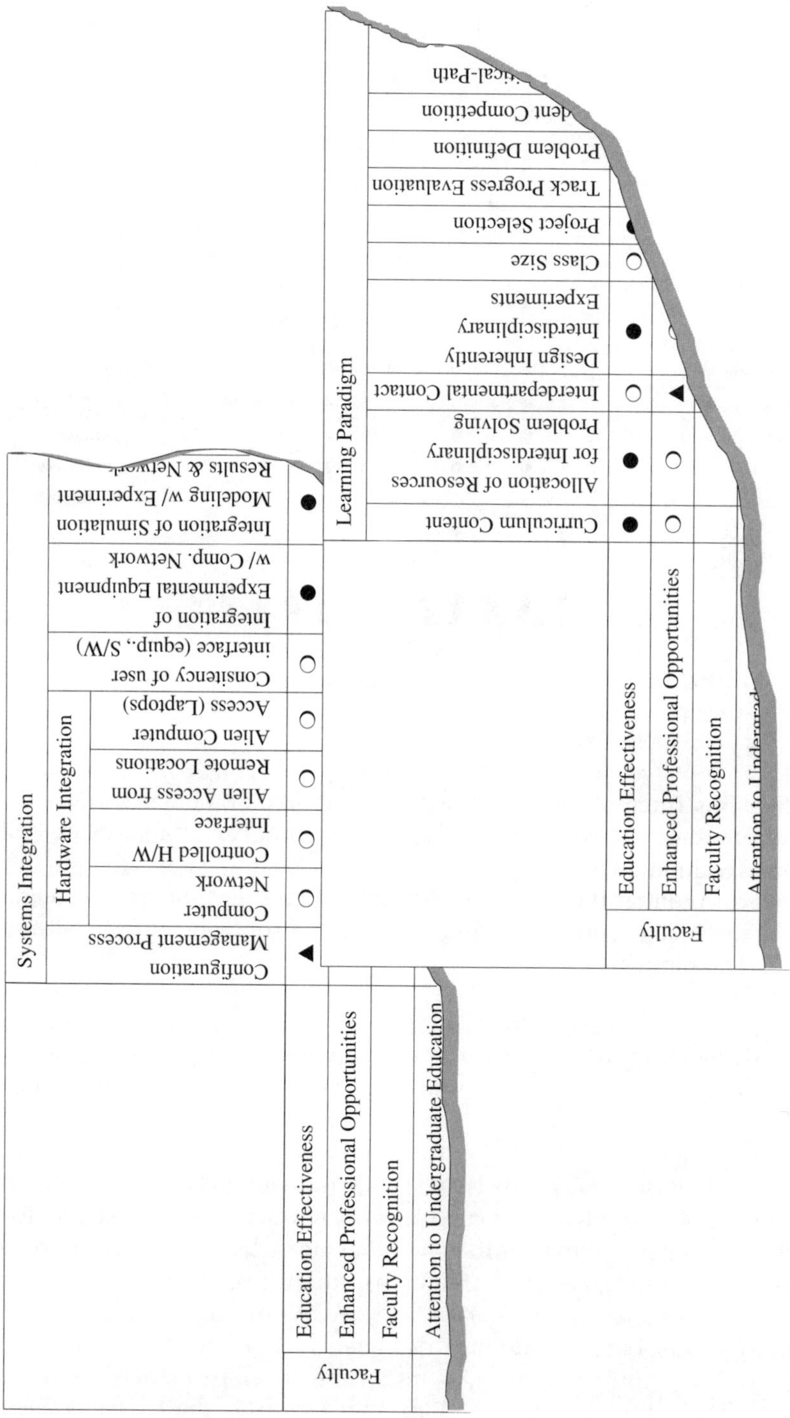

Figure 13-4 Illustration of Detailed QFD Charts analyzing the development of an Integrated Teaching Laboratory for University of Colorado at Boulder, College of Engineering and Applied Science.

implementing the ITL that included specific responsibility assignments for all aspects of the ITL, including funding and construction.

13.2.4 Remarks

The QFD analysis also supported the marketing and design phases of the ITL. As a result of the efforts of the faculty, the university has raised the necessary funds prior to each deadline, and the design phase is exactly on schedule. The faculty team has commented that QFD solidified the vision that gave the team focus and the necessary direction to implement an integrated plan of action. First year, "hands-on" design courses are currently being taught in an interdisciplinary approach and the response from the students — and the faculty — show that motivation is not a problem! The program director said that students and teachers are willingly volunteering to be a part of the program even though the work is harder and the hours are longer, because the process and results are fun and satisfying.

QFD in an educational setting should not be a surprise. The diverse interests in a university need methods for interaction and communication, and QFD (applied in a manner using many of the attributes of the Integrated QFD Approach™) provides disciplined focus to bring about results.

13.3 Aluminum extrusion process: Westinghouse Electric Corporation

The authors have stressed that the power of QFD comes from the use of QFD with statistical tools and methods. In the area of process improvement QFD provides the analysis to conduct designed experiments to evaluate the best process techniques. These two techniques — QFD and DOE — combined to improve and control the process of tight tolerance aluminum extrusion for Westinghouse Electric Corporation.[52]

13.3.1 Background

The extrusion improvement study was initiated in response to quality, cost, and schedule problems associated with producing high-precision waveguide aluminum extrusions. These waveguide elements are used in slotted planar array antennas of sophisticated radar systems. The precision of the waveguide extrusion is a prime factor in the performance of the antenna array. The waveguide elements used for this type of antenna are finned thin-walled aluminum extrusions whose function is to provide structural integrity and electrical performance.

13.3.2 Purpose

The object of the QFD study was to improve and control the current process of tight tolerance aluminum extrusion. The complex antenna assemblies rely on the extrusion process for product uniformity and cost. Radio frequency (RF) transmission is sensitive to surface deviations within waveguides, and extrusions eliminated the need for dip brazing or welding. Aluminum extruded waveguide provides structural integrity while minimizing weight. Extrusions, are, however, susceptible to twist, bow, and cross-sectional variability. Changes in the complex process can produce positive or negative results and therefore must be done in a controlled environment. The methods commonly used in the extrusion industry typically rely on an iterative "correction" procedure of both die geometry and press practices which is similar to a one-factor-at-a-time experiment.

13.3.3 QFD Results

A cross-functional team consisting of various experts from manufacturing disciplines as well as the customer was formed to brainstorm new die designs and process procedures, benchmark world-class extruders, and evaluate existing process parameters. This team-building concept worked to solidify the partnership between the customer, the primary supplier and subsupplier, and define and use their value-added capabilities.

The systems QFD analysis matrix presented in Figure 13-5 produced customer priorities of specific product quality characteristics. Functional requirements were weighted by the customer, then mapped by the team to the product specifications. Relationships between customer needs and product specifications identified the key specifications and led to development of separate matrices for the die design press and press practices as presented in Figure 13-6. The die design matrix enumerated die features that contributed to the key specifications. The press practices matrix identified press practices which affected the same specifications. These two final matrices were then used to identify areas for focused improvement and control.

The QFD analysis brought together information from experts from the customer, supplier, and subsupplier. The process of QFD analysis opened communication between the three groups to prioritize, plan, and strategize on approaches for a common set of performance criteria. For instance, the customer had no criteria that governed the placement of the weld seam — only that it did not impact on electrical and/or structural performance. The weld seams had been placed in the corners by the die designers reflecting typical visual requirements that in this case did not exist. The QFD helped the team prioritize the quality characteristics that were most important to the customer. This was followed with a mapping of their subprocess that affected the key characteristics and eventually a logical selection of factors for experimentation.

A statistical experiment was designed using Taguchi experimental design criteria to investigate the main effects of selected parameters of the die geometries and press practices on the quality of extrusion as defined by the QFD.

The QFD methodology combined with the experimental design provided a quantifiable and repeatable procedure for analyzing various combinations of

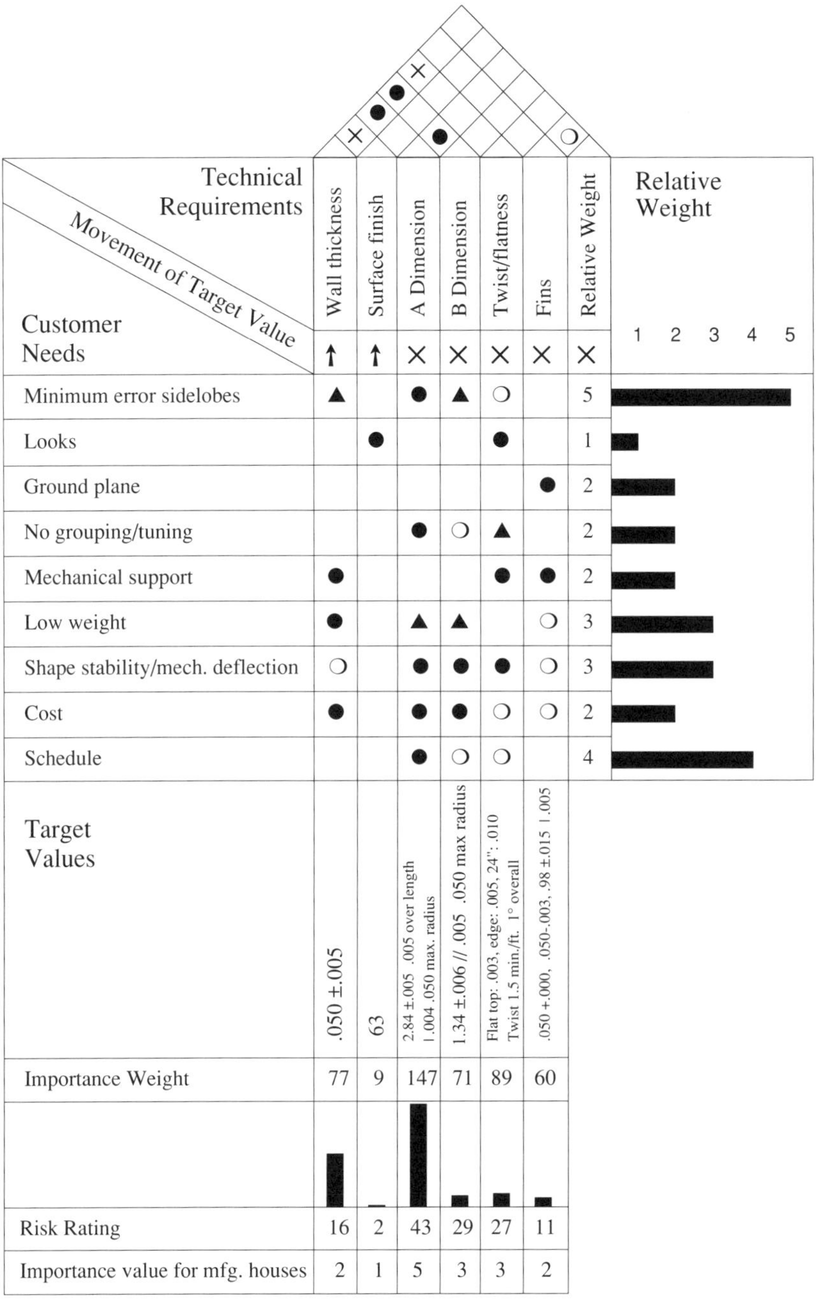

Figure 13-5 High-precision waveguide extrusion — System level QFD matrix.

	Design/Technical Requirements	Relative Importance	Press "Practices"/Operational Requirements: Drop Height	Die Alignment	Container Temp.	Container Wear
Twist	Twist (max. 1.5 min./ft.)	4	▲	●		●
	Twist < 1° overall	4		❍	❍	
tness	Flat over 24" (.010 max.)	3		▲	▲	

	Design/Technical Requirements	Relative Importance	Die Design: Bearing surface flatness, squareness	Length of port leg	Bearing length	Weld chamber	Under-cut geometry
Twist	Twist (max 1.5 min./ft.)	4	●	●	●	❍	
	Twist < 1° overall	4	●	●	●		
tness	Flat over 24" (.010 max.)	3	●	●	●		●
	Process Parameter Relative Weight		486	351	342	225	144
	Performance Targets		.0002 over 2.5" cut	Based on press size, PSI, and billet size	History CNC repeat, 2X wall thickness	75" deep	Shear edge

Figure 13-6 High-precision waveguide extrusion manufacturing process and operational QFD matrices for the die design and press operational requirements.

die geometries and press practices to produce tight tolerance products with minimum variation. The process is generic and can be used in many process-improvement applications to help accelerate change in the extrusion industry.

13.3.4 Remarks

QFD can be used as a tool for mapping requirements anywhere in the design process. As the Westinghouse example shows, long-accepted practices and processing techniques often do not incorporate shop practices that have changed, nor reflect variation that has developed in processing. Involving the customer and the suppliers helps to define better requirements and brainstorm solutions that can profit all entities involved in the process. QFD analysis also provides focus for the application of DOE, saving time and money.

13.4 Using QFD as an analysis tool in environmental compliance

Sometimes the use of QFD analysis to solve a specific problem leads to unanticipated benefits. Because QFD often requires *system* thinking and definition of interactions, technical requirements can be developed that do more than merely solve a design problem, but, also lead to overall improvements.

13.4.1 Background

More and more companies are not only turning to QFD as a requirements analysis method, but as a proactive problem-aversion method because of its ability to prioritize, plan, and implement. At the Westinghouse Electric Corporation in Baltimore, concern with volatile organic compounds (VOCs), used in painting, and upcoming regulatory requirements prompted the company to analyze its painting process to determine if compliance with the new regulations could be accomplished.[53]

13.4.2 Purpose

QFD was used to support the identification and reduction of VOCs in a painting operation while maintaining quality. The goal was to prove to the State of Maryland that Westinghouse would be in compliance with state regulations regarding volatile emissions by January 1, 1993.

13.4.3 QFD Results

Figures 13-7 and 13-8 illustrate how Integrated QFD Approach™ elements combine in a waste minimization example. The matrix presented in

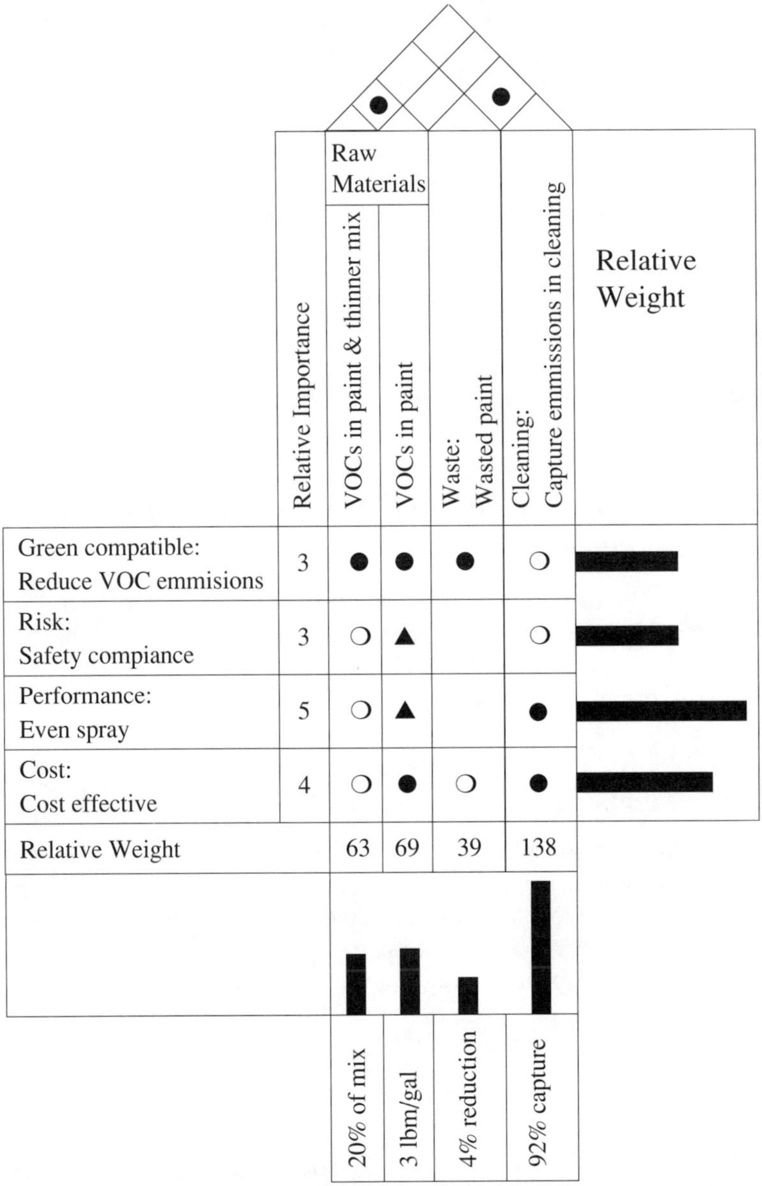

Figure 13-7 Environmentally conscious design QFD analysis to reduce Volatile Organic Compounds (VOCs) in a painting environment.

Figure 13-7 contains only parts of the full analysis, but provides an illustration of ranking and comparison of various key elements involved. The targets were developed by an integrated product team in order to meet emission reduction and performance requirements, and in this case to avoid a potentially substantial fine if compliance requirements were not met. The results

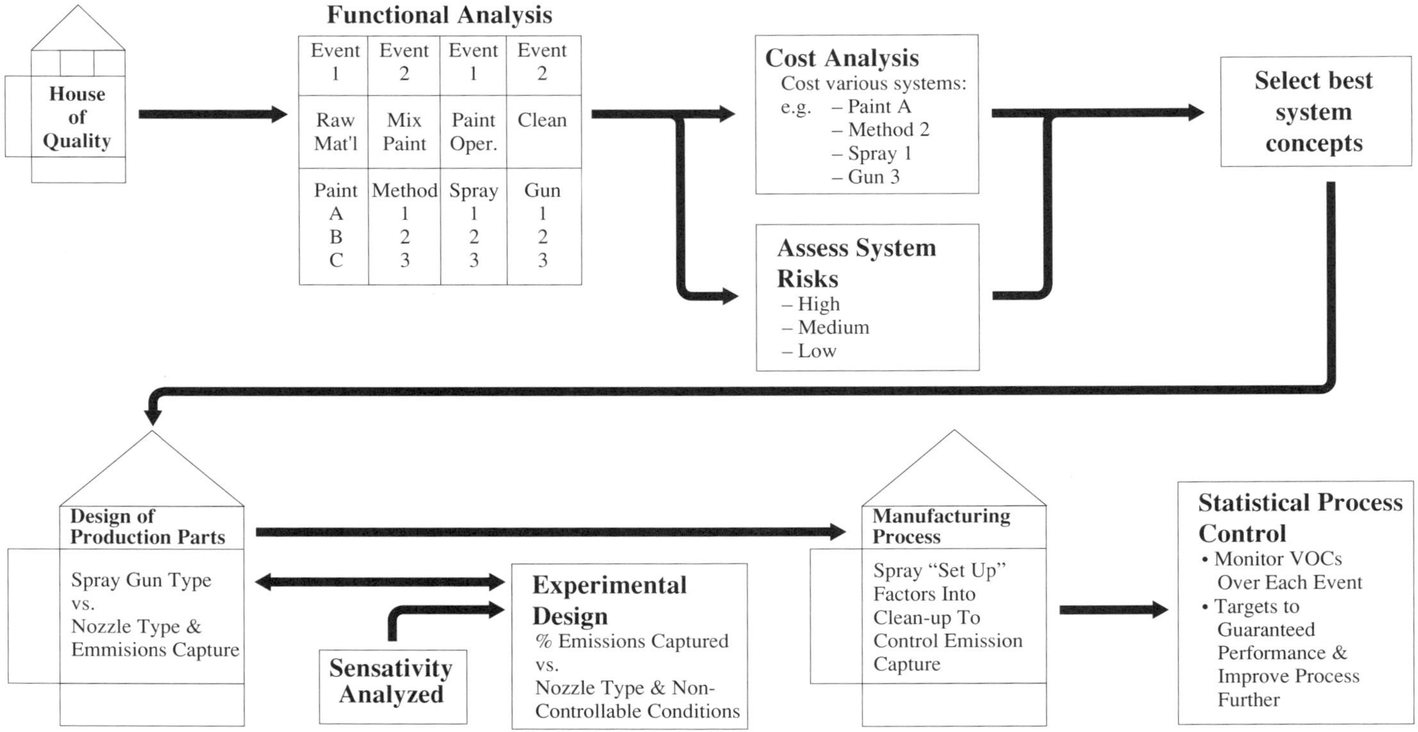

Figure 13-8 Environmentally Conscious Design Integrated QFD Approach™ for waste minimization and streamlined compliance to reduce volatile organic compounds (VOCs) in a painting environment.

of the QFD analysis led the team to change the paint spray gun nozzles and recapture 92% of spray emissions, which also brought the process into compliance with regulations. In addition, both paint and paint thinner (incorporating VOCs) were reduced, providing ongoing savings beyond the scope of the original problem of capturing VOC emissions.

The system QFD matrix requirements presented in Figure 13-8 show how the process was further analyzed and optimized through a functional analysis of the product and process being examined. Figure 13-8 incorporates elements of the Integrated QFD Approach™ such as Functional Analysis, DOE, and SPC. For example, in the manufacturing process, paint comes in as a raw material, it is mixed, applied through a spray gun and cleaned from the painting operation. By analyzing various options of paint types with methods of mixing, spraying, and cleaning, the product can be systematically optimized for cost performance and risk reduction.

Once the concept that best meets the performance targets established in the initial System Matrix is established, a further analysis is performed on the subsystem products. For example, the types of spray guns required to optimize the amount of VOC emissions captured are analyzed not only for their individual performance, but also for their performance over various anomalous conditions through the process of DOE. After defining the subsystem, the affected manufacturing processes are further analyzed in a QFD matrix. System validation is provided by statistically gathering data (SPC) to assure performance repeatability.

13.4.4 Remarks

Westinghouse not only met the compliance requirements for VOC emissions for the painting operation, but also reduced material usage and streamlined the process without degradation of quality. This is an example of the power of QFD when combined with DOE and SPC. The cost of conducting the QFD analysis and purchasing new equipment was offset substantially by the savings in material, not to mention avoidance of fines for failing to comply with the new emission standards.

13.5 Martin Marietta Aerospace Group: QFD and waste reduction

Regulations on hazardous and nonhazardous waste will continue to increase and restrictions will proliferate in the foreseeable future. Because of this reality, companies are looking to reduce the waste generated at their facilities as waste reduction becomes a far better economic solution than cleanup or storage of hazardous material. QFD analysis can play an important part in focused development of substitute materials and process improvement to reduce usage of materials that cannot be replaced by less toxic or nontoxic substances.

13.5.1 *Background*

Over the past several years Martin Marietta has taken steps throughout the corporation to comply with the Montreal Protocol by eliminating ozone-depleting chlorofluorocarbons (CFCs) and other solvents such as 1,1,1-trichloroethane (TCA) from all cleaning processes. One of the actions taken by Martin Marietta Aerospace Group (MMAG) was to analyze the hydrocarbon/ester/nonionic surfactant (HENS) semi-aqueous cleaning technology in its Electronics Manufacturing Facility (EMF) for reduction of these chemicals.[54]

13.5.2 *Purpose*

MMAG used a QFD analysis to identify the critical factors and requirements for selecting a cleaning solvent, and other factors in the soldering and assembly of printed wire assemblies. The objective was to use non-CFC biodegradable cleaning agents within the EMF assembly process so that effluents from the facility would meet local, state, and federal requirements. Solvents such as freon TMS and TCA were to be completely eliminated.

13.5.3 *Results*

In order to establish the customer needs related to the selection of a cleaning solvent, a survey was conducted that involved inputs from 13 industry and Department of Defense (DoD) electronics technologists.

The results of the survey were integrated as customer needs of a QFD top-level matrix and are presented in Figure 13-9. Two important conclusions were drawn based upon the priority ranking of the needs:

- "Meeting EPA requirements" edged out "Flux Solubility Parameters" as the most important aspect in selecting the cleaning solvent. "Zero Ozone Depletion Level (ODP)" also made the top third, which indicated a high degree of concern about the environment by the electronics assembly community.
- "Flash Point" and "Odor" did not appear to be a concern, because there are safe, adequate measures available in terms of equipment technology and methods for storage and handling of solvents.

The results of the analysis of the technical requirements are presented in the lower half of Figure 13-9. Some of the QFD team's conclusions were as follows:

- The Cleaner (solvent) Selection, Equipment, Solder Paste, and CFC Elimination requirements were deemed as the four most important areas for concentration.
- Device Stand-off Height (a mechanical requirement in the automated cleaning process) was considered secondary and systems test was of least importance.

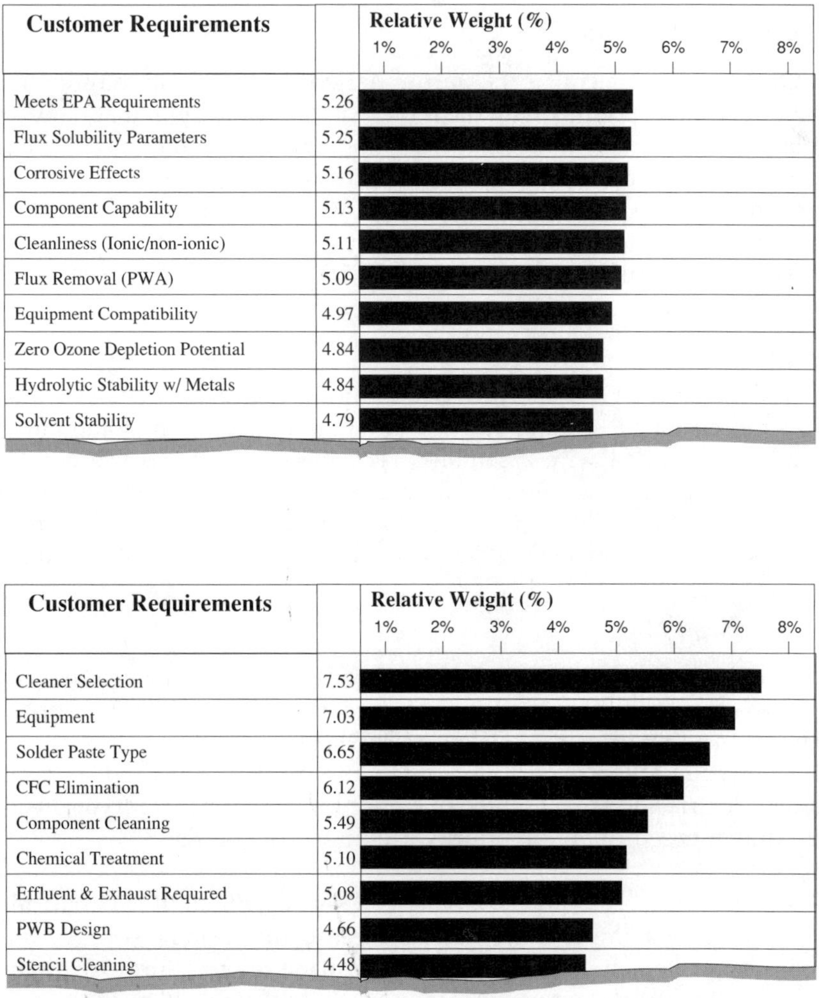

Figure 13-9 QFD evaluation results of automated semiaqueous cleaning of printed wire assemblies.

- A high degree of concern for the environment was reflected by the participants as a whole in that the treatment process and effluent/exhaust discharge quality characteristics scored high enough to make the top third of critical technical requirements to control.

The results were further analyzed using DOE, in order to select the optimum process characteristics and solvent. In addition to the DOE, comparison tests were run to determine if the new cleaning process was technically equal to the TCA solvent degreaser. Based upon the results of the experimental design, the new process result could be predicted to a statistical level of accuracy of 98.7%, and was a substitute for the old TCA process.

13.5.4 *Remarks*

QFD analysis synthesized the desires of the customer for environmental soundness in the cleaning of the printed wire assemblies and assisted MMAG in defining a new cleaning and soldering process that met the environmental criteria without sacrificing quality in the cleanliness of the parts. This study provided the support for a change in the process. QFD also affirmed some of the intuitive considerations of the team with regard to changes in the process.

13.6 *An example of RSM™ application to environmental processes*

Business has seen a shift of emphasis over the last 20 years from compliance to environmental strategy. No longer is it enough just to comply with regulations. Environmental performance affects shareholders, communities, and competitiveness. Companies have discovered that green marketing also means more greenbacks. Environmental managers can be found in all phases of the business structure from manufacturing to purchasing to finance.

Against this backdrop of an ever-expanding environmental industry affecting the way products are designed, any methodology of quality must also take into consideration environmental issues of waste reduction or minimization, pollution control, waste cleanup and environmentally conscious design and manufacturing. Although this book cannot address these issues in the detail necessary for full understanding of how to use QFD in environmental cleanup, the example presented here is given to show how even an area as highly regulated as the environment can make use of the basic principles of QFD.

The authors present a brief example of the applicability of RSM™ in environmental strategy and cleanup.

13.6.1 *Background*

In the environmental world, strategy is a somewhat elusive component that may be generated from more than one source. Companies often have environmental strategies as part of their company mission or long-term objectives. Governmental entities regulating the environment or industry may have strategies that run parallel to, or even conflict with, corporate strategy. In addition, community leaders, concerned groups, or individuals may also have strategies that impact government and corporate strategies. As a result, there has been a great deal of attention paid to the fact that a lot of money has been spent on very little toxic and hazardous waste cleanup. How is it possible to reconcile these varied and potentially gridlocking interests?

13.6.2 *Purpose*

An example of how RSM™ can help create a coherent strategy and resolve conflicting requirements or desires can be seen by using the Remedial

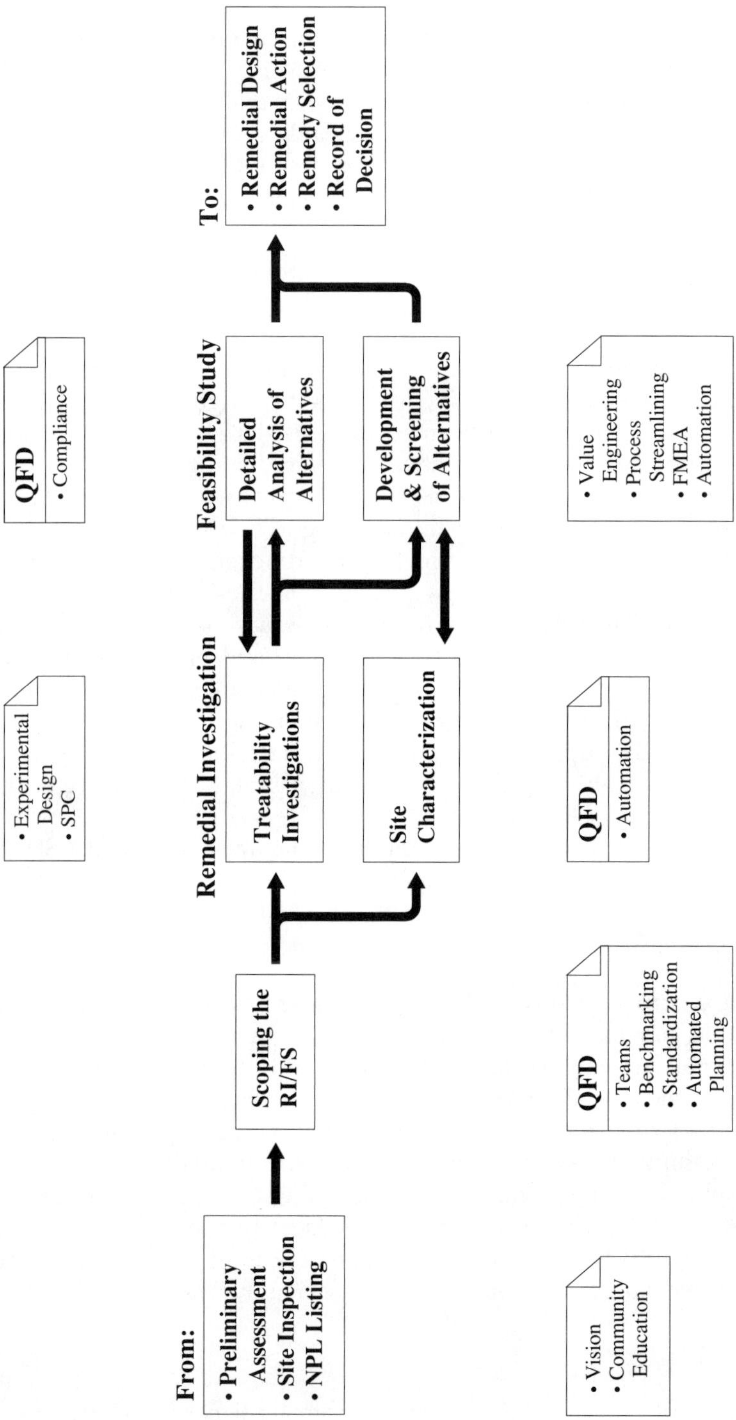

Figure 13-10 Application of QFD and quality techniques to the Remedial Investigation/Feasibility Study (RI/FS).

Investigation/Feasibility Study (RI/FS) scoping model employed at Superfund sites. This model, shown in Figure 13-10 shows the interplay between the elements of cleanup at a Superfund site. It is not important to comprehend the individual details of the complicated process in order to understand the application of RSM™ to the model in developing coherent strategy and implementing that strategy through the process. The important concept to understand in Figure 13-10 is the use of the strategy to create the vision for the process at the time of Preliminary Assessment and Site Inspection. By utilizing a QFD approach among the various parties involved at the site, a single unified vision, or strategy, can be developed and pursued. The authors have indicated where the various elements of RSM™ come into play during the process and are utilized to maintain feedback and consensus with the overall strategy.

13.6.3 QFD "Results": a strawman QFD as applied to the remedial investigation/feasibility study process

To illustrate how to deal with the varying interests at a Superfund Site, the authors have created Figure 13-11 — a "strawman" example of utilizing the Integrated QFD Approach™ matrices in a hypothetical scoping of the RI/FS. The most important idea contained in Figure 13-11 is the visualization of tracing customer needs all the way through to doing the actual data collection. What this model is meant to accomplish is to surface the conflict by tracing the requirements of the parties all the way through to actual data collection to support technical analysis and the eventual cleanup solution or solutions.

13.6.4 Remarks

The authors have seen first-hand that much of the work in environmental cleanup, however well-intentioned, is lacking basic systems thinking and organization. The application of methodology such as RSM™, incorporating Systems Engineering and Concurrent Engineering principles as well as sound strategy, could go far in accelerating cleanup and reducing waste.

13.7 3M hand-held wire marker: QFD and the new product

One of the most effective methods of creating a successful new product is to know that what is being made is what the customer wants. QFD analysis is at its best when used in conjunction with direct customer feedback early in the requirements development stage. 3M, well known for creativity and developing new products, uses QFD extensively in developing new products, and also makes extensive use of the customer as part of the process.[55]

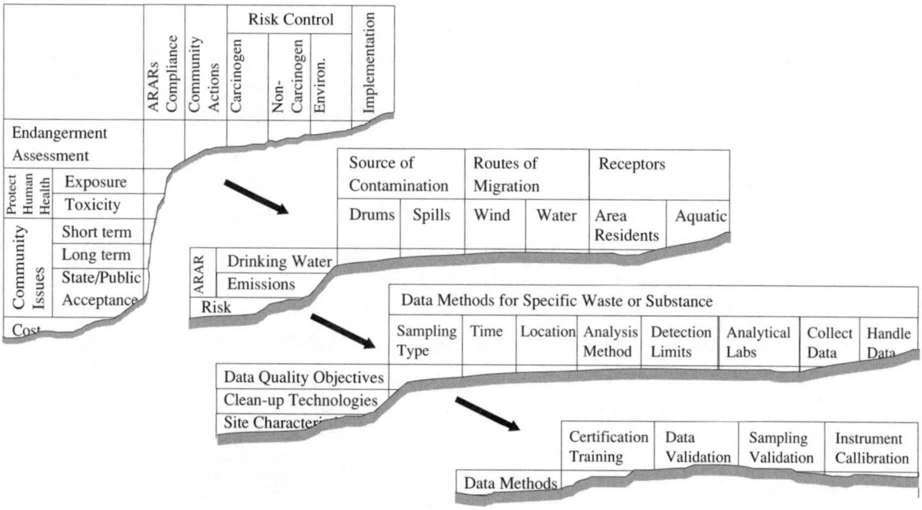

Figure 13-11 An example flowdown of QFD matrices as applied to the Remedial Investigation/Feasibility Study (RI/FS).

13.7.1 Background

The 3M Austin Center in Texas was given the task of creating a new generation of products called wire markers that are used to identify electrical wiring for use in a variety of applications. 3M had been selling a low-cost wire marker for several years, but after initial discussions with some of their customers, 3M saw an opportunity for upgrading the product to better fit developing customer needs.

13.7.2 Purpose

3M had conducted some substantial marketing information on attributes for a new wire marker. The purpose of the QFD was to verify some of the marketing data through involvement of customers directly, and to create a prioritization of the customer needs to help direct the focus of the technology development.

13.7.3 QFD Results

The customers or end users in the field were invited into the 3M facility to discuss the needs for the development of a hand-held printer. The first day of the QFD analysis involved the customers directly in the process. The results of the first day's morning session provided a prioritized list of customer needs as illustrated in Figure 13-12. In the afternoon session, the customer provided insights into how the marker would be used in the field, what kind of physical environments the marker would be used in and the

		Relative Importance	Company Now	Competitor "A"	Planned (2 years)	Ratio of Improvement	Sales Point	Importance Weight	Relative Weight	Relative Weight
Companion Tool	Small enough to carry around	2	3	4	3	1		4.0	6	
	Light Weight	3	4	4	4	1		3.0	4	
Reliable	Last Long (2 years)	2	3	3	3	1		4.0	6	
	Well constructed	3	4	3	4	1		3.0	4	
Permanent	Stable Image (text)	2	2	3	4	2		8.0	11	
	Marker stays on wire	5	2	3	4	2	●	15.0	21	

Figure 13-12 3M QFD example of customer requirements for a hand-held wire marker.

marking material would be subjected to, and specific targets the team should consider when developing the marker. These are reflected in Figure 13-13.

The QFD analysis not only confirmed much of the previously collected marketing data, but solidified alliances with customers. Customers offered to be beta test sites and to use new marker designs in worst-case environments. Customers also agreed to track the product performance as a means of supporting the integrated product development. The beta site information would substantiate the robustness of the system.

The next 2 days were spent developing the rest of the design concept QFD matrix, developing technical target values, and evaluating concepts against the requirements. As of the writing of this book, the design group had created prototypes and was preparing for customer focus sessions prior to production.

13.7.4 Remarks

The detailed results of the QFD analysis could not be revealed here based on the competition-sensitive nature of the new product, but the results show that the involvement of the customers during the initial stages of development not only confirmed the need for the product, but created alliances with the customers for the product testing. This is one of the key strategies of the 1990s — new alliances between customers and manufacturers that begin when the product is still in the concept-development stage.

		Keep It Simple			Marker Performance			Portability		
		Fast to Enter Information	Minimum Learning Time	Works like a Well Known Product	Solvent Resistant	Not Affected by Heat	Fits All Wires	Weight	Control Volume	Design Shape
Companion Tool	Small enough to carry around	○		▲				○	●	●
	Light Weight	○						●	○	○
Technical Importance		256	188	193	394	394	394	231	231	231
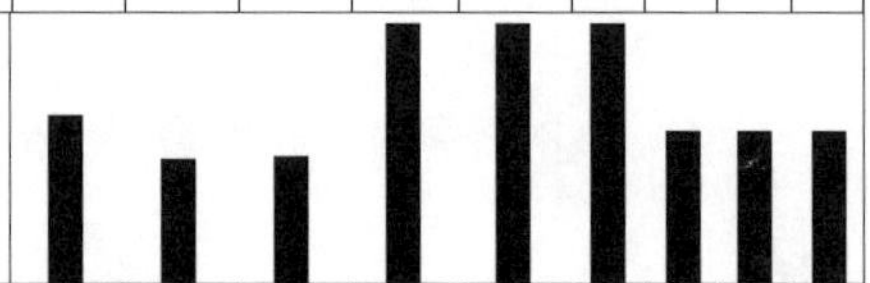										
Target Values		Same time as it takes to write info.	10 min. to use it, 20 min. to maintain it	X% task complete within time limit	Standard resistance/ 0/64 flag	105°C rating 0/64 flag	10-22/fits properly	Less that 1 lb.	25 cubic in.	Fits all pouch/carry holster
Concept "A"		○	●	●	▲	▲	▲	●	○	○
Concept "B"		●	○	○	▲	▲	▲	●	●	●
Concept "C"		▲	▲	▲	○	○	○	▲	▲	▲
Concept "D"		▲	▲	▲	●	●	○	▲	▲	▲

Figure 13-13 3M QFD example of technical requirements for a hand-held wire marker.

13.8 Conclusion

This chapter has provided a look at the application of QFD in some radically different scenarios. The authors are also using QFD, RSM™, and the Integrated QFD Approach™ with financial institutions, medical device manufacturers, and computer data storage companies to name but a few of the additional types of application of these techniques. The reader should recognize that QFD is truly a powerful tool when used properly, and that the "proof is in the puddin'" when it comes to applying QFD to actual problem-solving situations.

chapter fourteen

Conclusion

I will not steep my speech in lies;
the test of any man lies in action
Pindar, (518–438 B.C.) *Olympian Odes*

With regard to excellence, it is not enough to know,
but we must try to have and use it.
Aristotle (384–322 B.C.), *Nicomachean Ethics*

Just Do It.
Modern admonition.

14.0 The right method at the right time

The authors' purpose in writing this book has not been to only describe the uses of QFD, but to incite the reader to use it — and use not only QFD, but the other methodologies described in this book in the context of QFD application. In its narrowest application, the QFD matrix by itself is just a method for prioritizing various elements. But when QFD analysis is used in conjunction with the methodology of RSM™, the Integrated QFD Approach™, Systems Engineering, Concurrent Engineering, Process Improvement, and Business Re-engineering it acts as a catalyst for change, an idea generator, a communication facilitator and an integrator of systems thinking.

Back in Chapter 2 the authors talked about the perception of quality, and that there have been many quality experts over the past 40 years who have propounded excellent concepts and principles for achieving quality. Deming was mentioned as having his name attached to TQM (although it has been said that Deming disliked the term and never used it himself). One of the goals of this book has been to demonstrate that the principles of quality need to be *understood*, but the *execution* of those principles is even more important. QFD helps to implement the principles of quality.

Many readers may be familiar with Deming's Fourteen Points for quality. These points were developed and refined by Deming over the years to provide insight and discussion into what Deming saw as key principles for achieving quality. QFD provides the means to implement Deming's Fourteen Points. Figure 14-1 lists Deming's Fourteen Points and shows how QFD techniques operate to implement the principle.[56] The first point, "Create a constancy of purpose toward improvement of product and service, with the aim to become competitive, stay in business, and provide jobs," should be the obvious goal of every business organization. But how is this goal achieved? QFD, especially when linked with strategic planning and integrated planning such as in RSM™, develops a customer-focused strategic plan with product, operational goals, and measurement systems that can be tracked to a long-term mission. The use of QFD can create the "constancy of purpose" Deming talks about by focusing the strategy of the company on the critical customer needs. The remainder of the Fourteen Points is similarly satisfied

Deming Points	Implementation Support Offered by Applying QFD Techniques
1. Create a constancy of purpose toward improvement of product and service, with the aim to become competitive, stay in business, and provide jobs.	Development of a customer-focused strategic plan with product and operational goals and measurement systems that can be tracked to a long-term customer focused mission.
2. Adopt the new philosophy. We are in a new economic age, created by Japan. Western management must awaken to the challenge, must learn their responsibilities, and take on leadership for change.	Provide a framework for team decision making and organizational empowerment through well defined, prioritized integrated plans including expected outcomes, and roles and responsibilities. Provides focus on meeting the customer needs and beating the competition.
3. Cease dependency on inspection to achieve quality. Eliminate the need for inspection on a mass basis by building quality into the product in the first place.	Customer focused performance targets that can be measured throughout the product life-cycle.
4. End the practice of awarding business on the basis of price tag. Instead, minimize total cost. Move toward a single supplier for any one item on a long-term relationship of loyalty and trust.	Provides a method to translate company goals to supplier expectations and performance measures for selection and subcontractor management.
5. Improve constantly and forever the system of production and service, to improve quality and productivity, and thus constantly decrease costs.	Provides measures for constant performance improvement and the interaction of product, service, and operational characteristics for continual system enhancement.

Figure 14-1 QFD supports the implementation of the Deming Principles.

Deming Points	Implementation Support Offered by Applying QFD Techniques
6. Institute training on the job.	Provides opportunity to educate all team members on critical issues of the product, process, and operations on what they are, how they function and on how these activities and goals relate to the customer.
7. Institute leadership. The aim for leadership should be to help people, machines and gadgets to do a better job. Supervision of management is in need of overhaul, as well as supervision of production workers.	Provides a mechanism for company personnel to define and prioritize their activities and translate them into integrated plans. Provides means for delegation, self-management and organizational empowerment.
8. Drive out fear so that everyone may work effectively for the company.	Provides a creative environment for brainstorming new methods and designs in an objective, non-judgemental environment.
9. Break down barriers between departments. People in research, design, sales, and production must work as a team to foresee problems of production and in use that may be encountered with the product or service.	Provides maximum involvement of the company personnel during the most exciting, innovative stages of product development and operational streamlining such that individuals recognize their contribution and perform their activities in an atmosphere of mutual respect. Operationally defines the ultimate customers needs and expectations so that everyone understands their contribution to the organization.
10. Eliminate slogans, exhortations and targets for the workforce that ask for zero defects and new levels of productivity.	Establishes system performance targets for a product, service of operation. Performance based on process not individuals.
14. Put everybody in the organization to work to accomplish the transformation. The transformation is everybody's job.	Provides the roadmap and rationale for change.

Figure 14-1 (continued)

by QFD and QFD-related mechanisms that have been discussed earlier in this book.

Is QFD the only way to meet Deming's Points? Hardly. But the "point" here is that the reader should come away from reading this book with the idea that it is just as important to know how to implement as it is to understand principles.

14.1 *It may be better to try and fail than never to have tried at all — but its best to have tried and succeeded!*

To have given something "the ol' college try" is a moving sentiment, but a business, and its employees, should want to do more than just try. QFD provides the vehicle for not only trying new applications to meet today's competitive business demands, but also succeeding in the attempt as well. The authors have created in this book the means for beginning a QFD pilot program that can be adapted to varying levels of sophistication and effort. If you can visualize the Mercedes, but only have the keys to the Saturn, the answer is not to walk until you can get the Mercedes keys! Use the keys to the Saturn to get to your destination.

The authors end by relating the story of the little boy and the starfish. One day, after a terrible storm the night before, a little boy went down to the ocean to walk along the beach. Upon arriving at the beach he saw his other friends playing, and people enjoying the sun. But he also saw that the huge waves from the storm had washed many starfish high on the beach away from where the tide could return them. His concern for the dying starfish compelled him to start running along the beach, picking up starfish and returning them to the surf. An old man watched the boy scurrying back and forth, grabbing starfish and running to the surf. He walked up to the little boy and said:

"Son, what in the world are you doing?"

"I'm trying to save the starfish," replied the winded youth, sweat beading on his brow.

"But son, there are far too many starfish on the beach for you to save them all. Don't you think you are wasting your time?" asked the old man.

The little boy frowned in thought for a few moments, and then replied:

"You are right that there are too many starfish for me to save them all, but the ones I've put back in the ocean do not think I've wasted my time!"

The point of the story is that the little boy was making some progress, but he really should have marshalled his resources (his friends), prioritized his customer's needs (big starfish vs. little starfish), and got his processes under control (how many of what size starfish could he and his friends carry at a time?).

Sentimentality is nice, but results are better! Use QFD to get the results you need.

Just do it.

References

1. *Proceedings of the 7th Annual NASA/Contractors Conference on Quality and Productivity*, Summary Report, April 1991, p. 3.
2. *Business Week*, 1992 Special Bonus Issue, p. 86.
3. On October 13, 1993, Bell Atlantic Corp. and Tele-Communications, Inc. announced the merger of the communications and cable giants, which pundits said would usher in the age of interactive communications faster than previously expected. *Business Week*, "Bell-Ringer!" pp. 32-36, October 25, 1993. The deal was dead by March of 1994, but the threat of the Bell-TCI merger has prompted similar, though not as large, mergers expected to open up the "information superhighway" jusy as quickly.
4. *Business Week*, June 9, 1987.
5. Smith, Fred, Chairman of Federal Express, in *Proceedings of the 7th Annual NASA/Contractors Conference on Quality and Productivity*, Summary Report, April 1991, p. 2.
6. Blackburn, Joseph D., ed., *Time-Based Competition*, Business One/Irwin, Homewood, 1991, p. 201.
7. *American Heritage Dictionary*, Second College Edition, 1985, p. 1013.
8. Ernst & Young Improvement Consulting Group, *Total Quality, An Executive's Guide for the 1990s*, Business One/Irwin, Homewood, 1990, p. 4.
9. *Proceedings of the 7th Annual NASA/Contractors Conference on Quality and Productivity*, Summary Report, April 1991, p. 1.
10. "Reengineering: The Hot New Managing Tool," *Fortune*, August 23, 1993, p. 41.
11. "Some Manufacturers Drop Efforts to Adopt Japanese Techniques," *Wall Street Journal* May 7, 1993.
12. "Reengineering: The Hot New Managing Tool," *Fortune*, August 23, 1993, p. 43.
13. "Reengineering: The Hot New Managing Tool," *Fortune*, August 23, 1993, p. 43.
14. "Leading the World-Class Company," Quality '92, Special Edition, *Fortune*, October 1992.
15. Kanter, Rosabeth Moss, *The Change Masters*, Simon & Schuster, New York, 1983, p. 156.
16. Eureka, William and Ryan, Nancy E., *The Customer-Driven Company*, ASI Press, Dearborn, 1988, p. 2.
17. King, Bob, *Better Designs in Half the Time*, Third Edition, Goal/QPC, Methuen, 1989, p. 2.
18. "The Customer-Driven Company," People Who Make A Difference, Supplement, *Fortune*, May 31, 1993, pp. 156–157.
19. "Quality Function Deployment Improves Furniture Design," *Quality Digest*, September 1992, p. 14.
20. "Quality Function Deployment Improves Furniture Design," *Quality Digest*, September 1992, p. 14.

21. "The Tough New Consumer," *Fortune*, Special Issue, Autumn/Winter 1993, p. 39.
22. "The Tough New Consumer," *Fortune*, Special Issue, Autumn/Winter 1993, p. 39.
23. "The Tough New Consumer," *Fortune*, Special Issue, Autumn/Winter 1993, p. 39.
24. Akao, Yoji, ed., *Quality Function Deployment*, Productivity Press, Cambridge, 1990.
25. Eureka, William E. and Ryan, Nancy E., *The Customer-Driven Company*, ASI Press, Dearborn, 1988.
26. Pugh, Stuart, *Total Design: Integrated Methods for Successful Product Engineering*, Addison-Wesley Publishing Company, Wokingham, 1990.
27. Pugh, Stuart, *Total Design: Integrated Methods for Successful Product Engineering*, Addison-Wesley Publishing Company, Wokingham, 1990, p. 5.
28. Blanchard, Benjamin S., *Systems Engineering Management*, John Wiley & Sons, Inc. New York, 1991, p. 28.
29. Mudge, Arthur E., CVS, *Value Engineering, A Systematic Approach*, J. Pohl Associates, Pittsburgh, 1990, p. 16.
30. Phadke, Madhav S., *Quality Engineering Using Robust Design*, Prentice Hall, Englewood Cliffs, 1989, p. 2.
31. Phadke, Madhav S., *Quality Engineering Using Robust Design*, Prentice Hall, Englewood Cliffs, 1989, p. 3.
32. Phadke, Madhav S., *Quality Engineering Using Robust Design*, Prentice Hall, Englewood Cliffs, 1989, p. 3.
33. Kiemele, Mark J. and Schmidt, Stephen R., *Basic Statistics: Tools for Continuous Improvement*, 3rd Ed., Air Academy Press, Colorado Springs, 1993, Chapter 8.
34. *American Heritage Dictionary*, Second College Edition, Houghton Mifflin Company, Boston, 1985, p. 1337.
35. The authors wish to thank Martin Marietta Astronautics Group for permission to use the Satellite Servicing System as a case example.
36. A minor debate exists as to whether the subject is "*system* engineering" in the singular, or "*systems* engineering" in the plural. The authors have chosen to use the plural term because it is often more descriptive of the actual function of the methodology, which is to analyze systems as part of a whole.
37. *American Heritage Dicitionary*, Second College Edition, Houghton Mifflin Company, Boston, 1985, p. 1234.
38. Chapman, William L., Bahill, A. Terry, and Wymore, A. Wayne, *Engineering Modeling and Desing*, CRC Press, Boca Raton, p. 10.
39. MIL-STD-499B, *Systems Engineering*, United States Department of Defense, May 15, 1991, Sec. 3.2.1.
40. Blanchard, Benjamin S., *System Engineering Management*, John Wiley & Sons, Inc., New York, 1991, p. 13.
41. Winner, Robert I., Pennell, James P., Bertrand, Harold E., and Slusarczuk, Marko G., *The Role of Concurrent Engineering in Weapons System Acquisiton*, IDA Report R-388, Institute for Defense Analysis, Alexandria, December 1988, p. 2.
42. Shina, Sammy G., *Concurrent Engineering and Design for Manufacture of Electronics Products*, Van Nostrand Reinhold, New York, 1991, p. 1.
43. Hammer, Michael and Champy, James, *Reengineering the Corporation*, AMACOM, New York, 1993, p. 32.
44. *American Heritage Dictionary*, Second College Edition, p. 987.
45. Hammer, Michael, *Harvard Business Review*, July-August 1990, pp. 104-112.

46. Hammer, Michael and Champy, James, *Reeingineering the Corporation*, HarperCollins, New York, 1993, p. 2.
47. Tomasko, Robert M., *Rethinking the Corporation*, AMACOM, New York, 1993, pp. 9–10.
48. Gene Hall, Jim Rosenthal, and Judy Wade, "How to Make Reengineering *Really* Work," *Harvard Business Review*, November-December 1993, p. 124.
49. Chapters 4, 10 and 11 also discuss strategy and problems to avoid in the utilization of any QFD analysis or application QFD as part of RSM™. This chapter identifies problems that can arise in a pilot program implementation of QFD.
50. This case study is presented by permission from Die Cut Technologies, Scott Flores, President. The authors wish to thank Mr. Flores for his participation in providing this information.
51. This case study was printed with permission from the University of Colorado at Boulder, College of Engineering and Applied Science. The authors wish to thank Dr. Richard Seebass, Dean of the College, and Dr. Jackie Sullivan, ITL Project Director, for providing the opportunity to use QFD in the ITL development and for providing follow-on reports on the success of the ITL.
52. This case study was printed with permission from Westinghouse Electric Corporation. The authors wish to thank Russ Calo of Westinghouse Electric Corporation for providing this information. The summary of the case study is based upon a paper written by Russ Calo, Westinghouse Electric Corporation, Steve Kramer, Westinghouse Electric Corporation, and John Kamauff, University of Virginia, titled "Total Quality Management in the Tight Tolerance Aluminum Extrusion Process."
53. This case study was published with permission from Westinghouse Electric Corporation, Baltimore, MD.
54. This case study appears with permission from Martin Marietta Astronautics Group.
55. This case study appears with permission from 3M. Specifics of the QFD matrix and the product requirements and target values have been omitted to preserve design confidentiality. The authors thank 3M for its permission to reference the results of the QFD analysis.
56. The Fourteen Points are taken from William W. Scherkenbach's, *The Deming Route to Quality and Productivity*, CEEPress Books, Washington, D.C., 1990.

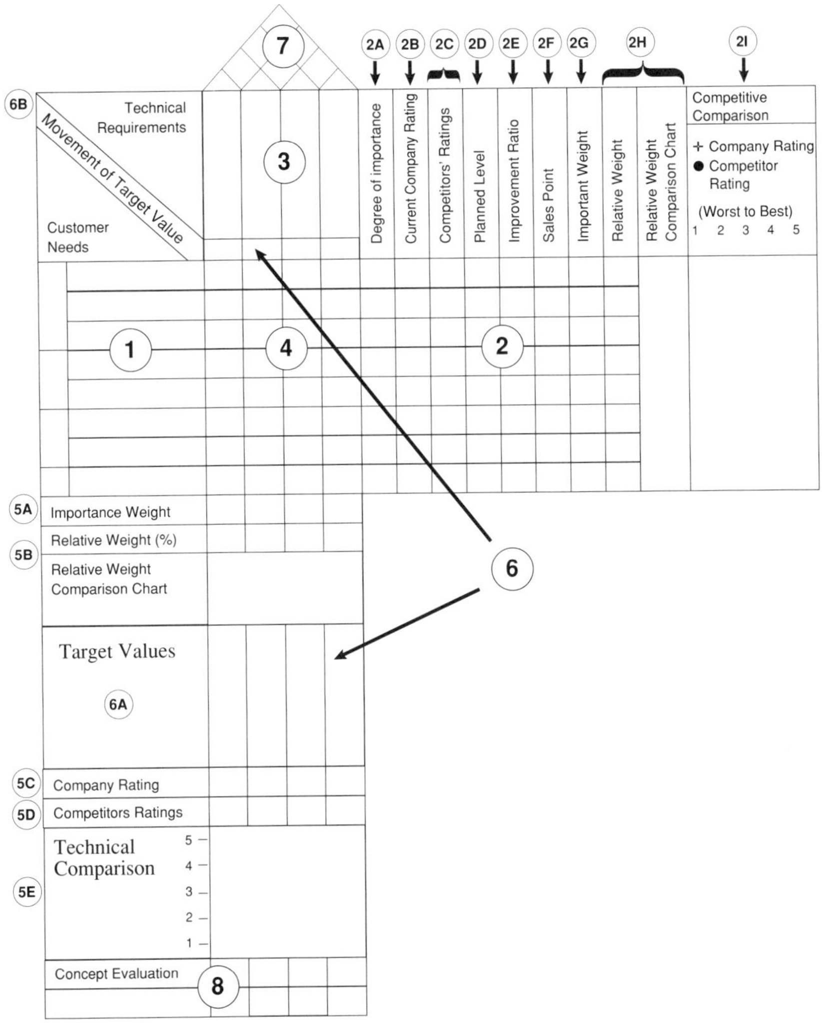

Appendix A Detailed Core QFD Matrix guide.

① Customer Requirements:

- Who are the primary customers of your product or service?
- Who are primary? Who are secondary?
- What do these customers want/expect from your product or service?

Steps:

1. Identify all customer groups.
   - Internal and external.
   - Possibility of conflicting needs.
   - May use end user as focus.
2. Prepare a plan for gathering and analyzing accurate information about the customer.
   - If possible, invite customer participation.
3. Perform a brainstorming session to establish Affinity and Tree diagrams.
4. Transfer Tree Diagram items to House of Quality.
5. Establish a set of definitions for key customer requirements.
   - Level of detail is tailored to available resources/team understanding.

(2A) Degree of Importance:

- How important is this need to your customer on a scale of 1 – 5 (5 being high)?
- Pick 5s first.
- When evaluating, ask would they give up something else for this? e.g. speed for cost?
- Or vice versa to establish a relative priority of importance.

Steps:

1. Identify relative priority based on information and perception of customer.
   - 1 to 5 scale most often used: Higher is better.
   - 1 to 10 scale used in cases where priorities are distinguishable.
   - Ratings have been determined based on statistical marketing analysis.

2. If possible, invite customer to review.

Appendix A 1.2

Appendix A (continued)

(2B) Current Company Rating:

- How well do you think your company meets this need?

Steps:

1. Identify how well your company meets the customer requirements.
2. Use customer inputs, business development data and research data as basis for rationale.
3. Use scale of 1 to 5, where 5 is best.

(2C) Competitors' Ratings:

- In one year, how well do you think your best competitors meet this need?
- or – How well do you think your best competitor will meet this need by the time you launch this new product, service or business operations?

Steps:

1. Identify how well their company meets the customer requirements.
2. Use customer inputs, business development data and research data as basis for rationale.
3. Use scale of 1 to 5, where 5 is best.

(2D) Planned Level:

- By the time you launch this new product, service or business operation where would you like your company to be relative to today?
- Do you have or should you get the resources to make this improvement taking into account how important this need is to the customer and how well your competitors provide this need?

Steps:

1. Set a time in the future for planning improvement prior to submitting or implementing the product such as:

(continued on next page)

Appendix A 1.3

Appendix A (continued)

- Proposal submittal.
- Product delivery.
- End of design phase or improvement phase.

2. Base rating of planned level of improvement within established time frame.
   - Use scale of 1 to 5, where 5 is best.
3. Focus on matching or surpassing competition.
4. Plans and resources for improvement should match ratings.

(2E) Improvement Ratio:

- Divide each number in the Planned Level column by its corresponding number in the Current Company Rating column.

$$\text{Improvement Ratio} = \frac{\text{Planned Level}}{\text{Current Company Rating}}$$

(2F) Sales Point:

- What gives your company a competitive advantage?

Steps:

1. Select attributes that would generate new business for additional products and services which provide "Excitement Quality", or when met would provide the company with a competitive advantage.
   - Items that help to capture the marketplace.
2. Rank the attributes by percieved value:
   - Major sales point ● = 1.5 value
   - Minor sales point ○ = 1.2 value
   - No sales point = 1.0 value

Appendix A 1.4

Appendix A (continued)

(2G) Importance Weight (Customer Requirements):

$$W = D \cdot R \cdot S$$

where:

W = Importance Weight

D = Degree of Importance (2A)

R = Importance Ratio (2E)

S = Sales Point Value (2F)

(2H) Relative Weight (Customer Requirements):

$$RW = \frac{W}{\Sigma W} \cdot 100$$

Where:

W = Importance Weight (2G)

ΣW = Sum of Importance Weights (sum of 2G column)

(2I) Competitive Comparison Chart:

- The use of symbols provides visual assistance in comparing company performance.

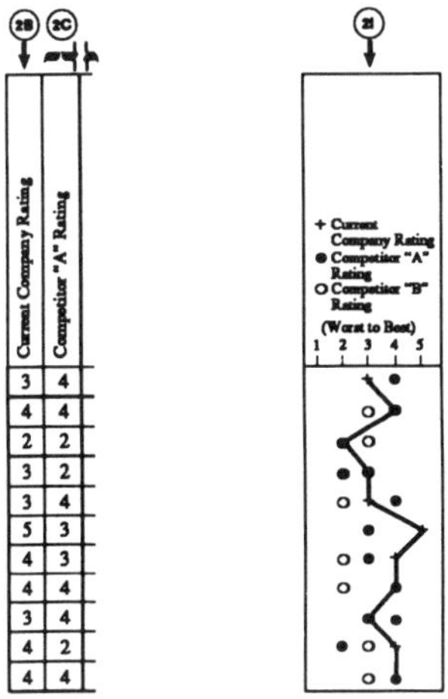

Appendix A 1.5

Appendix A (continued)

③ Technical Requirements:

- What can I control that allows me to meet my customer objectives?

Steps:

1. Develop a list internally to answer the following question:
   "If I control ___'a'___ then I am meeting my customers requirements ___'b'___."

   Where "a" is a possible technical requirement (skill, technique, technology, or machine), and "b" is a customer need, (something in section 1.
2. Develop requirements through a brainstorming session with the multifunctional team and create a hierarchy diagram to transfer to the matrix.

④ Correlation Matrix:

- Indicates strength of technical requirements to satisfy customer's need as indicated by the following symbols:
  - ● = 9 (Strong relationship)
  - O = 3 (Moderate relationship)
  - ▲ = 1 (Weak relationship)
  - = 0 (Blank – No relationship)
- "If I control (the technical requirement) it will (significantly (●), moderately (O), slightly (▲), or not (blank) impact my customer objective to (insert specific customer need) ."

(5A) Importance Weight:

$IW = \Sigma VR$

where:

ΣVR = Sum of Value Relationship column (4)

Appendix A 1.6

Appendix A (continued)

(5B) Relative Weight (Technical Requirements):

$$RW = \frac{W}{\Sigma W} \cdot 100$$

Where:

W = Importance Weight (5A)

ΣW = Sum of Importance Weights (sum of 5A row)

(5C) (5D) (5E) Competitive Rating (Technical Performance):

1. Identify how well you and competitors are capable of meeting the technical requirements.
2. Use scale of 1 to 5, where 5 is best.
3. Translate numerical values into symbols for visual comparison.

(6A) Target Values:

1. Assign specific target values for as many requirements as possible.
   - Define specific goals/ranges for designers and engineers.
   - Establish targets for trade studies and analysis.
2. Use historical data where possible.
3. These are starting points that can be revised based on detailed analysis.

(6B) Movement of Target Values:

1. Use symbols to indicate desired direction for each corresponding technical requirement target value:
   ⇑ – Increase
   ⇓ – Decrease
   × – Meet specified nominal value

Appendix A 1.7

Appendix A (continued)

⑦ Identify Technical Interactions:

1. Identify and compare the interaction of implementing technical requirements to the established target values:
   - ● = Strong positive
   - ○ = Positive
   - ✕ = Negative
   - ✱ = Strong Negative
2. Ask "If I try to improve this parameter or operation will it have a positive or negative effect on something else I am trying to control?"

⑧ Concept Evaluation

1. How well does this concept meet the technical requirements?
   ● Concept meets the requirements in all anomolous conditions and in some cases exceeds the target requirement.
   ○ Concept meets or nearly meets the requirements under most conditions.
   ▲ Concept cannot meet the target parameter under nearly all conditions. This could be a significant technical gap that could be a potential "show stopper."

Appendix A 1.8

Appendix A (continued)

Index

D

E

F

M

N

O

P

Q

R

S